New Perspectives in Optical Design

New Perspectives in Optical Design

Guest Editors

Xing Peng
Ci Song
Shuo Qiao

MDPI

Basel • Beijing • Wuhan • Barcelona • Belgrade • Novi Sad • Cluj • Manchester

Guest Editors

Xing Peng
College of Intelligent Science and Technology
National University of Defense Technology
Changsha
China

Ci Song
College of Intelligent Science and Technology
National University of Defense Technology
Changsha
China

Shuo Qiao
College of Intelligent Science and Technology
National University of Defense Technology
Changsha
China

Editorial Office
MDPI AG
Grosspeteranlage 5
4052 Basel, Switzerland

This is a reprint of the Special Issue, published open access by the journal *Photonics* (ISSN 2304-6732), freely accessible at: www.mdpi.com/journal/photonics/special_issues/UE93HA2BP5.

For citation purposes, cite each article independently as indicated on the article page online and using the guide below:

Lastname, A.A.; Lastname, B.B. Article Title. *Journal Name* **Year**, *Volume Number*, Page Range.

ISBN 978-3-7258-3344-3 (Hbk)
ISBN 978-3-7258-3343-6 (PDF)
https://doi.org/10.3390/books978-3-7258-3343-6

Contents

photonics

Communication

A New Method to Enhance the Light–Matter Interaction by Controlling the Resonance of Electrons

Xuefang Hu [1,2], Haoyang Mao [2], Sisi Yang [3,*], Changgui Lu [4], Xiangyue Zhao [4] and Mengjia Lu [5]

[1] College of Digital Technology and Engineering, Ningbo University of Finance & Economics, Ningbo 315175, China; huxuefang@nbufe.edu.cn
[2] Ningbo Yongxin Optics Co., Ltd., Ningbo 315040, China; 12230099@zju.edu.cn
[3] Ningbo Institute of Education Sciences, Ningbo 315099, China
[4] Advanced Photonics Center, School of Electronic Science & Engineering, Southeast University, Nanjing 210018, China; changguilu@seu.edu.cn (C.L.); zhaoxiangyuenwpu@126.com (X.Z.)
[5] School of Information Technology, Jiangsu Open University, Nanjing 210036, China; mjlu@seu.edu.cn
* Correspondence: 18892619125@163.com

Abstract: The manipulation of surface plasmon polaritons (SPPs) plays an essential role in plasmonic science and technology. However, the modulation efficiency and size of the device in the traditional method suffer from weak light–matter interaction. Herein, we propose a new method to enhance the light–matter interaction by controlling the resonance of electrons in a sandwich structure which is composed of an interdigital electrode, dielectric, and doped semiconductor. The numerical results show that the resonance of electrons occurs when their vibrational frequency under electrostatic field matches well with the oscillation frequency of the propagating SPPs. The intensity of the electric field is enhanced about 8%, which can be utilized to improve the modulation efficiency and minimize the footprint of device to a great extent. These findings pave a new way towards higher precision sensor and more compact modulator.

Keywords: light–matter interaction; modulation efficiency; resonance

check for updates

Received: 13 December 2024
Revised: 7 January 2025
Accepted: 20 January 2025
Published: 22 January 2025

Citation: Hu, X.; Mao, H.; Yang, S.; Lu, C.; Zhao, X.; Lu, M. A New Method to Enhance the Light–Matter Interaction by Controlling the Resonance of Electrons. *Photonics* **2025**, *12*, 95. https://doi.org/10.3390/photonics12020095

1. Introduction

Surface plasmon polaritons (SPPs) are modes of electromagnetic waves that propagate along a metal surface due to the interaction between light waves and surface charges [1–6]. In this mode, the light wave is localized at the interface between metal and dielectric, and there is a strong coupling between the light field and the free electrons on the metal surface [7–10]. Recently, the SPPs-based circuit elements including waveguides [11,12], modulators [13–16], and photodetectors [17,18] have been widely demonstrated to benefit from this strong couple mode [19,20]. However, this mode is difficult to be further enhanced due to the limited light–matter interaction. Thus, a long interaction region is usually adopted in a plasmon modulator to improve its modulation efficiency, leading to a large footprint of the device. Herein, it is highly desirable to propose a new method to enhance the light–matter interaction for the ultra-compact modulator.

The surface charges of metal are resonant with the incident light and produce a great enhancement at the interface of metal and dielectric in propagating SPPs [21–23]. Nevertheless, the motion of electrons can also be driven and controlled by the electrostatic field. In this condition, the resonance of electrons may happen when the vibration frequency of the electrons under an electrostatic field matches well with the oscillation frequency of the propagating SPPs [24]. The electric field intensity at the interface can be further enhanced if resonance occurs, which can be utilized to improve the modulation efficiency and reduce the footprint of a plasmon modulator owing to the enhanced light–matter interaction [25,26]. Although it is difficult to modulate the electron concentration and movement in metal by the electrostatic, the modulation in some materials with low electron concentration, such as heavily doped semiconductor, transparent conduction oxides, and two-dimensional materials [27–29], is feasible. Consequently, it is possible to achieve an enhanced light–matter interaction in these materials.

In this paper, we proposed a sandwich structure composed of interdigital electrode, dielectric, and doped semiconductor. The electromagnetic particle-in-cell (PIC) method has been developed based on the finite-difference time-domain (FDTD) method to analyze the electric field distribution and laws of the electron motion in doped semiconductor [30,31] (the PIC code is a central simulation tool for a wide range of physics studies, from semiconductors to cosmology or accelerator physics, and, in particular, to plasma physics Especially, it can vividly reflect the interaction between the electromagnetic fields and electrons, and the technique follows the motion of a large assembly of charged particles in their self-consistent electric and magnetic fields [32]). The numerical results showed that the resonance occurs when the vibration frequency of electrons under an electrostatic field is the same as that of SPPs, and the electric intensity is enhanced to about 8% as the periodic electrostatic field and the propagating SPPs are employed simultaneously. These findings pave a new way to enhance the light–matter interaction and have a potential application, such as improving the modulation efficiency, minimizing the device size, and optimizing the sensing precision, in optical modulators, detectors and sensors.

2. Model

In order to analyze the interaction between the electrostatic field and the propagating SPPs at the particle level, a hybrid structure composed of interdigital electrode, dielectric, and heavy doped semiconductor is designed in Figure 1. The polarized light is incident on the grating and stimulates the SPPs that propagate along the interface of silicon dioxide (SiO_2) and the heavily doped semiconductor, in which the electrons oscillate with the incident light. On the other hand, the electrons can also oscillate under the electrostatic field when the voltage is applied to the interdigital electrode individually. Herein, the oscillation frequency of the electrons can be modulated by the voltage dynamically, and the propagating SPPs can be modulated if the wavelength of SPPs is fixed. The resonance may occur when the vibration frequency of the electrons under an electrostatic field matches well with the oscillation frequency of propagating SPPs. The electric field intensity will be enhanced and the amplitude of electron motion will be enlarged if the resonance occurs in the simulation process, which can be utilized to enhance the light–matter interaction at nanoscale. (The enhanced electric field intensity can be observed during the experimental process). The PIC method is employed to analyze the motion of electrons in the doped semiconductor within one period of 1800 nm and the position ranges from −900 nm to 900 nm.

The main parameters for the simulation are listed in Table 1. These parameters serve as references, with their actual values being the product of the set value and the unit value. The value of ωr (frequency) is defined by the user [32].

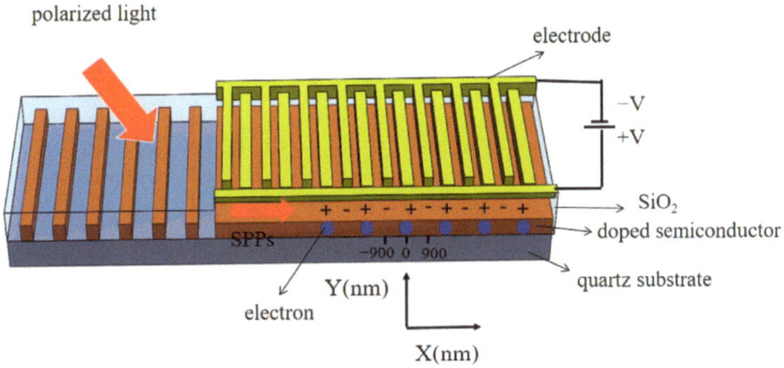

Figure 1. Schematic illustration of the interdigital electrode-dielectric-doped simiconductor structure, the electrons in doped simiconductor can be controlled by the voltage applied on electrode.

Table 1. The list of main parameters for the simulation.

Units of charge(e)	e
Units of mass(m)	m_e
Units of velocity(c)	c
Units of time(Tr)	${\omega_r}^{-1}$
Units of length(Lr)	c/ω_r
Units of electric field(Er)	$m_e c \omega_r / e$

3. Results and Discussion

3.1. The Characteristics of Electron Under the Electrostatic Field

The electrons are accelerated from one side to the center and then decelerated to another side because the symmetric external forces are applied on them when the opposite voltage is applied on the electrode, respectively. The distribution of the electric field intensity under the electrostatic field is shown in Figure 2a, the electric field is in the opposite direction at the side of the 0 position. The trajectory of a single electron located at the position of 585 nm is shown as the red line in Figure 2b. The amplitude ranges from −585 nm to 585 nm. The black line is the electron located at the position of 195 nm, and the amplitude of the electron is between −195 nm and 195 nm. As shown, the amplitude of the electron is closely related to the position of the particle, in other words, the amplitude of the electron depends on its original location under the electrostatic field.

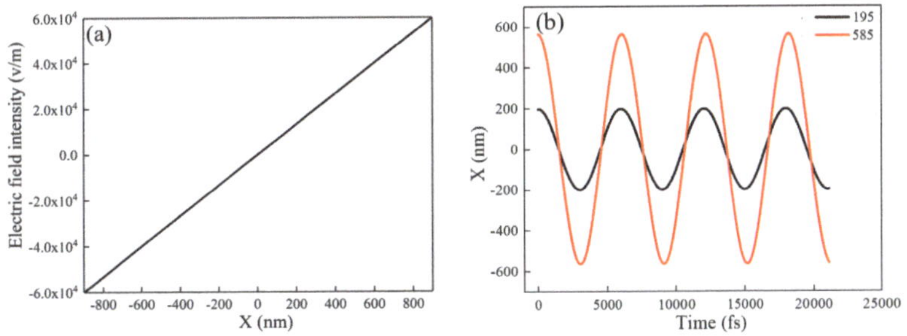

Figure 2. (**a**) The distribution of electric field intensity under the electrostatic field (**b**) the trajectory of electron at different position versus the time.

Figure 3a–c show the trajectory of a single electron under the electrostatic field with values of 10^4, 10^5, and 10^6 V/m, respectively, where the X-axis represents the time while the Y-axis represents the position of the electron. It can be observed that the oscillation frequency of the electron varies with the magnitude of the electrostatic field. The electron has a faster acceleration due to the greater force imposed by the electric field, leading to a shift in electron vibration frequency. It can be demonstrated that the vibration frequency of the electron can be controlled by the applied voltage dynamically.

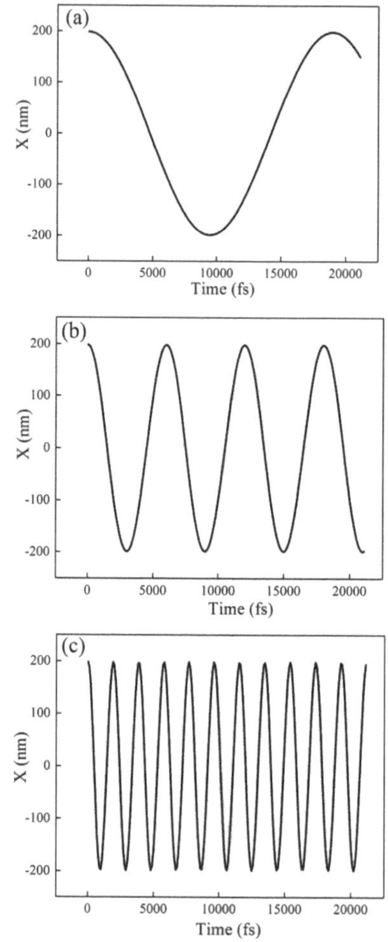

Figure 3. The trajectory of electron under different electrostatic fields: (**a**) 10^4, (**b**) 10^5, and (**c**) 10^6 V/m.

3.2. The Distribution of Electric Intensity Under the Propagating SPPs

The propagating SPPs we employed are in the form of A0 × cos(ωt + 8 × pi × x), where the A0 is set as 0.001. (The propagation loss is neglected in simulation as the main purpose of the model is to analyze the movement of electrons under various conditions). The distribution of the electric field at the time of 628.27 fs is shown in Figure 4, which represents a standard sine wave used to approximate the propagating SPPs.

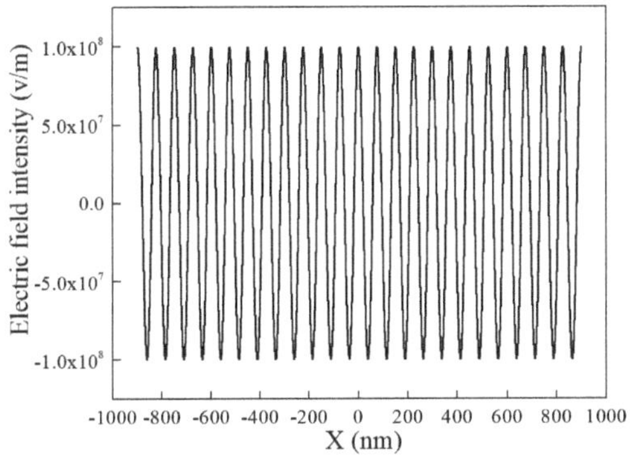

Figure 4. The electric field distribution at the time of 628.27 fs.

3.3. The Distribution of Electric Intensity Under the Electrostatic Field and Propagating SPPs Simultaneous

When the periodically distributed electrostatic field and the propagating SPPs are employed in the proposed model simultaneously, its electric field distribution is shown in Figure 5, when compared to Figures 2a and 4, it becomes apparent that the electric field distribution represents the superposition of both the electrostatic field and the propagating SPPs.

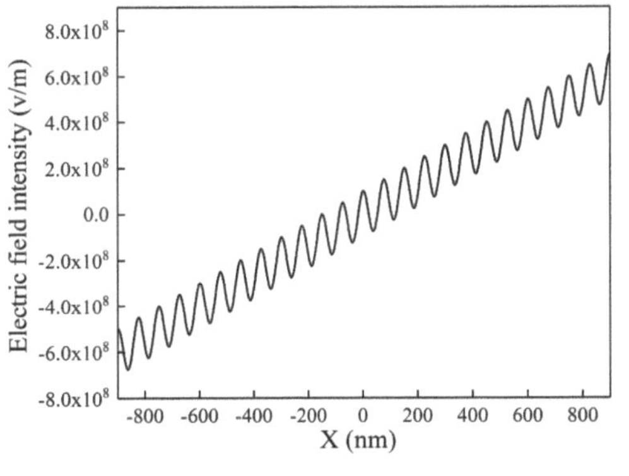

Figure 5. The electric field distribution at the time of 628.27 fs when the voltage is applied.

3.4. The Resonance of Electrons

3.4.1. Analysis of Amplitude

We fixed the frequency of the propagating SPPs and changed the magnitude of voltage, the numerical results show that the resonance occurs when the vibration frequency of electrons under an electrostatic field is the same as the oscillation frequency of SPPs, leading to an increase in the amplitude of the electron. As demonstrated in Figure 6, the oscillation amplitude of the same electron in the resonant state is significantly larger than in the normal state clearly demonstrating the resonance of the electron.

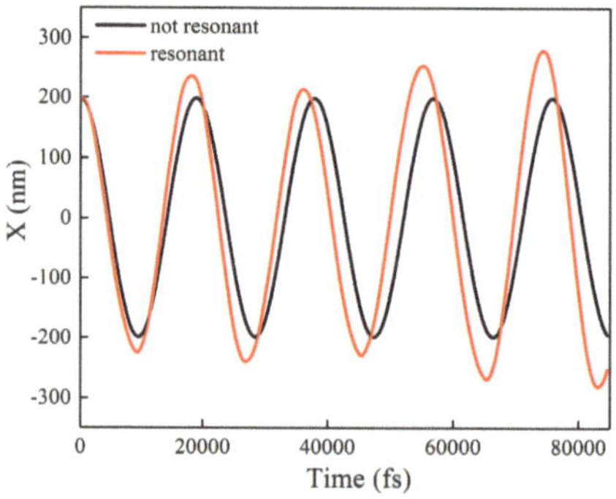

Figure 6. The trajectory of the electron under the resonant and normal state.

3.4.2. Analysis of Electric Field Intensity

The electric field intensity at the position of 500 nm versus different frequencies of propagating SPPs is depicted in Figure 7a. The electric field intensity varies with the frequency of SPPs, and its intensity reaches a maximum at a frequency of 4.3×10^{11} Hz while the electrostatic field is fixed at 10^5 V/m. At this frequency, the intensity is enhanced by approximately 8%. The electric intensity will not vary with the frequency of SPPs if it is just the superposition of externally applied electric field with SPPs, which clearly indicates that resonance has occurred and the enhancement of light–matter interaction is achieved by the proposed structure. Figure 7b shows the electric field intensity as a function of the position in both the resonant state and the normal state. It becomes evident that the electric field intensity at the resonant state is larger than the normal state and the electric field intensity is obviously enhanced. These findings pave the way for enhancing the light–matter interaction and have great potential applications in the fields of optical modulation and sensing.

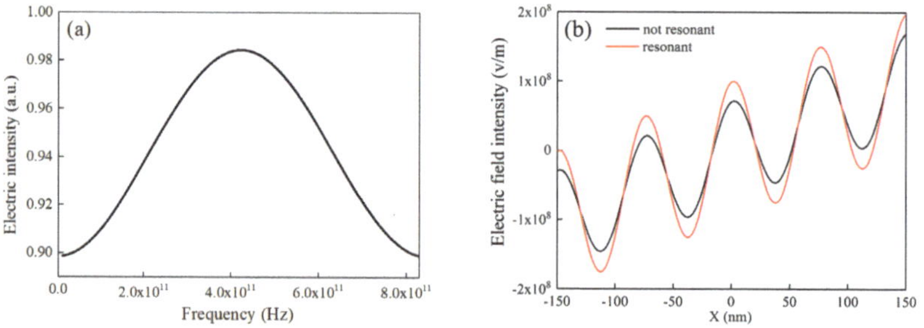

Figure 7. The electric field distribution (**a**) versus different SPPs frequency (**b**) at different state versus the position.

4. Conclusions

In conclusion, a hybrid structure composed of an interdigital electrode, dielectric, and doped semiconductor is proposed to enhance the light–matter interaction in this paper. The resonance of the electron occurs when the frequency of propagating SPPs is 4.3×10^{11} Hz while the electrostatic field is fixed at 10^5 V/m. The electric field and the amplitude reach their maximum, and the electric intensity is enhanced by approximately 8%. This effect can be harnessed to improve the modulation efficiency and minimize the footprint of the plasmonic modulator. These findings pave the way towards a higher precision sensor and a more compact modulator.

Author Contributions: C.L. and H.M. designed the study, X.H. and X.Z. finished the numerical simulation, M.L. and S.Y. interpreted the results and wrote the manuscript. All authors gave final approve for publication. All authors have read and agreed to the published version of the manuscript.

Funding: The work was financially supported by National Natural Science Foundation of China (NSFC) (12274075). Scientific Research Fund of Zhejiang Provincial Education Department (Y202352891).

Data Availability Statement: The data relative to the experiments discussed in this work are available upon reasonable request from the corresponding author Sisi Yang.

Conflicts of Interest: Authors Xuefang Hu and Haoyang Mao were employed by the company Ningbo Yongxin Optics Co., Ltd. The remaining authors declare that the research was conducted in the absence of any commercial or financial relationships that could be construed as a potential conflict of interest.

References

1. Pala, R.A.; Shimizu, K.T.; Melosh, N.A.; Brongersma, M.L. A Nonvolatile Plasmonic Switch Employing Photochromic Molecules. *Nano Lett.* **2008**, *8*, 1506–1510. [CrossRef] [PubMed]
2. Abeysinghe, D.C.; Nader, N.; Myers, J.; Hendrickson, J.R.; Cleary, J.W.; Walker, D.E.; Chen, K.-H.; Liu, Y.; Mou, S. Edge Doping Effect to the Surface Plasmon Resonances in Graphene Nanoribbons. *J. Phys. Chem. C* **2019**, *123*, 19820–19827. [CrossRef]
3. Behnia, S.; Fathizadeh, S.; Javanshour, E.; Nemati, F. Light-Driven Modulation of Electrical Current through DNA Sequences: Engineering of a Molecular Optical Switch. *J. Phys. Chem. B* **2020**, *124*, 3261–3270. [CrossRef] [PubMed]
4. Bouillard, J.-S.G.; Dickson, W.; O'connor, D.P.; Wurtz, G.A.; Zayats, A.V. Low-Temperature Plasmonics of Metallic Nanostructures. *Nano Lett.* **2012**, *12*, 1561–1565. [CrossRef] [PubMed]
5. Chang, W.-S.; Lassiter, J.B.; Swanglap, P.; Sobhani, H.; Khatua, S.; Nordlander, P.; Halas, N.J.; Link, S. A Plasmonic Fano Switch. *Nano Lett.* **2012**, *12*, 4977–4982. [CrossRef]
6. Chen, T.; Tong, F.; Enderlein, J.; Zheng, Z. Plasmon-Driven Modulation of Reaction Pathways of Individual Pt-Modified Au Nanorods. *Nano Lett.* **2020**, *20*, 3326–3330. [CrossRef] [PubMed]
7. Cox, J.D.; de Abajo, F.J.G. Single-Plasmon Thermo-Optical Switching in Graphene. *Nano Lett.* **2019**, *19*, 3743–3750. [CrossRef]
8. Cox, J.D.; de Abajo, F.J.G. Nonlinear Graphene Nanoplasmonics. *Acc. Chem. Res.* **2019**, *52*, 2536–2547. [CrossRef] [PubMed]
9. Du, W.; Zhao, J.; Zhao, W.; Zhang, S.; Xu, H.; Xiong, Q. Ultrafast Modulation of Exciton–Plasmon Coupling in a Monolayer WS2–Ag Nanodisk Hybrid System. *ACS Photonics* **2019**, *6*, 2832–2840. [CrossRef]
10. Emboras, A.; Niegemann, J.; Ma, P.; Haffner, C.; Pedersen, A.; Luisier, M.; Hafner, C.; Schimmel, T.; Leuthold, J. Atomic Scale Plasmonic Switch. *Nano Lett.* **2016**, *16*, 709–714. [CrossRef]
11. Agrawal, A.; Susut, C.; Stafford, G.; Bertocci, U.; McMorran, B.; Lezec, H.J.; Talin, A.A. An Integrated Electrochromic Nanoplasmonic Optical Switch. *Nano Lett.* **2011**, *11*, 2774–2778. [CrossRef] [PubMed]
12. Kress, S.J.P.; Antolinez, F.V.; Richner, P.; Jayanti, S.V.; Kim, D.K.; Prins, F.; Riedinger, A.; Fischer, M.P.C.; Meyer, S.; McPeak, K.M.; et al. Wedge Waveguides and Resonators for Quantum Plasmonics. *Nano Lett.* **2015**, *15*, 6267–6275. [CrossRef] [PubMed]
13. Hossein, K.; Vahed, H. A structure of electro-absorption hybrid plasmonic modulator using silver nano-ribbon. *Opt. Quantum Electron.* **2023**, *55*, 894.
14. Chauhan, D.; Sbeah, Z.; Dwivedi, R.P.; Nunzi, J.-M.; Thakur, M.S. An investigation and analysis of plasmonic modulators: A review. *J. Opt. Commun.* **2024**, *45*, 483–511. [CrossRef]
15. Zurak, L.; Wolf, C.; Meier, J.; Kullock, R.; Mortensen, N.A.; Hecht, B.; Feichtner, T. Modulation of surface response in a single plasmonic nanoresonator. *Sci. Adv.* **2024**, *10*, eadn5227. [CrossRef] [PubMed]
16. Gao, X.; Gu, Z.; Ma, Q.; Cui, W.Y.; Cui, T.J.; Chan, C.H. Reprogrammable spoof plasmonic modulator. *Adv. Funct. Mater.* **2023**, *33*, 2212328. [CrossRef]
17. Wang, L.; Zhang, Y.; Wang, B.; Wei, Y.; Zhang, B.; Meng, L.; Liu, T.; Wang, B.; Han, B.; Jiang, Z.; et al. High-performance infrared Ge-based plasmonicphotodetector enhanced by dual absorption mechanism. *APL Photonics* **2020**, *5*, 096104. [CrossRef]
18. Ma, Z.; Kikunaga, K.; Wang, H.; Sun, S.; Amin, R.; Maiti, R.; Tahersima, M.H.; Dalir, H.; Miscuglio, M.; Sorger, V.J. Compact Graphene Plasmonic Slot Photodetector on Silicon-on-Insulator with High Responsivity. *ACS Photonics* **2020**, *7*, 932–940. [CrossRef]
19. Liu, J.; Khayrudinov, V.; Yang, H.; Sun, Y.; Matveev, B.; Remennyi, M.; Yang, K.; Haggren, T.; Lipsanen, H.; Wang, F.; et al. InAs-Nanowire-Based Broadband Ultrafast Optical Switch. *J. Phys. Chem. Lett.* **2019**, *10*, 4429–4436. [CrossRef]
20. Luo, Y.; Wang, Y.; Liu, M.; Zhu, H.; Chen, O.; Zou, S.; Zha, J. Colloidal Assembly of Au–Quantum Dot–Au Sandwiched Nanostructures with Strong Plasmon–Exciton Coupling. *J. Phys. Chem. Lett.* **2020**, *11*, 2449–2456. [CrossRef]

21. Menghrajani, K.S.; Nash, G.R.; Barnes, W.L. Vibrational Strong Coupling with Surface Plasmons and the Presence of Surface Plasmon Stop Bands. *ACS Photonics* **2019**, *6*, 2110–2116. [CrossRef] [PubMed]
22. Min, B.K.; Nguyen, V.-T.; Kim, S.J.; Yi, Y.; Choi, C.-G. Surface Plasmon Resonance-Enhanced Near-Infrared Absorption in Single-Layer MoS_2 with Vertically Aligned Nanoflakes. *ACS Appl. Mater. Interfaces* **2020**, *12*, 14476–14483. [CrossRef] [PubMed]
23. Oshita, M.; Takahashi, H.; Ajiki, Y.; Kan, T. Reconfigurable Surface Plasmon Resonance Photodetector with a MEMS Deformable Cantilever. *ACS Photonics* **2020**, *7*, 673–679. [CrossRef]
24. Singh, D.; Malik, H.K. Terahertz emission by multiple resonances under external periodic electrostatic field. *Phys. Rev. E* **2020**, *101*, 043207. [CrossRef] [PubMed]
25. Xiong, X.; Zhou, Y.; Luo, Y.; Li, X.; Bosman, M.; Ang, L.K.; Zhang, P.; Wu, L. Plasmon-Enhanced Resonant Photoemission Using Atomically Thick Dielectric Coatings. *ACS Nano* **2020**, *14*, 8806–8815. [CrossRef] [PubMed]
26. Maleki, M.J.; Soroosh, M.; AL-Shammri, F.K.; Alkhayer, A.G.; Mondal, H. Design and simulation of a compact subwavelength graphene-based switch for surface plasmon polariton transmission in integrated optoelectronic circuits. *Plasmonics* **2024**, 1–13. [CrossRef]
27. Haddadan, F.; Soroosh, M. Design and simulation of a subwavelength 4-to-2 graphene-based plasmonic priority encoder. *Opt. Laser Technol.* **2023**, *157*, 108680. [CrossRef]
28. Saeed, M.; Ghaffar, A.; Rehman, S.U.; Naz, M.Y.; Shukrullah, S.; Naqvi, Q.A. Graphene-based plasmonic waveguides: A mini review. *Plasmonics* **2022**, *17*, 901–911. [CrossRef]
29. Sharma, T.; Zhang, Z.; Wang, J.; Cheng, Z.; Yu, K. Past, present, and future of hybrid plasmonic waveguides for photonics integrated circuits. *Nanotechnol. Precis. Eng.* **2024**, *7*, 045001. [CrossRef]
30. Ding, W.J.; Lim, J.Z.J.; Do, H.T.B.; Xiong, X.; Mahfoud, Z.; Png, C.E.; Bosman, M.; Ang, L.K.; Wu, L. Particle simulation of plasmons. *Nanophotonics* **2020**, *9*, 3303–3313. [CrossRef]
31. Zhou, J.; Hu, M.; Zhang, Y.; Zhang, P.; Liu, W.; Liu, S. Numerical analysis of electron-induced surface plasmon excitation using the FDTD method. *J. Opt.* **2011**, *13*, 035003. [CrossRef]
32. Derouillat, J.; Beck, A.; Pérez, F.; Vinci, T.; Chiaramello, M.; Grassi, A.; Flé, M.; Bouchard, G.; Plotnikov, I.; Aunai, N.; et al. Smilei: A collaborative, open-source, multi-purpose particle-in-cell code for plasma simulation. *Comput. Phys. Commun.* **2018**, *222*, 351–373. [CrossRef]

photonics

MDPI

Communication

Bifunctional Electromagnetic Manipulation of Surface Waves Using Metasurfaces Under One Circularly Polarized Incidence

Min Kang *, Lixing Chen, Shuaipeng Qin, Liang Ma, Aoxiang Rui and Shiqing Li *

Department of Applied Physics, Zhejiang University of Technology, Hangzhou 310023, China
* Correspondence: 211122090021@zjut.edu.cn (M.K.); sql@zjut.edu.cn (S.L.)

Abstract: The ability to freely manipulate the wavefronts of surface plasmon polaritons (SPPs) or surface waves (SWs), particularly with multifunctional integration, is of great importance in near-field photonics. However, conventional SPP control devices typically suffer from low efficiency and single-function limitations. Although recent works have proposed metasurfaces that achieve bifunctional SPP manipulation, their implementation relies on the excitations of circularly polarized (CP) light with different helicities. Here, we propose a generic approach to designing bifunctional SPP meta-devices under single-helicity circularly polarized incidence. Constructed using carefully selected and arranged meta-atoms that possess both structural resonance and a geometric phase, this kind of meta-device can exhibit two distinct SPP manipulation functionalities in both co- and cross-polarized output channels under one CP incidence. As proof of this concept, we designed a bifunctional meta-device in the microwave regime and numerically demonstrated that it can convert a normally incident left circularly polarized (LCP) beam into SWs, exhibiting both a focused wavefront in the co-polarized output channel and a deflected wavefront in the cross-polarized output channel. Our findings substantially enrich the capabilities of metasurfaces to manipulate near-field electromagnetic waves, which can find many applications in practice.

Keywords: bifunctional; metasurface; surface waves; co-polarized; cross-polarized

check for updates

Received: 17 December 2024
Revised: 9 January 2025
Accepted: 17 January 2025
Published: 20 January 2025

Citation: Kang, M.; Chen, L.; Qin, S.; Ma, L.; Rui, A.; Li, S. Bifunctional Electromagnetic Manipulation of Surface Waves Using Metasurfaces Under One Circularly Polarized Incidence. *Photonics* **2025**, *12*, 91. https://doi.org/10.3390/photonics12010091

1. Introduction

Surface plasmon polaritons (SPPs), which are elementary excitations arising from the coupling of photons and free-electron oscillations at dielectric/metal interfaces, have garnered significant attention over the past few decades due to their wide range of applications in sub-diffraction-limit imaging, biological and chemical sensing, on-chip photonic circuits, and more [1,2]. In low-frequency regimes, where natural SPPs do not occur, spoof SPPs on structured metals offer equally intriguing possibilities for a variety of applications [3–7]. The ability to control these surface waves (SWs, including SPPs and spoof SPPs) is crucial for advancing near-field photonics.

During the last two decades, metasurfaces, which can locally provide abrupt phase shifts at subwavelength intervals to tailor the phase of the incident waves, have been proposed and have shown strong capabilities to manipulate EM waves with flat configurations, subwavelength sizes, and high efficiencies [8–12]. In response to linearly or circularly polarized (CP) waves, the phase responses of meta-atoms on these metasurfaces are dictated by either structural resonance [13,14] or the Pancharatnam–Berry (PB) mechanism [14,15]. Many fascinating wavefront manipulation effects have been discovered, including anomalous reflection/refraction [16,17], the photonic spin Hall effect [18–20], meta-holograms [21–23], flat lenses [24,25], complex beam generation [26–30], spatial light modulation with multifunctionality [31–33], and many others [34–36]. Additionally, metasurfaces have also been proposed for converting PWs into SWs [37–40] and manipulating the wavefront of SWs [41–43]. However, despite the great successes already achieved with metasurfaces, the existing designs for achieving SPP excitation and wavefront manipulation still rely on separate devices for each functionality, limiting their integration into compact systems. Recently, polarization-controlled bifunctional metasurfaces for near-field SW manipulations have been proposed and experimentally demonstrated in both the microwave and optical regimes.

In this paper, we propose a high-efficiency bifunctional meta-device capable of simultaneously exciting SWs and manipulating their wavefronts under single-helicity circularly polarized incidence. Unlike previous works that require separate devices or operated under different helicities for each functionality, our approach integrates both functionalities into one metasurface under a single CP incidence. To validate the concept, we designed a microwave meta-device and numerically characterized its performance. By engineering a series of meta-atoms with tailored reflection phases, we constructed a meta-device with pre-designed phase profiles, enabling the conversion of an incident LCP plane wave into both a focused surface wave (SW) beam and a deflected SW beam in the co- and cross-polarized output channels (Figure 1). Our results pave the way for the realization of ultra-compact devices that integrate diverse functionalities for near-field manipulations, with potential applications in near-field sensing, imaging, and on-chip photonics.

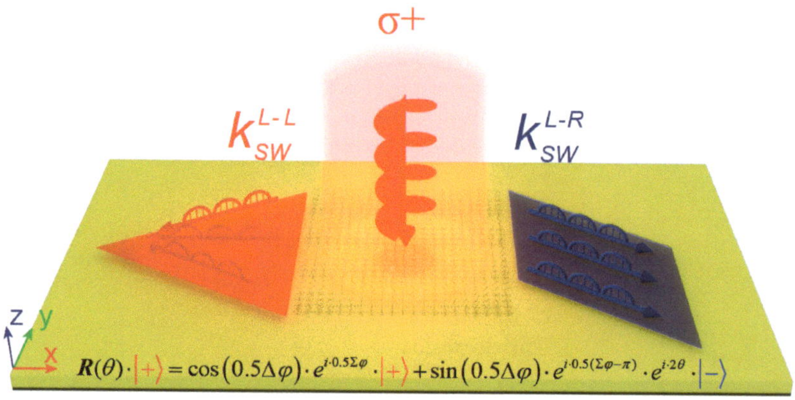

Figure 1. Schematic principle of proposed metasurfaces. Schematic of the designed bifunctional metasurface for near-field SW manipulations in both co- and cross-polarized output channels, as shined using single-helicity circularly polarized (CP) waves. The output SWs can achieve an arbitrary functionality, such as SWs with a focusing wavefront in the co-polarized output channel and SWs with a deflecting wavefront in the cross-polarized output channel.

2. The Physical Concept

We describe our strategy for designing a meta-device that exhibits bifunctional surface wave (SW) control, with distinct functionalities implemented in the co- and cross-polarized output channels under LCP wave incidence. We need to impose independent spatial functional wavefronts onto each output CP component. As discussed in Ref. [35], by introducing geometric phases and structural resonance, the electromagnetic energy carried by the co- and cross-polarized reflected fields can be fully phase-modulated with independent wavefronts, and the output can be described using the following equation:

$$R(\theta)\cdot|+\rangle = cos(0.5\Delta\varphi)\cdot e^{i\cdot 0.5\Sigma\varphi}\cdot|+\rangle + sin(0.5\Delta\varphi)\cdot e^{i\cdot 0.5(\Sigma\varphi-\pi)}\cdot e^{i\cdot 2\theta}\cdot|-\rangle \quad (1)$$

in which $R(\theta)$ represents the Jones matrix of each element with location (x, y), θ represents the rotation angle of the meta-atom derived from the rotation matrix $M(\theta) = \begin{bmatrix} cos\theta & sin\theta \\ -sin\theta & cos\theta \end{bmatrix}$, $\Delta\varphi = \varphi_{xx} - \varphi_{yy}$ is the phase difference between two orthogonal linear polarizations, and $\sum\varphi = \varphi_{xx} + \varphi_{yy}$ is the phase summation of these two polarizations. The Jones matrix of the concerned reflective metasurface can be expressed as follows, $R(0) = \begin{bmatrix} r_{xx} & 0 \\ 0 & r_{yy} \end{bmatrix}$, where $r_{xx} = R_{xx}e^{i\varphi_{xx}}$ and $r_{yy} = R_{xx}e^{i\varphi_{yy}}$ represent the reflection coefficient of the x-pol and y-pol incident, respectively. The normal wave corresponds to the mirror reflection of the incident EM waves, while the anomalous wave results from the mutual interference of scattered light with the meta-atoms. Ignoring the absorption loss, the efficiencies of the normal and anomalous waves are $R_n = \cos^2(0.5\Delta\varphi)$ and $R_a = \sin^2(0.5\Delta\varphi)$, respectively. Obviously, the condition $0.5\Delta\varphi = \varphi_{xx} - \varphi_{yy} = \pi/4$ can ensure that the output LCP and RCP channels share an equal efficiency of 50%.

It can be observed in Equation (1) that there are theoretically two CP output components resulting from the LCP wave incidence. The co-polarized component retains the same CP state as the input $|+\rangle$, while the other exhibits the opposite polarization state $|-\rangle$, representing the cross-polarized output. The phase profile imprinted onto the co-polarized component, denoted as $\Phi_1 = 0.5\sum\varphi$, indicates that the wavefront of the co-polarized output can be directly tuned through phase summation of the two orthogonal linear polarizations, which is further adjusted according to the dimensions of the meta-atom. Meanwhile, the phase profile introduced in the cross-polarized field, which can be viewed as a geometric phase assisted by the propagation phase, is represented as $\Phi_2 = 0.5(\sum\varphi - \pi) + 2\theta$. The phase pattern of the co-polarized components ϕ_2 can be further determined according to the chirality-assisted phase, indicating that the cross-polarized components require an additional degree of freedom to decouple the inherent consistency between the co- and cross-polarized output channels. Moreover, the geometric phase influences only the cross-polarized field, $\phi_{PB} = \pm 2\theta$ (with $\pm$ corresponding respectively to the LCP and RCP incident waves). In other words, the rotation angle associated with the geometric phase can be expressed as $\theta = \frac{1}{2}(\Phi_2 - \Phi_1 + \frac{\pi}{2})$.

In order to convert normally incident PWs into SWs with a specific wavefront, the metasurface needs to exhibit a specific phase profile. For example, if the phase profile is given by $\Phi(x, y) = \xi_x x + k_{SW}\left(\sqrt{y^2 + F^2} - F\right)$ or $\Phi(x, y) = \xi_x x + \xi_y |y|$, where $\xi_x > k_0$, with k_0 being the free-space wave vector, the generated SWs will acquire a wavefront of a focusing or Bessel beam. Now, our strategy is very clear: we need to design a metasurface, of which the co-polarized and cross-polarized output channels are imparted with the required phase profiles to generate SWs with a particular wavefront. Based on the analysis above, it can be inferred that we first need to design a series of meta-atoms so that their $0.5\sum\varphi$ values cover the entire 2π range. Meanwhile, by assigning different initial phases to the meta-atoms in each column, the resulting functions are combined to generate SWs with focusing beam profiles in the co-polarized output waves, while SWs with deflecting beam profiles in the cross-polarized output waves can be independently realized through additional rotation of the meta-atoms.

3. The Meta-Atom Designs

We choose the microwave regime for proving our scheme. The meta-atoms we designed are depicted in Figure 2a, which is a sandwich structure consisting of a metallic Jerusalem cross and a ground metallic plane, separated by a 1.5 mm thick dielectric spacer ($\varepsilon_r = 3 + 0.01i$). The ground metallic plane blocks all transmissions through the system and, more importantly, couples with the metallic crosses to create magnetic resonances at the frequencies dictated by the geometrical parameters [44,45]. Due to the presence of the metallic ground plate, these meta-atoms can completely reflect EM waves polarized along two cross directions (denoted as the x and y axes in the following) but with different reflection phases, φ_{xx} and φ_{yy}. Due to the resonance mechanism, the reflection phases φ_{xx} and φ_{yy} are primarily related to the dimensions of the meta-atom in the x-direction and the y-direction, respectively. Figures 2c and 2d, respectively, show the simulated $0.5\Delta\varphi$ and $0.5\Sigma\varphi$ values as a function of $l_x(l_x = l_1 + b_1)$ and $l_y(l_y = l_2 + b_2)$ at 10 GHz. To achieve bi-functional manipulation of both the co-polarized and cross-polarized output channels, a database of meta-atoms (with a period of 7 mm and a dielectric layer thickness of 1.5 mm) is established by varying the length l_x and the width l_y, as shown in these figures, to provide the necessary phase distributions for the required functionalities. Based on the analysis in the previous section, it is evident that under the input of the LCP wave, $0.5\Delta\varphi$ governs the energy ratio of the LCP and RCP output components, and $0.5\Sigma\varphi$ determines the phase imprinted onto the LCP output component. For simplicity, a series of meta-atoms at a working frequency of 10 GHz, as shown by the purple lines in Figure 2c, is chosen for constructing the metasurface, which possess the same $0.5\Delta\varphi = \pi/4$, ensuring that the output LCP and RCP channels share an equal amount of energy.

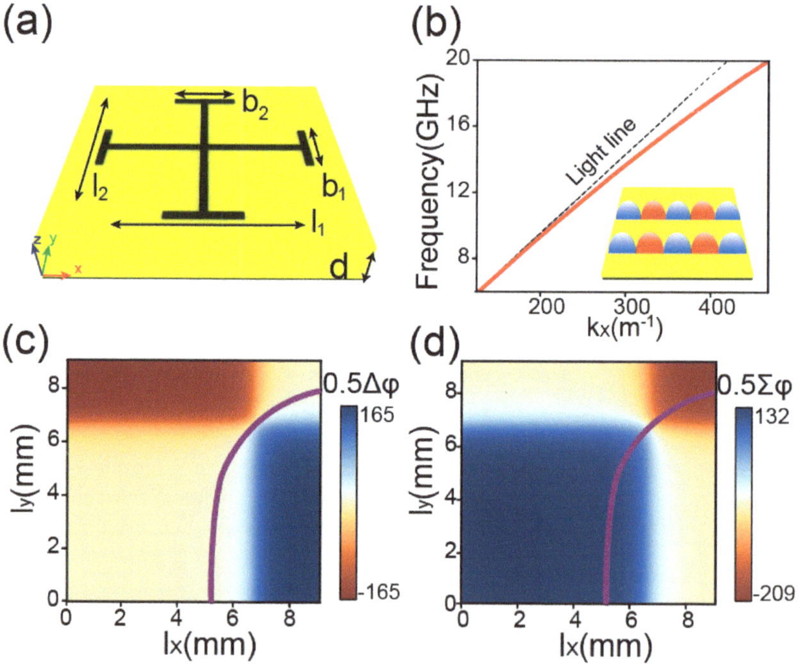

Figure 2. Characterization of the designed meta−atoms and plasmonic metal. (**a**) Geometry of the designed meta−atom (sized 7×7 mm^2) composed of the metallic Jerusalem cross structure array and a flat metal mirror separated by a 1.5 mm thick dielectric spacer. (**b**) Finite element method (FEM) −simulated dispersion relation (red line) of the eigen SWs supported by the plasmonic metal, as depicted in the inset. (**c**,**d**) Pseudocolor maps of simulated (**c**) $0.5\Delta\varphi$ and (**d**) $0.5\Sigma\varphi$ for the structure of (**a**) in a parameter space spanned by $l_x(l_x = l_1 + b_1)$ and $l_y(l_y = l_2 + b_2)$ at 10 GHz. The purple lines indicate $0.5\Delta\varphi = \pi/4$.

It is worth noting that some of the functionalities to be implemented here are related to the control of the SWs; therefore, we design a "plasmonic metal" supporting the spoof SWs in the microwave regime, which is a metallic ground plane with a 1.5 mm thick dielectric layer put on top of it (see the inset in Figure 2b). Figure 2b depicts the FEM-simulated dispersion relation of the SW modes supported by such a system, which exhibits an eigen wave vector $k_{SW} = 1.0245k_0$ with k_0 being the free-space wave vector at the frequency of 10 GHz. For a real metal at microwave frequencies, its absorption power (P_{ab}) can be calculated using $P_{ab} = \frac{1}{2}\sqrt{\pi u_0 f/\sigma}\int |H_t|^2 ds$ [46], where H_t (obtained in the PEC approximation) is the tangential component of the magnetic field on the metal's surface, σ is the metal's conductivity, f is the wave frequency, and s represents the metal's surface. Using this formula, we calculate the absorption power (P_a, caused by the metal) in the unit cell for the SSPPs and then evaluate the propagation length using $L = 1/2\alpha$, where $\alpha = P_a/2P_0$ and P_0 is the SSPP's power. The propagation length of the SSPP is found to be $L_{SSPP} = 113.64\lambda_0$.

4. Meta-Device Realizations: Microwave Simulations

With both the physical concept and the properties of our meta-atoms fully understood, we now use these meta-atoms to demonstrate a bifunctional meta-device that integrates both the focusing and deflecting effects of the SWs under LCP wave incidence at 10 GHz. The SW focusing beam can converge the electromagnetic waves to a single point, achieving high directivity. It is widely used in fields such as optical imaging and near-field microscopy, where precise control over the wave concentration is essential. On the other hand, the deflecting beam redirects the electromagnetic waves to a specific angle, offering fine control over the wave propagation. This is commonly applied in radar systems, beam steering, and other precision-guided technologies.

To achieve the bifunctional integration of SW focusing and deflection, the meta-device needs to exhibit the following chiral-dependent phase distributions under LCP wave incidence:

$$\begin{cases} \varphi^+(x,y) = 0.5\Sigma\varphi = -\xi_1 x + \xi_1\left(\sqrt{y^2+F^2}\right) \\ \varphi^-(x,y) = 0.5\Sigma\varphi + \sigma\cdot 2\theta(x,y) = -k_{sw} - k_{sw}sin\theta\cdot y \end{cases} \quad (2)$$

where $\xi_1 = 1.071k_0$, $k_{SW} = 1.0245k_0$, $\theta = 60^{\circ}$, and F = 200 mm in this case that we are concerned with. According to our analyses presented in Section 1, as depicted in Figure 1, such a meta-device can convert a normally incident LCP beam ($\sigma = +$) into two separate beams: the co-polarized component is converted into an SW, which is then focused onto a point at a distance F away from the device's center, while the cross-polarized ($\sigma = -$) component is converted into an SW, which is reflected in an oblique direction at $\theta = 60^{\circ}$. We now employ the meta-atoms designed in Sec. 3 to construct this meta-device. To obtain the phase profile $\varphi^+(x,y)$ range coverage as accurately as possible, a library of 84 meta-atoms is meticulously established based on the cells indicated by the purple lines in Figure 2d.

The desired co- and cross-polarized phase profiles in Equation (2) are depicted in Figure 3a, respectively, with linear and parabolic phase profiles in the y-direction under LCP wave incidence. For the co-polarized output component, to achieve the required phase gradient ξ_1 in the x-direction, each column contains a set of four different meta-cells with an equal spacing in $0.5\Sigma\varphi$, arranged periodically. Specifically, the first column follows the sequence ..., cell 1, cell 2, cell 3, cell 4, ..., and the second row follows the pattern ..., cell 5, cell 6, cell 7, cell 8, ... To obtain the desired parabolic phase profile in the y-direction, the $0.5\Sigma\varphi$ value of the meta-atom between each set (i.e., ..., cell 9, cell 5, cell 1, cell 5, cell 9, ...) follows a parabolic distribution. In general, the desired phase distribution of the co-polarized component determines the required sizes (l_x and l_y) of the meta-atoms at different positions. Then, by integrating the required phase distribution for the cross-polarized output component with the relationship $\phi_{PB} = 2\theta$, we can unambiguously determine the required orientation angles $\theta(x,y)$ of positions at different locations, i.e., $\theta(x,y) = \left(-k_{SW} - k_{sw}sin\theta \cdot y + \xi_1 x - \xi_1\left(\sqrt{y^2+F^2}\right)\right)/2$. Ultimately, such a strategy guides us in designing our bifunctional metasurface, with the structural details displayed in Figure 3b–d.

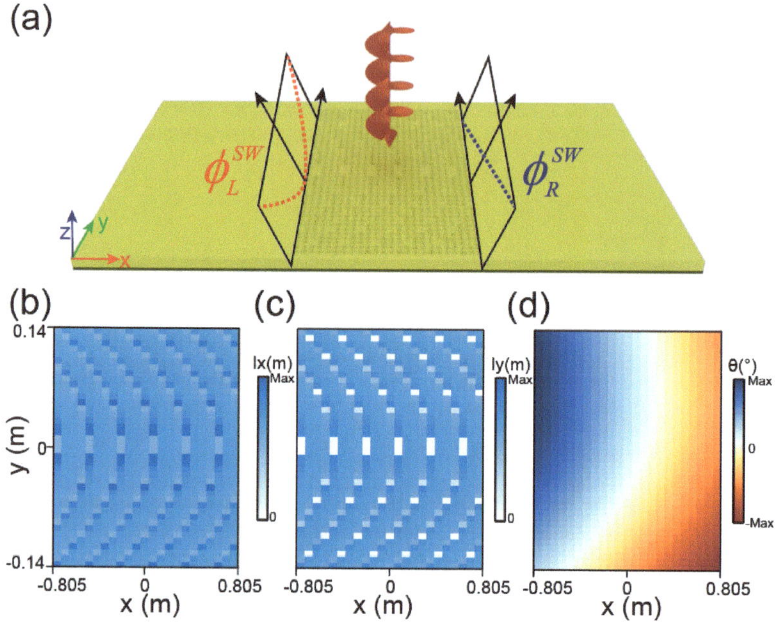

Figure 3. Structural details of the designed bifunctional meta−device. (**a**) The proposed meta-device for generating SWs with both hyperbolic and linear phase profiles dictated under LCP wave illumination. (**b**–**d**) Distributions of (**b**) $l_x(x,y)$, (**c**) $l_y(x,y)$, and (**d**) $\theta(x,y)$ of the designed bifunctional meta-device.

We next employed FEM simulations to validate our theoretical predictions. Figure 4a shows the simulated Re[Ez] field distributions on a reference plane 1 mm above the whole device under LCP wave incidence. The simulated patterns clearly demonstrate that the incident LCP wave is first converted into SWs, which are then focused onto a focal point in the co-polarized output channel and deflected in the oblique direction in the cross-polarized output channel. Moreover, to characterize the performance of SW focusing further, Figure 4b presents the near-field $|E_y|^2/|E_{y0}|^2$ distribution of the SWs at a numerically calculated focal length F = 196 mm, plotted against the y-coordinate, showing the excellent performance of the SW focusing beam. The other functionality, SW deflection, is also characterized by the Re[Ez] field on the xy plane with z = 1 mm, as shown in Figure 4c. This result exhibits excellent agreement with the theoretical prediction, with the SWs being directed in the $\theta = 60^{\circ}$ direction. Notably, the desired co- and cross-polarized wave vector in Equation (2) with $\xi_1 = 1.071k_0 = 1.04k_{SW}$ and k_{SW}. The eigen wave vector of the SW supported by the "plasmonic metal" is $k_{SW} = 1.0245k_0$ at this frequency. Although the wave vectors are not completely matched, the working efficiency of both output channels remains high. The FEM-simulated efficiencies of these two functionalities (i.e., SW focusing and SW deflection) are estimated as 40% and 37%, respectively. They are lower than the ideal value of 50%, which can be primarily attributed to the following reason: only spoof SWs with transverse-magnetic (TM) polarization exist on the designed "plasmonic metal", while the incident CP wave contains both TM and transverse-electric (TE) components [47].

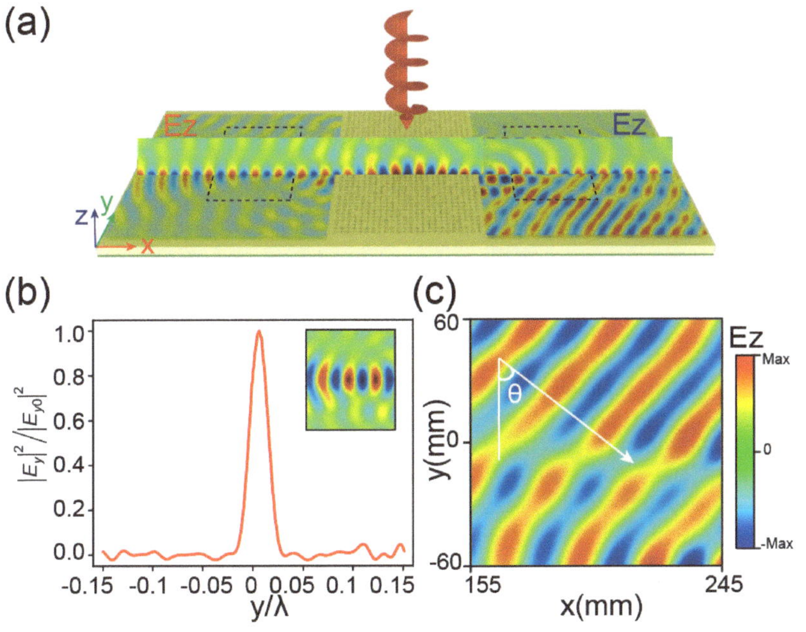

Figure 4. Bifunctional meta−device for near−field SW manipulations. (**a**) FEM−simulated near-field Re[*Ez*] patterns in the whole system, as the meta−device is illuminated by an LCP wave. (**b**) The near-field $|E_y|^2/|E_{y0}|^2$ distribution of the SWs at the focal length position versus the y coordinate with *F* = 196 mm. The inset shows part of the SW focusing in the co−polarized output channel. (**c**) The simulated near-field Re[*Ez*] profiles of the SW deflection in the cross−polarized output channel. Here, the frequency is fixed at 10 GHz.

5. Conclusions

In summary, we have proposed a novel approach that utilizes a single metasurface to simultaneously achieve bifunctional near-field manipulation of the SWs in reflection mode, under the excitation of an input circularly polarized wave with a specific helicity. By integrating two distinct mechanisms (the geometric phase and resonant mechanisms), the metasurface enables phase modulation for both the co-polarized and cross-polarized channels, thereby facilitating wavefront manipulation of the SWs. As a proof of concept, we carefully designed a meta-device and performed simulations that effectively validated the proposed scheme. Based on this design approach, the meta-device is capable of converting an incident LCP wave into focusing SWs in the co-polarized channel and deflecting SWs in the cross-polarized channel at 10 GHz. This method not only provides a compact solution for SW manipulation but also integrates both functionalities into one metasurface under a single CP incidence, which distinguishes it from previous designs that have typically required separate devices to realize different functionalities or operate diverse functionalities under different helicities. Our results open up a new avenue for the manipulation of surface waves using a single, ultra-compact meta-device, which combines small feature sizes with multifunctionality, making it well suited to future integrated optics applications.

Author Contributions: Conceptualization, S.L. and M.K.; methodology, M.K.; simulation, M.K., L.C., S.Q., L.M. and A.R. All authors have read and agreed to the published version of the manuscript.

Funding: This research was funded by the National Natural Science Foundation of China, grant number 12104401, and the Natural Science Foundation of Zhejiang Province, grant number LZ25A040002.

Institutional Review Board Statement: Not applicable.

Informed Consent Statement: Not applicable.

Data Availability Statement: The data that support the findings of this study are available from the corresponding author upon reasonable request.

Conflicts of Interest: The authors declare no conflicts of interest.

References

1. Raether, H. *Surface Plasmons on Smooth and Rough Surfaces and on Gratings*; Springer: Berlin/Heidelberg, Germany, 1988.
2. Barnes, W.L.; Dereux, A.; Ebbesen, T.W. Surface plasmon subwavelength optics. *Nature* **2003**, *424*, 824–830. [CrossRef] [PubMed]
3. Pendry, J.B.; Martín-Moreno, L.; Garcia-Vidal, F.J. Mimicking surface plasmons with structured surfaces. *Science* **2004**, *305*, 847–848. [CrossRef]
4. Maier, S.A.; Andrews, S.R.; Martín-Moreno, L.; García-Vidal, F.J. Terahertz surface plasmon-polariton propagation and focusing on periodically corrugated metal wires. *Phys. Rev. Lett.* **2006**, *97*, 176805. [CrossRef]
5. Gan, Q.; Fu, Z.; Ding, Y.J.; Bartoli, F.J. Ultrawide-bandwidth slow-light system based on THz plasmonic graded metallic grating structures. *Phys. Rev. Lett.* **2008**, *100*, 256803. [CrossRef] [PubMed]
6. Lockyear, M.; Hibbins, A.; Sambles, J. Microwave surface-plasmon-like modes on thin metamaterials. *Phys. Rev. Lett.* **2009**, *102*, 73901. [CrossRef] [PubMed]
7. Zhang, H.C.; Liu, S.; Shen, X.; Chen, L.H.; Li, L.; Cui, T.J. Broadband amplification of spoof surface plasmon polaritons at microwave frequencies. *Laser Photonics Rev.* **2015**, *9*, 83–90. [CrossRef]
8. Sun, S.; He, Q.; Xiao, S.; Xu, Q.; Li, X.; Zhou, L. Gradient-index meta-surfaces as a bridge linking propagating waves and surface waves. *Nat. Mater.* **2012**, *11*, 426–431. [CrossRef] [PubMed]
9. Ni, X.; Emani, N.K.; Kildishev, A.V.; Boltasseva, A.; Shalaev, V.M. Broadband light bending with plasmonic nanoantennas. *Science* **2012**, *335*, 427. [CrossRef]
10. Yu, N.; Genevet, P.; Kats, M.A.; Aieta, F.; Tetienne, J.-P.; Capasso, F.; Gaburro, Z. Light propagation with phase discontinuities: Generalized laws of reflection and refraction. *Science* **2011**, *334*, 333–337. [CrossRef] [PubMed]
11. Kowerdziej, R.; Ferraro, A.; Zografopoulos, D.C.; Caputo, R. Soft-Matter-Based Hybrid and Active Metamaterials. *Adv. Opt. Mater.* **2022**, *10*, 2200750. [CrossRef]
12. Lio, G.E.; Ferraro, A.; Zappone, B.; Parka, J.; Schab-Balcerzak, E.; Umeton, C.P.; Riboli, F.; Kowerdziej, R.; Caputo, R. Unlocking Optical Coupling Tunability in Epsilon-Near-Zero Metamaterials Through Liquid Crystal Nanocavities. *Adv. Opt. Mater.* **2023**, *12*, 2302483. [CrossRef]
13. Li, S.; Dong, S.; Yi, S.; Pan, W.; Chen, Y.; Guan, F.; Guo, H.; Wang, Z.; He, Q.; Zhou, L.; et al. Broadband and high-efficiency spin-polarized wave engineering with PB metasurfaces. *Opt. Express* **2020**, *28*, 15601–15610. [CrossRef] [PubMed]
14. Bomzon, Z.; Biener, G.; Kleiner, V.; Hasman, E. Space-variant Pancharatnam-Berry phase optical elements with computer-generated subwavelength gratings. *Opt. Lett.* **2002**, *27*, 1141–1143. [CrossRef] [PubMed]
15. Shitrit, N.; Bretner, I.; Gorodetski, Y.; Kleiner, V.; Hasman, E. Optical Spin Hall Effects in Plasmonic Chains. *Nano Lett.* **2011**, *11*, 2038–2042. [CrossRef] [PubMed]
16. Sun, S.; Yang, K.Y.; Wang, C.M.; Juan, T.-K.; Chen, W.T.; Liao, C.Y.; He, Q.; Xiao, S.; Kung, W.-T.; Guo, G.-Y.; et al. High efficiency broadband anomalous reflection by gradient meta-surfaces. *Nano Lett.* **2012**, *12*, 6223–6229. [CrossRef] [PubMed]
17. Grady, N.K.; Heyes, J.E.; Chowdhury, D.R.; Zeng, Y.; Reiten, M.T.; Azad, A.K.; Taylor, A.J.; Dalvit, D.A.R.; Chen, H.-T. Terahertz Metamaterials for Linear Polarization Conversion and Anomalous Refraction. *Science* **2013**, *340*, 1304–1307. [CrossRef]
18. Yin, X.; Ye, Z.; Rho, J.; Wang, Y.; Zhang, X. Photonic Spin Hall Effect at Metasurfaces. *Science* **2013**, *339*, 1405–1407. [CrossRef] [PubMed]
19. Luo, W.; Xiao, S.; He, Q.; Sun, S.; Zhou, L. Photonic Spin Hall Effect with Nearly 100% Efficiency. Adv. *Opt. Mater.* **2015**, *3*, 1102–1108. [CrossRef]
20. Luo, W.; Sun, S.; Xu, H.X.; He, Q.; Zhou, L. Transmissive Ultrathin Pancharatnam-Berry Metasurfaces with Nearly 100% Efficiency. *Phys. Rev. Appl.* **2017**, *7*, 044033. [CrossRef]
21. Zheng, G.; Mühlenbernd, H.; Kenney, M.; Li, G.; Zentgraf, T.; Zhang, S. Metasurface Holograms Reaching 80% Efficiency. *Nat. Nanotechnol.* **2015**, *10*, 308–312. [CrossRef]
22. Chen, W.T.; Yang, K.; Wang, C.; Huang, Y.-W.; Sun, G.; Chiang, I.-D.; Liao, C.Y.; Hsu, W.-L.; Lin, H.T.; Sun, S.; et al. High-Efficiency Broadband Meta-Hologram with Polarization-Controlled Dual Images. *Nano Lett.* **2014**, *14*, 225–230. [CrossRef]
23. Deng, Z.-L.; Deng, J.; Zhuang, X.; Wang, S.; Li, K.; Wang, Y.; Chi, Y.; Ye, X.; Xu, J.; Wang, G.P.; et al. Diatomic Metasurface for Vectorial Holography. *Nano Lett.* **2018**, *18*, 2885–2892. [CrossRef] [PubMed]

24. Chen, X.; Huang, L.; Mühlenbernd, H.; Li, G.; Bai, B.; Tan, Q.; Jin, G.; Qiu, C.-W.; Zhang, S.; Zentgraf, T. Dual-polarity plasmonic metalens for visible light. *Nat. Commun.* **2012**, *3*, 1198. [CrossRef]
25. Ding, X.; Monticone, F.; Zhang, K.; Zhang, L.; Gao, D.; Burokur, S.N.; de Lustrac, A.; Wu, Q.; Qiu, C.-W.; Alù, A. Ultrathin Pancharatnam-Berry metasurface with maximal cross-polarization efficiency. *Adv. Mater.* **2015**, *27*, 1195–1200. [CrossRef] [PubMed]
26. Li, S.; Tsakmakidis, K.L.; Jiang, T.; Shen, Q.; Zhang, H.; Yan, J.; Sun, S.; Shen, L. Unidirectional guided-wave-driven metasurfaces for arbitrary wavefront control. *Nat. Commun.* **2024**, *15*, 5992. [CrossRef]
27. Genevet, P.; Yu, N.; Aieta, F.; Kats, M.A.; Blanchard, R.; Gaburro, Z.; Capasso, F. Ultra-thin plasmonic optical vortex plate based on phase discontinuities. *Appl. Phys. Lett.* **2012**, *100*, 013101. [CrossRef]
28. Wang, Z.; Dong, S.; Luo, W.; Jia, M.; Liang, Z.; He, Q.; Sun, S.; Zhou, L. High-efficiency generation of Bessel beams with transmissive metasurfaces. *Appl. Phys. Lett.* **2018**, *112*, 191901. [CrossRef]
29. Pfeiffer, C.; Grbic, A. Controlling vector Bessel beams with metasurfaces. *Phys. Rev. Appl.* **2014**, *2*, 044012. [CrossRef]
30. Mehmood, M.Q.; Mei, S.; Hussain, S.; Huang, K.; Siew, S.Y.; Zhang, L.; Zhang, T.; Ling, X.; Liu, H.; Teng, J.; et al. Visible-frequency metasurface for structuring and spatially multiplexing optical vortices. *Adv. Mater.* **2016**, *28*, 2533–2539. [CrossRef] [PubMed]
31. Huang, Y.; Xiao, T.; Chen, S.; Xie, Z.; Zheng, J.; Zhu, J.; Su, Y.; Chen, W.; Liu, K.; Tang, M.; et al. All-optical controlled-NOT logic gate achieving directional asymmetric transmission based on metasurface doublet. *Opto-Electron. Adv.* **2023**, *6*, 220073. [CrossRef]
32. Zhang, F.; Pu, M.; Li, X.; Ma, X.; Guo, Y.; Gao, P.; Yu, H.; Gu, M.; Luo, X. Extreme-Angle Silicon Infrared Optics Enabled by Streamlined Surfaces. *Adv. Mater.* **2021**, *33*, 2008157. [CrossRef] [PubMed]
33. Zhang, F.; Guo, Y.; Pu, M.; Chen, L.; Xu, M.; Liao, M.; Li, L.; Li, X.; Ma, X.; Luo, X. Meta-optics empowered vector visual cryptography for high security and rapid decryption. *Nat. Commun.* **2023**, *14*, 1946. [CrossRef] [PubMed]
34. Dong, S.; Li, S.; Ling, X.; Hu, G.; Li, Y.; Zhu, H.; Zhou, L.; Sun, S. Broadband spin-unlocked metasurfaces for bifunctional wavefront manipulations. *Appl. Phys. Lett.* **2022**, *120*, 180404. [CrossRef]
35. Yuan, Y.; Sun, S.; Chen, Y.; Zhang, K.; Ding, X.; Ratni, B.; Wu, Q.; Burokur, S.N.; Qiu, C.-W. A fully phase-modulated metasurface as an energy-controllable circular polarization router. *Adv. Sci.* **2020**, *7*, 2001437. [CrossRef] [PubMed]
36. Yu, N.; Genevet, P.; Aieta, F.; Kats, M.A.; Blanchard, R.; Aoust, G.; Tetienne, J.-P.; Gaburro, Z.; Capasso, F. Flat optics: Controlling wavefronts with optical antenna metasurfaces. *IEEE J. Sel. Top. Quantum Electron.* **2013**, *19*, 4700423.
37. Huang, L.; Chen, X.; Bai, B.; Tan, Q.; Jin, G.; Zentgraf, T.; Zhang, S. Helicity dependent directional surface plasmon polariton excitation using a metasurface with interfacial phase discontinuity. *Light Sci. Appl.* **2013**, *2*, e70. [CrossRef]
38. Pors, A.; Nielsen, M.G.; Bernardin, T.; Weeber, J.C.; Bozhevolnyi, S.I. Efficient unidirectional polarization-controlled excitation of surface plasmon polaritons. *Light Sci. Appl.* **2014**, *3*, e197. [CrossRef]
39. Sun, W.; He, Q.; Sun, S.; Zhou, L. High-efficiency surface plasmon meta couplers: Concept and microwave-regime realizations. *Light Sci. Appl.* **2016**, *5*, e16003. [CrossRef]
40. Li, S.; Wang, Z.; Dong, S.; Yi, S.; Guan, F.; Chen, Y.; Guo, H.; He, Q.; Zhou, L.; Sun, S. Helicity-Delinked Manipulations on Surface Waves and Propagating Waves by Metasurfaces. *Nanophotonics* **2020**, *9*, 3473–3481. [CrossRef]
41. Wang, Z.; Li, S.; Zhang, X.; Feng, X.; Wang, Q.; Han, J.; He, Q.; Zhang, W.; Sun, S.; Zhou, L. Excite spoof surface plasmons with tailored wavefronts using high-efficiency terahertz metasurfaces. *Adv. Sci.* **2020**, *7*, 2000982. [CrossRef]
42. Li, S.; Tu, W.; Zhang, H.; Yan, J.; Shen, L. Surface wave control via unidirectional surface magnetoplasmon waveguide arrays. *Opt. Mater. Express* **2024**, *14*, 996–1007. [CrossRef]
43. Dong, S.; Zhang, Y.; Guo, H.; Duan, J.; Guan, F.; He, Q.; Zhao, H.; Zhou, L.; Sun, S. Highly Efficient Wave-Front Reshaping of Surface Waves with Dielectric Metawalls. *Phys. Rev. Appl.* **2018**, *9*, 4032. [CrossRef]
44. Sievenpiper, D.; Broas, R.F.J.; Alexopoulos, N.G.; Yablonovitch, E. High-Impedance Electromagnetic Surfaces with a Forbidden Frequency Band. *Microw. Theory Tech. IEEE Trans.* **1999**, *47*, 2059. [CrossRef]
45. Hao, J.M.; Zhou, L.; Chan, C.T. An effective-medium model for high-impedance surfaces. *Appl. Phys. A* **2007**, *87*, 281–284. [CrossRef]
46. Kong, J.A. *Electromagnetic Wave Theory*; EMW Publishing: Boston, MA, USA, 2008.
47. Duan, J.; Guo, H.; Dong, S.; Cai, T.; Luo, W.; Liang, Z.; He, Q.; Zhou, L.; Sun, S. High-efficiency chirality-modulated spoof surface plasmon meta-coupler. *Sci. Rep.* **2017**, *7*, 1354. [CrossRef]

photonics

Article

A Simple Polarization-Based Fringe Projection Profilometry Method for Three-Dimensional Reconstruction of High-Dynamic-Range Surfaces

Xiang Sun [1,2,*], Zhenjun Luo [1], Shizhao Wang [1], Jianhua Wang [1], Yunpeng Zhang [1] and Dandan Zou [1]

1 School of Electrical and Automation Engineering, East China Jiaotong University, Nanchang 330013, China; 2023028080800021@ecjtu.edu.cn (Z.L.); 8362@ecjtu.edu.cn (S.W.); 1398@ecjtu.edu.cn (J.W.); 2023028080800005@ecjtu.edu.cn (Y.Z.); ddzou@ecjtu.edu.cn (D.Z.)

2 Shanghai Engineering Research Center of Ultra-Precision Optical Manufacturing, School of Information Science and Technology, Fudan University, Shanghai 200438, China

* Correspondence: 3313@ecjtu.edu.cn

Abstract: Three-dimensional (3D) reconstruction of high-dynamic-range (HDR) surfaces plays an important role in the fields of computer vision and image processing. Traditional 3D measurement methods often face the risk of information loss when dealing with surfaces that have HDR characteristics. To address this issue, this paper proposes a simple 3D reconstruction method, which combines the features of non-overexposed regions in polarized and unpolarized images to improve the reconstruction quality of HDR surface objects. The optimum fringe regions are extracted from images with different polarization angles, and the non-overexposed regions in normally captured unpolarized images typically contain complete fringe information and are less affected by specular highlights. The optimal fringe information from different polarized image groups is gradually used to replace the incorrect fringe information in the unpolarized image, resulting in a complete set of fringe data. Experimental results show that the proposed method requires only 24~36 images and simple phase fusion to achieve successful 3D reconstruction. It can effectively mitigate the negative impact of overexposed regions on absolute phase calculation and 3D reconstruction when reconstructing objects with strongly reflective surfaces.

Keywords: high dynamic range; three-dimensional reconstruction; polarized image; unpolarized image; phase fusion; absolute phase

Received: 7 December 2024
Revised: 23 December 2024
Accepted: 30 December 2024
Published: 31 December 2024

Citation: Sun, X.; Luo, Z.; Wang, S.; Wang, J.; Zhang, Y.; Zou, D. A Simple Polarization-Based Fringe Projection Profilometry Method for Three-Dimensional Reconstruction of High-Dynamic-Range Surfaces. *Photonics* **2025**, *12*, 27. https://doi.org/10.3390/photonics12010027

1. Introduction

Due to advantages such as high speed, non-contact capability, and simple hardware configuration, visual 3D reconstruction is widely applied in various fields, including defect detection, industrial part quality control, and reverse engineering [1–4]. Existing structured light methods are generally suitable for measuring the morphology of diffuse surfaces. However, when measuring HDR surfaces, it is challenging to balance high-reflectivity and low-response areas due to the limitations of the camera's intensity response range under a single exposure. Reconstructing the 3D shape of HDR objects is hindered by challenges in processing overexposed regions. Overexposure typically occurs in HDR scenes, leading to the loss of detail in certain areas of the image, which severely affects the accuracy of the reconstruction results. Traditional reconstruction techniques often fail to effectively restore the geometric features of these areas, limiting their applicability in practical applications. Therefore, the measurement of HDR objects is one of the main challenges in 3D reconstruction [5]. To overcome this challenge, various studies have proposed corresponding solutions [6]. For example, Hu et al. [7] proposed an efficient adaptive fringe projection method that uses unaligned gray code patterns to eliminate phase errors caused by low-frequency fringe patterns and employs the images captured during the coordinate mapping process for optimal projection intensity calculation and 3D reconstruction. Sun et al. [8] proposed a method that effectively suppresses highlights in single texture-rich images based on wavelet transform and fusion strategy while preserving texture details. This method effectively suppresses highlights in texture-rich visual detection images with extensive highlight regions. Zhu et al. [9] proposed a method for HDR surface 3D reconstruction that utilizes a shared demodulated phase unwrapping mechanism and a multi-indicator fusion strategy. By sharing the decoded results of misaligned gray codes (MGCs), the method unwraps the wrapped phases of fringe images with varying intensities. Ultimately, a fusion model is established to generate the optimal phase map based on exposure quality, phase smoothness, and pixel efficiency. Other solutions for measuring HDR objects include the use of polarized light [10] and multiple cameras [11]; however, these methods rely on additional auxiliary hardware to enhance system performance, which may lead to increased costs and computational complexity due to the inclusion of multi-camera or polarized camera setups. Sun et al. [12] proposed a prior-guided restoration method for removing specular highlights in Fringe Projection Profilometry (FPP) images, enabling accurate 3D shape reconstruction. However, this method is effective only for small-area highlights and performs poorly in regions with large-area overexposure.

Multi-exposure fusion significantly enhances image quality under various lighting conditions by expanding dynamic range, improving detail, reducing noise, and increasing color accuracy. Consequently, some scholars have proposed using multiple exposures for the 3D reconstruction of HDR surfaces. For instance, Li et al. [13] proposed an exposure mapping fusion method for precise 3D reconstruction of HDR surfaces. Unlike existing methods based on image intensity, this new approach utilizes exposure maps for 3D reconstruction, thereby avoiding the issue of intensity saturation caused by insufficient camera measurement range; at the same time, the fusion of frequency components from multiple exposure maps significantly improves the signal-to-noise ratio. Wang et al. [14] maximize the projection stripe intensity to ensure a good signal-to-noise ratio, capturing two sets of stripes in advance to estimate the exposure time interval and selecting appropriate exposure times along with high-speed projection techniques. Suitable pixels are extracted from the stripe sequence for 3D reconstruction. Wang et al. [15] utilize the intensity responses of the RGB channels in color images to project a uniform blue image onto the object's surface, dividing it into three regions with different reflectance. It applies a crosstalk coefficient function to obtain the intensity of overexposed areas and calculates the optimal exposure time through the camera's linear photometric response. Ultimately, the blue stripe patterns corresponding to the three sets of optimal exposure times are fused into a combined HDR image to compute the absolute phase. Tang et al. [16] proposed a structured light technique based on automatic exposure, which determines the exposure time and captures phase-shift images by analyzing the relationship between phase errors caused by noise and intensity modulation. Ultimately, it extracts the brightest unsaturated pixels to obtain the composite phase-shift image. Additionally, some researchers [17,18] use a color camera to divide a color image into three monochrome images with different intensities to achieve this objective.

Polarization can significantly eliminate image highlights and is widely used as an efficient solution for detecting high-reflectivity targets. Lee et al. [19] improved a microscope system using linear polarizers, adjusting the polarizers to 90 degrees and 270 degrees to eliminate most specular components. To obtain HDR images, some researchers have used polarization cameras [20] and PFA cameras [21] to achieve this. However, in 3D reconstruction systems, using these two types of cameras not only increases the cost of the experiments, but also complicates the calibration process of the 3D reconstruction system. Some researchers have used CNN to obtain HDR images [22–25]. Beak et al. [26] proposed a polarization BRDF (pBRDF) model that describes the mixed polarization of specular and diffuse reflection. Based on this model, a new inverse rendering method is introduced, which jointly optimizes the pBRDF and normals to address the ill-posed problem in inverse rendering. Specular reflections hinder the use of 3D laser scanners, while phase-shifting-based 3D systems are more suitable for inspecting low-curvature surfaces. Therefore, Morel et al. [27] proposed an automatic 3D detection system based on polarization analysis, which obtains surface normal information by analyzing the polarization state of reflected light. Using an improved relaxation algorithm, the object shape is reconstructed based on the polarization data. This method has a high computational complexity, especially when the data volume is large, the object shape is complex, or high precision is required. To solve the errors in depth perception caused by high-reflectivity surfaces, Huang et al. [28] designed a polarization-structured light 3D sensor to address these issues. Although the results are very promising, its practical application is limited due to high technical requirements and development costs. Huang et al. [29] proposed a method that can estimate the degree of linear polarization (DOLP) in the scene, using DOLP to differentiate between targets and selectively reconstruct them, thereby enhancing the advantages of 3D reconstruction for the targets. However, this method is not suitable for the reconstruction of HDR objects.

In the current application of polarization techniques in HDR imaging and 3D reconstruction, many methods still face significant limitations, such as reliance on expensive equipment [20,21], high computational complexity [27], or challenges in processing HDR scenes [29]. The method proposed demonstrates several notable advantages in addressing these issues. First, the method efficiently combines non-overexposed regions of polarized and unpolarized images. Unlike traditional methods that rely solely on polarized or HDR images, the proposed approach can accurately extract surface details in the presence of specular highlights, while minimizing the loss of surface information due to overexposure. By utilizing both types of images simultaneously, the method effectively reduces the impact of overexposed areas on the result. Second, the optimization of fringe information is another key contribution of this method. By selecting the best regions from multiple sets of polarized images and applying them to the erroneous fringe data in the unpolarized image, the method ensures that the fringe information in the final image is more accurate. Finally, the implementation of the method is relatively simplified, as it does not depend on specialized polarization cameras or complex hardware configurations. This makes it highly feasible and cost-effective for practical applications. By focusing on the combination of the advantages of polarized and unpolarized images, the method reduces the need for complex hardware, simplifies system design, and improves overall implementation efficiency. In summary, the proposed method successfully overcomes the limitations of traditional approaches in overexposed region reconstruction by efficiently combining polarized and unpolarized imaging, providing a more practical, efficient, and cost-effective solution for 3D reconstruction of high-dynamic-range surfaces.

2. Principle of HDR Object Reconstruction Method

2.1. Reasons for Overexposed Areas in HDR Object Measurement

When measuring HDR objects, if the intensity of the stripe projection is too weak, the camera will be unable to fully capture stripe information from low-reflection areas. On the other hand, if the stripe projection intensity is too strong, it may cause large overexposed regions in highly reflective areas. Therefore, the relationship between the camera's exposure time and the stripe projection intensity should generally be determined before acquiring the stripe images. When determining the camera's exposure time, the brightness of the projector should allow the camera to effectively capture stripe information from the non-overexposure regions (R_F). Meanwhile, due to the characteristics of HDR objects, highly reflective areas may result in the formation of overexposed regions (R_G), as shown in Figure 1. The overexposure area will result in incomplete stripe information within that area, making accurate phase calculation impossible.

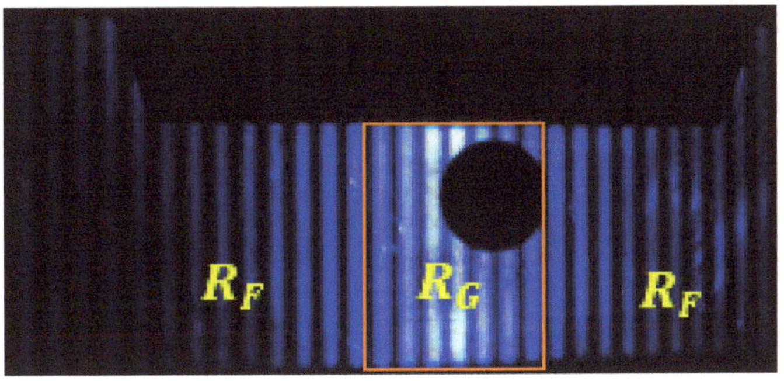

Figure 1. The overexposed region R_G and non-overexposed region R_F in the unpolarized image (from a set of images).

2.2. Analysis of Polarization Characteristics

Light waves can be categorized into linear polarization, circular polarization, and elliptical polarization. Linear polarization refers to the electric field vector vibrating along a specific direction, while circular polarization involves the electric field vector rotating in a circular path in space, and elliptical polarization is a combination of the two [30,31]. Linear polarizers effectively filter different polarization states of light, particularly scattered and reflected light. This means that only light waves aligned with the polarizer can pass through, enhancing the contrast of the stripes and making image details clearer, which aids in better identifying and analyzing the contours of the stripes. In many cases, smooth surfaces like plastic or metal can produce strong reflections that interfere with the clarity of the stripes. By placing a polarizer between the light source and the imaging device, the impact of reflected light from smooth surfaces can be significantly reduced, making the stripe patterns more distinct and minimizing ambient light noise, thereby improving the accuracy of stripe recognition. The relevant characteristics of polarization provide the theoretical basis for the proposed method.

In a fixed exposure time, phase information is not recorded in overexposed areas. In the absence of phase information, when a linear polarizer is placed in front of the camera, the measured intensity I at a specific image location or pixel can be expressed as a function of the polarizer orientation angle φ:

$$I(\varphi) = I_U + I_A \cos[2(\theta - \varphi)] = I_U\{1 + P\cos[2(\theta - \varphi)]\} \quad (1)$$

In the equation, θ represents the orientation angle of the major axis of the polarization ellipse and I_U is the total intensity at each pixel divided by 0.5. $P = I_A/I_U$ is the degree of linear polarization at that pixel. The reference axes for angles φ and θ can be chosen arbitrarily. Images are captured at three different angles of the polarizer: φ = 0, φ = 45, and φ = 90 [32,33]. From these three images, the complete information about the light's polarization state can be obtained, and the following expressions can be used to recover I_U, I_A, and θ for each pixel in the image:

$$\begin{cases} I_U = (I_0 + I_{90})/2 \\ I_A = \sqrt{(I_{45} - I_U)^2 + (I_{45} - I_U)^2} \\ \theta = \arctan[(I_{45} - I_U)/(I_{45} - I_U)]/2 \end{cases} \tag{2}$$

Here, the indices 0, 45, and 90 indicate the rotation angles of the polarizer when capturing each specific image. Since θ and $\theta + \pi$ are indistinguishable for phase-blind sensors, the meaningful range for θ is limited to π, with θ ranging from 0 to π. Therefore, when acquiring the set of polarization images, the rotation angle of the polarizer should be varied within this range to obtain the corresponding polarization images.

The absolute phase map indicates that the fringe image obtained at 0 degrees of polarization can further divide the overexposed areas in the unpolarized image into overexposed and non-overexposed regions; these overexposed regions are determined based on the pixel information in the absolute phase diagram, as shown in Figure 2a. As the polarization angle increases, a specific angle is reached at which the overexposed area in Figure 2a transitions to a non-overexposed region, as shown in Figure 2b. In general, two rounds of polarization are sufficient to capture the complete global fringe information of the object; however, if the overexposed areas in the unpolarized image are too large, multiple-angle polarization images may be needed to obtain the full global fringe information. The relevant theory provides the theoretical basis for applying linear polarizers in front of cameras and projectors to capture the fringe information in overexposed regions, with the experimental setup illustrated in Figure 3.

Figure 2. (**a**) is the overexposed region R_G and non-overexposed region R_F in the brighter polarization set images, and (**b**) is the non-overexposed region R_F in the darker polarization set images.

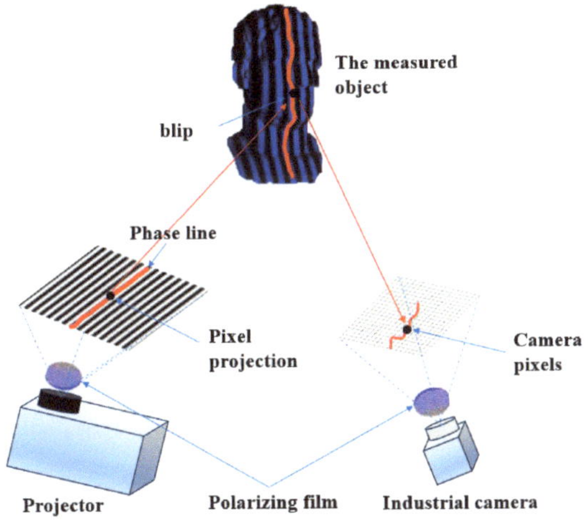

Figure 3. Experimental design diagram.

2.3. Phase Expansion and Calculation

In the three-frequency four-step phase-shift method, the intensity of the projection pattern is represented by Equation (3).

$$I(x,y) = A(x,y) + B(x,y)cos\left[\varphi(x,y) + \frac{2\pi n}{N}\right] \quad (3)$$

In the equation, $A\ (x, y)$ represents the background intensity of the fringe image, $B\ (x, y)$ is the modulation amplitude, $\varphi(x,y)$ is the phase value, n is the phase-shift index, N is the number of phase steps, and (x, y) denotes the pixel position in the projected image. In the four-step phase-shifting method, $n = 0, 1, 2, 3$ and $N = 4$, resulting in phase-shifts of 0, $\pi/2$, π, and $3\pi/2$ for the four fringe images. The intensity is represented by Equation (4).

$$\begin{cases} I_0(x,y) = A(x,y) + B(x,y)cos[\varphi(x,y)] \\ I_1(x,y) = A(x,y) - B(x,y)sin[\varphi(x,y)] \\ I_2(x,y) = A(x,y) - B(x,y)cos[\varphi(x,y)] \\ I_3(x,y) = A(x,y) + B(x,y)sin[\varphi(x,y)] \end{cases} \quad (4)$$

Based on the fringe projection images captured by the camera, the related values can be calculated using the formula shown in Equation (5):

$$\begin{cases} \varphi(x,y) = \tan^{-1}\frac{\sum_{n=1}^{N} I_n(x,y)\sin(2\pi n/N)}{\sum_{n=1}^{N} I_n(x,y)\cos(2\pi n/N)} \\ A(x,y) = \frac{\sum_{n=1}^{N} I_n(x,y)}{N} \\ B(x,y) = \frac{2}{N}\sqrt{[\sum_{n=1}^{N} I_n(x,y)\sin(2\pi n/N)]^2 + [\sum_{n=1}^{N} I_n(x,y)\cos(2\pi n/N)]^2} \end{cases} \quad (5)$$

From Equations (4) and (5), the phase $\varphi(x,y)$ in the four-step phase-shifting method can be obtained, as shown in Equation (6):

$$\varphi(x,y) = \tan^{-1}\frac{I_3(x,y) - I_1(x,y)}{I_0(x,y) - I_2(x,y)} \quad (6)$$

From the above equations, the range of $\varphi(x,y)$ is determined to be from $-\pi$ to π. To obtain a globally unambiguous absolute phase, phase unwrapping is performed. The phase unwrapping algorithm used in this paper is the multi-frequency phase unwrapping method. Based on the aforementioned equations, the phase information at three different frequencies is represented as $\varphi_1(x,y)$, $\varphi_2(x,y)$, and $\varphi_3(x,y)$. This involves calculating the phase difference between the first and second sets, as shown in Equation (7):

$$\varphi_{12}(x,y) = \begin{cases} \varphi_1(x,y) - \varphi_2(x,y) & if\ \ \varphi_1(x,y) > \varphi_2(x,y) \\ \varphi_1(x,y) + 2\pi - \varphi_2(x,y) & else \end{cases} \tag{7}$$

Similarly, the phase difference between the second and third sets can be calculated, as shown in Equation (8):

$$\varphi_{23}(x,y) = \begin{cases} \varphi_2(x,y) - \varphi_3(x,y) & if\ \ \varphi_2(x,y) > \varphi_3(x,y) \\ \varphi_2(x,y) + 2\pi - \varphi_3(x,y) & else \end{cases} \tag{8}$$

By combining Equations (7) and (8), the final phase difference can be obtained, as shown in Equation (9):

$$\varphi_{123}(x,y) = \begin{cases} \varphi_{12}(x,y) - \varphi_{23}(x,y) & if\ \ \varphi_{12}(x,y) > \varphi_{23}(x,y) \\ \varphi_{12}(x,y) + 2\pi - \varphi_{23}(x,y) & else \end{cases} \tag{9}$$

The phase unwrapping algorithm increases the measurement range by reducing the signal-to-noise ratio, so the synthesized phase map $\varphi_{\ 123}(x,y)$ is used as a reference phase to assist in the unwrapping process. The synthesized wavelength λ_{ed} and the smaller wavelength λ_d define a scaling factor. This scaling factor is applied to the phase at the carrier frequency to determine the fringe order, ultimately leading to the final absolute phase $\varphi_{\ ed}(x,y)$, as shown in Equation (10):

$$\begin{cases} K_d(x,y) = Round[\frac{(\lambda_{ed}/\lambda_d)\varphi_{\ 123}(x,y) - \varphi_1(x,y)}{N}] \\ \varphi_{\ ed}(x,y) = \varphi_1(x,y) + 2\pi K_d(x,y) \end{cases} \tag{10}$$

In the equation, $K_d\ (x,y)$ represents the integer fringe order, which plays a crucial role in phase unwrapping by helping to determine whether there is a 2π ambiguity in the phase and ensuring that the final absolute phase $\varphi_{\ ed}(x,y)$ can be correctly recovered.

2.4. Phase Fusion

Images captured by a camera cannot intuitively determine the precise location of overexposure, nor can they directly identify whether stripe information is missing in the captured images. The overexposed areas may vary slightly in images taken at different frequencies and with different phase shifts. Therefore, the best method is to assess the regions of unsuccessful phase retrieval based on the final phase retrieval results. This method is also applied to the brighter group of polarized images, where the assessment is based on the unsuccessful phase retrieval regions in the images captured without a polarizer. For the remaining darker polarized images, successful phase retrieval areas are selected step-by-step from larger regions based on the brighter group's images, until the measured object area has no unsuccessful phase retrieval regions. Finally, all successfully retrieved phase areas are merged into a single image. The phase fusion process will gradually replace the phase information in the overexposed areas with the correct phase information. To better understand, we assume that the final phase incorrect region for a unpolarized image group of an object is defined as $(y_1 : y_1^1,\ x_1 : x_1^1)$. The final phase incorrect region for a brighter polarized image group is $(y_2 : y_2^1,\ x_2 : x_2^1)$. The final phase information for the darker polarized image group in the region $(y_2 : y_2^1,\ x_2 : x_2^1)$ is correct. Subsequently, the phase information of the darker polarized image group in the region $(y_2 : y_2^1,\ x_2 : x_2^1)$ will replace the phase information of the brighter polarized image group in that region. Finally, the fused phase information from the brighter polarized image group in the region $(y_1 : y_1^1,\ x_1 : x_1^1)$ will replace the phase information of the unpolarized image group in that region. In this way, a complete global phase information can be obtained.

If we denote the phase information of the first image at position (x, y) as $I_1\ (x, y)$, and the pixel values of the second image at the same position as $I_2\ (x, y)$, we can express the replacement of pixel information in a region using the formula shown in Equation (11):

$$I_2(x,y) = f(I_1\ (x,y)) \tag{11}$$

where f is a replacement function. If the replacement is for a region R, as shown in Equation (12)

$$I_2(x,y) = \begin{cases} f(I_1(x,y)) & if\ \ (x,y) \in R \\ I_2(x,y) & else \end{cases} \tag{12}$$

this indicates that within the region R, the phase information of the pixels from the first image will replace the phase information of the pixels from the second image. The logic of the second formula is applied twice in this paper to replace the phase information of the overexposed region with the correct phase information. The flowchart of the proposed method is shown in Figure 4:

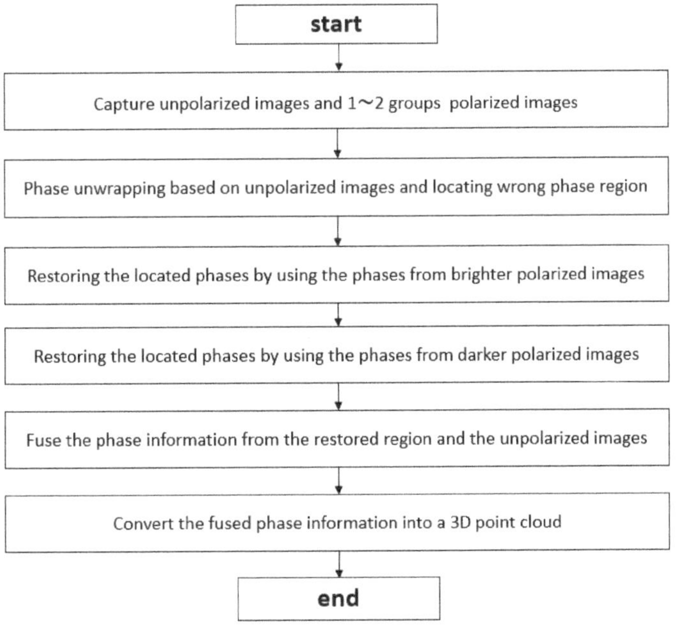

Figure 4. The flowchart of the method presented in this paper.

3. Experiment

3.1. Color Selection of Structured Light in the Three-Frequency Four-Step Method

This paper adopts the three-frequency four-step phase-shifting method for three-dimensional reconstruction. In the stripe projection profilometry system, the generated images containing sinusoidal fringe information are imported into the projector, which then projects the grating pattern onto the surface of the measured object. The camera captures the fringe information on the object's surface, and finally, MATLAB 2021b is used for data processing to obtain the phase information. Blue light has a shorter wavelength, so it scatters more strongly when interacting with small particles or surfaces [34]. This is in accordance with the principle of Rayleigh scattering, which states that shorter wavelengths scatter more than longer wavelengths. Therefore, using blue stripes can effectively reduce the scattering of ambient light. Improving the accuracy and reliability of reflection signals is beneficial for 3D reconstruction of HDR surfaces. Additionally, blue stripes exhibit significant differences in reflection characteristics across various materials (such as metal, plastic, and fabric), which helps make the stripe information clearer. For this reason, the color of structured light in this experiment is blue.

3.2. Construction of Experimental System

As shown in Figure 5, this paper presents an FPP system, where the structured light generator (projector) is a Tengju MIN-TJ20U with a resolution of 1280 × 720, and the camera's CMOS sensor is a Daheng MER-132-43U3C with a resolution of 1292 × 964. The entire system has been well calibrated using the Camera Calibration Toolbox in MATLAB [35,36].

This paper uses the four-step phase-shifting method and the three-frequency heterodyne method to calculate the absolute phase, with the stripe patterns projected by the projector having periods of 30, 32, and 35.

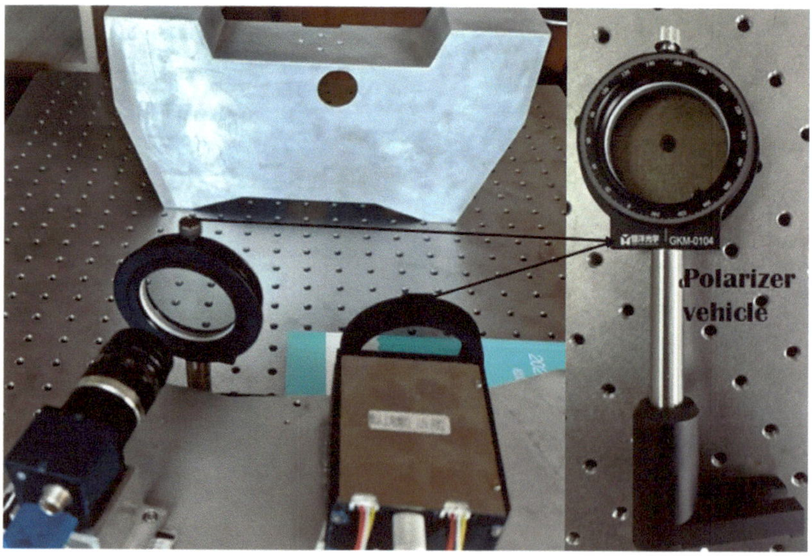

Figure 5. The physical diagram of the system built in this paper.

3.3. Three-Dimensional Reconstruction of HDR Surfaces

The metal plate used in this experiment is shown in Figure 5. First, before testing the 3D reconstruction system built in this paper, the camera exposure time and projector intensity need to be determined. The method for determining light intensity involves projecting a blue striped pattern onto the object's surface. With the camera's exposure time fixed at t = 20,000 μs, the criterion for judging light intensity is that the stripe information in the non-overexposed regions of the object can be accurately and completely captured by the camera. After determining the light intensity, a set of unpolarized stripe images is taken. Subsequently, with the camera exposure time, projection brightness, and positions of other objects unchanged, polarizers are placed in front of the camera and projector, as shown in Figure 5. When capturing the brighter set of polarized images, two polarizer angles of 0 degrees are used. For the darker set of polarized images, only the polarizer in front of the camera needs to be rotated. The number of polarized image sets is determined based on the requirement that the polarized images can complement the stripe information of the overexposed regions in the unpolarized images, as shown in Figure 6. In the polarized images, only the regions with successful phase unwrapping are selected for phase fusion.

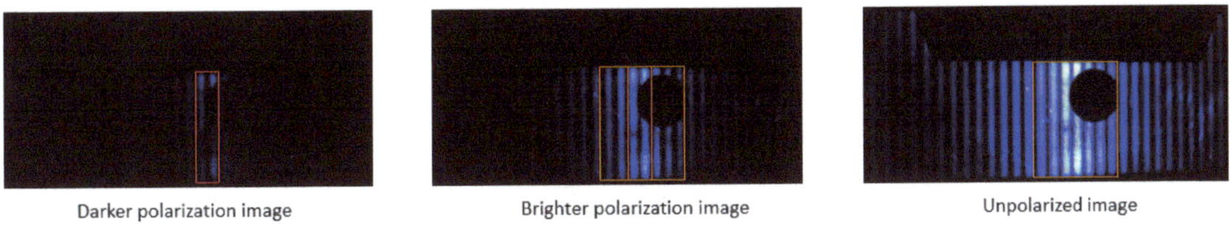

Figure 6. The sum of the non-overexposed regions in the polarized image set can cover the overexposed region in the unpolarized image.

After obtaining the required sets of unpolarized and polarized images, the blue stripe image obtained by traditional methods [37] is shown in Figure 7a; this method does not include a polarizer. The calculated absolute phase is illustrated in Figure 7b, and the 3D point cloud generated by traditional methods is displayed in Figure 7c. In the overexposed regions, pixel saturation leads to a loss of detail, making the stripe information unreliable. This results in incorrect calculations of the absolute phase and the 3D reconstruction of the point cloud. Figure 7 visually illustrates this issue.

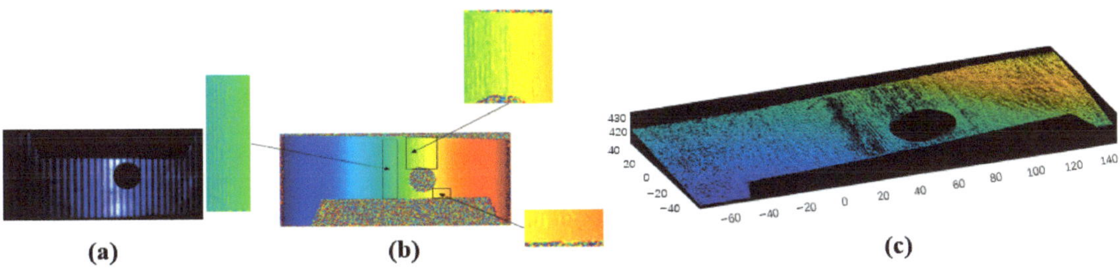

Figure 7. The unpolarized image of the metal plate is shown in (**a**), the calculated absolute phase is presented in (**b**), and the generated 3D point cloud is displayed in (**c**).

In the polarized image set, we first compute the final phase map from the brighter polarized images to identify the successful demodulation areas within the overexposed regions of the unpolarized images. Then we calculate the final phase using the darker polarized images to determine the successful demodulation areas within the overexposed regions of the brighter polarized images. After identifying these regions, to reduce the noise introduced by the polarizer, we use Fourier filtering to remove it and then calculate the fringe information for each cycle and fuse the successfully demodulated phase information. The specific steps involve capturing the blue stripe images from the unpolarized and polarized image sets, and then eliminating the overexposed areas while retaining the correct phase information. Finally, we combine the retained phase information through pixel-to-pixel stitching to create a complete global phase map. This process is roughly illustrated in Figure 8. From Figure 8, we obtain the complete phase information for each cycle, leading to the final phase map of the metal plate, as shown in Figure 9a. As can be seen, the final phase information obtained is accurate, and the 3D reconstructed point cloud further supports the feasibility of the results, as illustrated in Figure 9b. Finally, a comparative analysis of the point cloud numbers in the fusion region was conducted. In this region, the point cloud count obtained using the unpolarized image was 42,785, while the method proposed yielded 52,129 point clouds. This indicates that the proposed method effectively addresses the negative impact of overexposed areas on 3D reconstruction when dealing with strongly reflective metal objects.

To demonstrate that the fused region of the proposed method does not affect the system's accuracy, Additional experiments were conducted using an unpolarized system. These experiments involved capturing fringe images at shooting angles different from those used in the previous experiments in this paper. At these angles, the overexposed regions in the fringe images reconstructed by the proposed method should be correctly identified as non-overexposed areas, with the fringe information in the reconstructed overexposed regions being accurate and complete. No specific requirements are set for other regions. This is illustrated in Figure 10.

The point cloud map obtained from the additional experimental images, after processing with traditional methods, is shown in Figure 11.

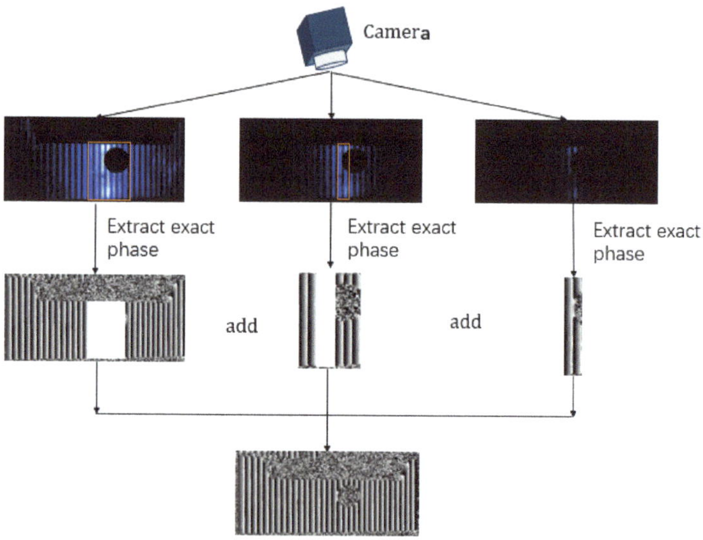

Figure 8. The fusion process of fringe information.

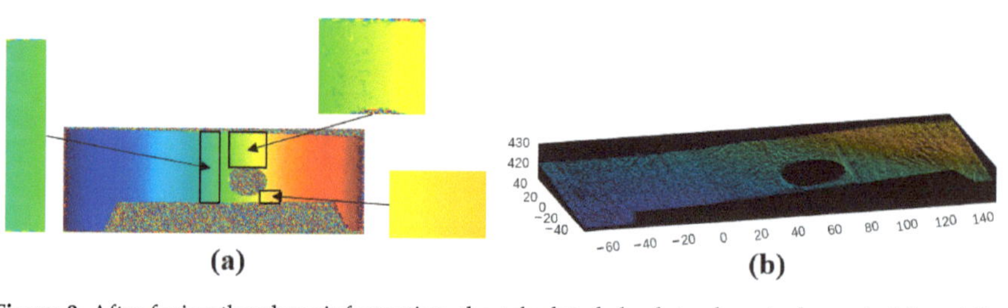

Figure 9. After fusing the phase information, the calculated absolute phase is shown in (**a**), and the generated 3D point cloud is displayed in (**b**).

Figure 10. Fringe image diagram.

From Figure 11, it can be seen that the point cloud map obtained by the system built in this paper, using traditional methods, correctly and completely reconstructs the overexposed region, which is identified as a plane. To further demonstrate that the proposed method does not affect the reconstruction accuracy, a detailed comparison was made between the point cloud map of the fused region and the point cloud map in Figure 11. Since the point cloud map corresponds to a plane, a better comparison of the corresponding region is achieved by comparing the point cloud map around the circular hole attachment, as shown in Figure 12. In Figure 12, (a) shows the point cloud map obtained from the experimental group images processed by traditional methods, (b) shows the point cloud map obtained using the proposed method, (c) highlights the region without fringe information, which corresponds to the fused region of the proposed method, and (d) shows a comparison of the same region of the two point cloud maps.

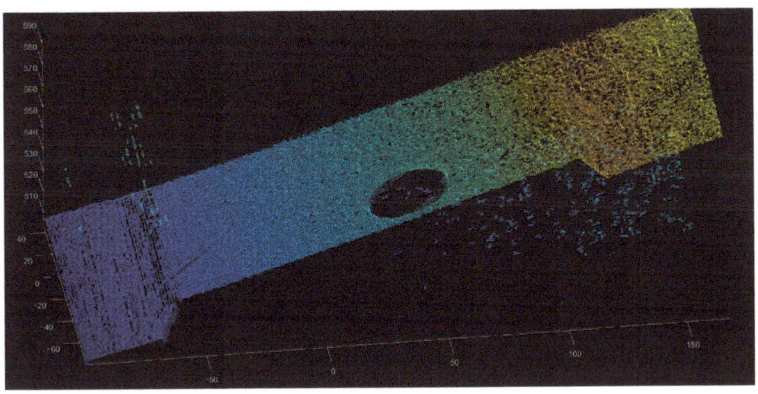

Figure 11. The point cloud maps obtained from the additional experiments.

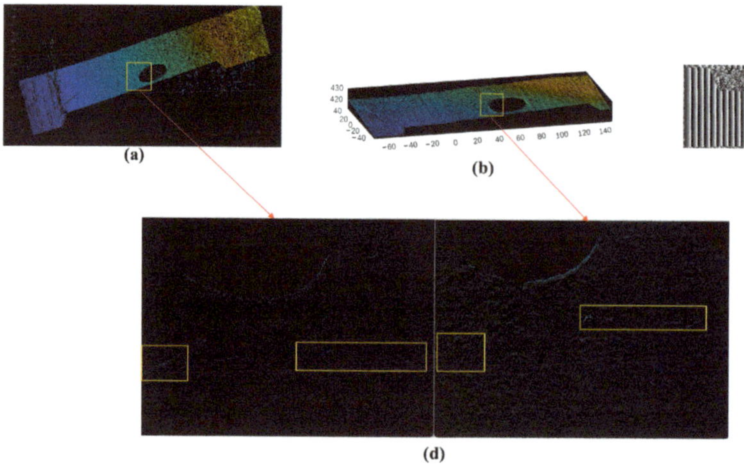

Figure 12. (**a**) is the point cloud map obtained from the experimental group images processed using traditional methods, (**b**) is the point cloud map obtained using the proposed method, (**c**) shows the region without fringe information, which corresponds to the fused region of the proposed method, and (**d**) provides a comparison of the same region from both point clouds.

From Figure 12d, it can be seen that due to differences in the shooting angles, there are some subtle variations in the details. However, in the more noticeable scratched region within the box in Figure 12d, the point cloud map of the fused region using the proposed method shows nearly identical details to the point cloud map of the surface scratches of the metal plate, as directly measured by the system built in this paper. This demonstrates that the proposed method has almost no impact on the accuracy of the system developed in this study.

To better demonstrate the feasibility of the proposed method, experiments were conducted on two additional objects with strong reflective characteristics: a kettle with an HDR surface and a metal standard part, as shown in Figure 13.

Figure 13. Physical drawings of water bottles and standard part.

When capturing the unpolarized image of the kettle, its shape resembles a cylinder, allowing us to relate the reflection of the kettle to that of a cylinder. When the light source is fixed in relation to the cylinder and parallel to its axis, the overexposed area will form a straight line along the height of the cylinder. This occurs because the side surface of the cylinder reflects light in a way that produces a linear characteristic of highlights at certain viewing angles. Consequently, the obtained unpolarized image is shown in Figure 14a. The overexposed area in the unpolarized image can lead to poor results for the absolute phase and 3D point cloud, as illustrated in Figure 14b,c. In capturing the polarized image set, we found that the overexposed area of the kettle is relatively small. By acquiring just one set of polarized images that meet the conditions, we can obtain the stripe information of the overexposed area in the unpolarized image. This indicates that the proposed method requires only 24 images to complete the 3D reconstruction of the overexposed area and the overall object when dealing with objects with smaller overexposed regions. The resulting absolute phase and 3D point cloud images are shown in Figures 14d and 14e, respectively. In the fusion region, the point cloud count obtained using the unpolarized image was 3024, whereas the method proposed resulted in 4366 point clouds. These results demonstrate that, under certain conditions, the involvement of only one set of polarized images in phase fusion does not affect the accuracy of the absolute phase and 3D point cloud.

The testing of metal standard parts can verify the feasibility of the proposed method, without affecting the accuracy of the 3D reconstruction system built in this paper. The unpolarized image of the metal standard part is shown in Figure 15a. The overexposed area can lead to poor results in the absolute phase map and 3D point cloud, as illustrated in Figure 15b,c. The absolute phase map obtained using the proposed method is shown in Figure 15d, and the final 3D point cloud is presented in Figure 15e. These images allow for a straightforward assessment of the correctness of the absolute phase map and the 3D point cloud. In the fusion region, the point cloud count obtained using the unpolarized image was 4469, whereas the method proposed resulted in 6677 point clouds, and the maximum z-direction distance of this region was compared, as shown in Table 1. The standard part has a width of 20.5 mm, and the resulting 3D point cloud further confirms the accuracy of the 3D reconstruction.

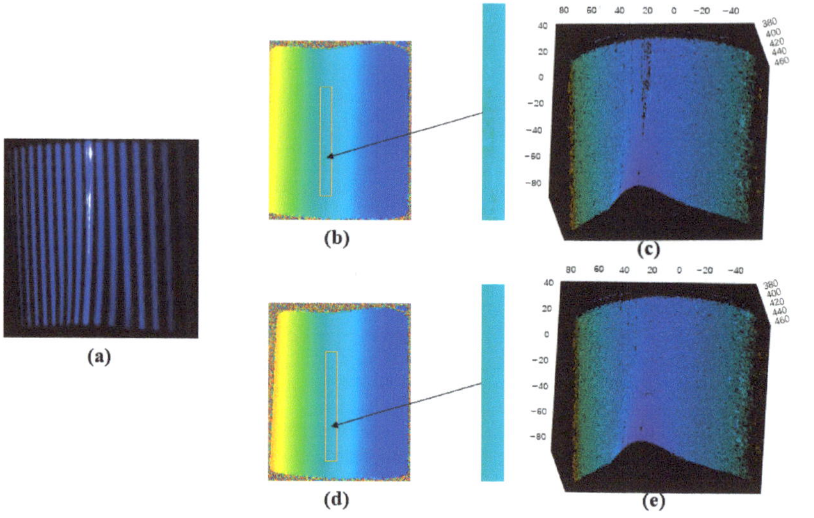

Figure 14. The unpolarized image of the kettle is shown in (**a**), the calculated absolute phase is presented in (**b**), and the generated 3D point cloud is displayed in (**c**); after fusing the phase information, the resulting absolute phase and 3D point cloud images are shown in (**d**) and (**e**), respectively.

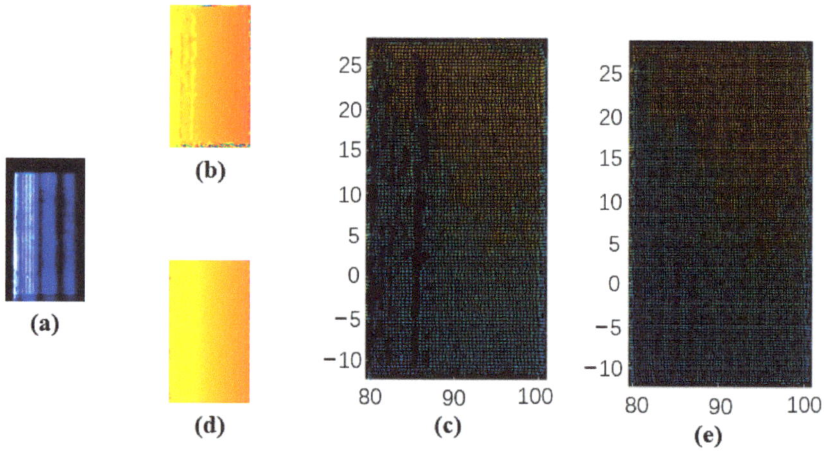

Figure 15. The unpolarized image of the standard part is shown in (**a**), the calculated absolute phase is presented in (**b**), and the generated 3D point cloud is displayed in (**c**); after fusing the phase information, the resulting absolute phase and 3D point cloud images are shown in (**d**) and (**e**), respectively.

As Table 1 shows, the average maximum z-direction distance of the proposed method is 0.1248 mm, while the reference value is 0.01 mm, and the relative error between the reference and the average value is 0.1148 mm. These errors mainly stem from measurement errors and plane fitting errors. In comparison, the original system's average maximum z-direction distance is 1.0581 mm, with a relative error of 1.0481 mm. Compared to the original system, the z-direction distortion for the proposed method is reduced by 88.2%.

Table 1. Comparison of the maximum z-direction distance in the fusion region of a standard part.

Experiment Number	Z-Coordinate (mm)		Z-Direction Distance (mm)	
	Original System	Proposed Method	Original System	Proposed Method
1	393.9603 395.0693	394.3923 394.5182	1.1090	0.1259
2	393.6461 394.6184	394.2794 394.4031	0.9723	0.1236
3	394.0147 395.1795	394.8597 394.9853	1.1648	0.1255
4	394.1236 395.2085	394.5039 394.6327	1.0849	0.1288
5	393.8975 394.8040	394.5889 394.7086	0.9064	0.1196
6	394.1406 395.2120	394.5238 394.6551	1.0714	0.1313
7	394.0340 395.1319	394.4489 394.5676	1.0979	0.1188

The proposed method has been demonstrated through the above experiments to not affect the accuracy of the 3D reconstruction system. The higher the accuracy of the system, the higher the accuracy when applying this method. Therefore, comparing the accuracy with the related literature may not be very meaningful. The primary focus of this paper is not on the accuracy of 3D reconstruction, but rather on addressing the impact of overexposed areas in HDR objects during their 3D reconstruction. Thus, the discussion centers on the convenience and ease of operation of the proposed method. The advantages of this approach include simplicity in calculations, strong practicality, and a reduced number of required images. Currently, no scholars have integrated multiple polarized image sets with unpolarized images, so to demonstrate the advantages of the proposed method, we compare it only with multi-exposure fusion methods [13–16], as shown in Table 2.

Table 2. Comparison of the proposed method with multi-exposure fusion methods.

Method	Li [13]	Wang [14]	Wang [15]	Tang [16]	The Proposed Method
Number of cameras	2	1	1	1	1
Number of captured images	-	$12 \times 4 + 4 \times 2 = 56$	$13 \times 3 + 1 = 40$	-	24 or 36
Number of image acquisition groups	5~9	4	3	5~6	2~3

Li's method [13] utilized two cameras and fused the main frequency components of multiple exposed images with exposure times ranging from 5 to 9. Wang Jianhua's method [14] used a wide range of four exposures to ensure the effectiveness of the fringes, involving four different exposure times and two four-level phase-shifting fringe patterns. Wang Zhangying's method [15] conducted phase measurements with three exposures. Additionally, Tang's method [16] employed 5 to 6 exposure times for automatic exposure techniques to estimate surface reflectivity. In contrast, the proposed method does not require calculating the surface reflectivity of the object; it only needs to determine an optimal exposure time. This method requires only 24 images when the overexposed region in the unpolarized image is small, and only 36 images even when the overexposed region is larger, and is the simplest to implement. The results indicate that our approach utilizes fewer exposure instances and fringe patterns compared to the other four methods, making it simpler in terms of computation and experimental setup. This not only reduces overall implementation costs but also enhances the feasibility of the technology in practical applications.

After comparing the implementation costs of the proposed method with other methods, in order to demonstrate the advantages of the proposed method in terms of 3D reconstruction results, we compare the point cloud images from References [14,15] with the point cloud image obtained using the proposed method. In Figure 16, (a) shows the fringe pattern, point cloud, and local magnified image obtained using the method from Reference [14], (b) shows the fringe pattern, point cloud, and local magnified image obtained using the method from Reference [15], and (c) shows the fringe pattern, point cloud, and local magnified image obtained using the proposed method.

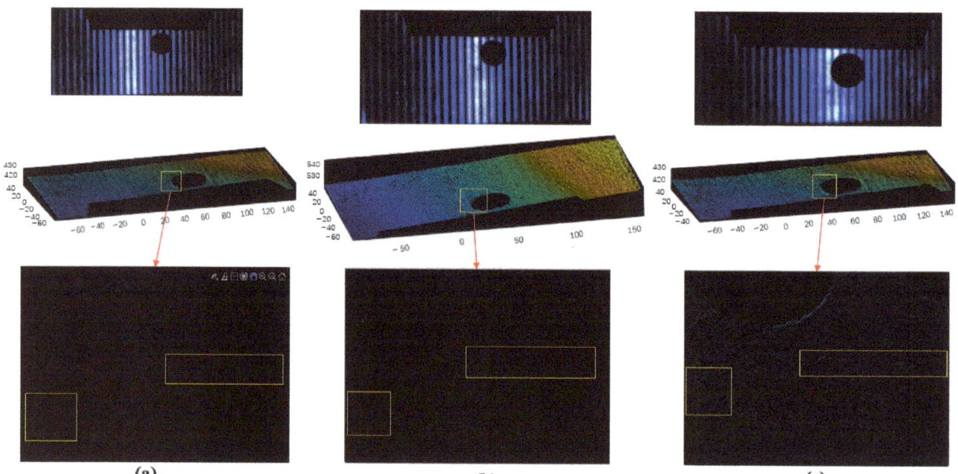

Figure 16. (**a**) shows the fringe pattern, point cloud, and local magnified image obtained using the method from Reference [14], (**b**) shows the fringe pattern, point cloud, and local magnified image obtained using the method from Reference [15], and (**c**) shows the fringe pattern, point cloud, and local magnified image obtained using the proposed method.

Although the proposed method requires the fewest images, its overall 3D reconstruction results are quite similar to those obtained from the other two methods. However, when we zoom in on the point cloud images in the region of fusion from the three methods, it becomes evident that the proposed method offers advantages in reconstructing finer details compared to the other two methods. This is because the proposed method uses a polarizer to capture fringe information in overexposed areas, while the other two methods rely on changing the exposure time to obtain fringe information in these regions. The change in exposure time may affect the camera's ability to capture detailed fringe information. In contrast, the polarizer filters out scattered light, noise, and light that is not aligned with the polarization direction, allowing the camera to capture more detailed fringe information in overexposed areas. As a result, the fringe information obtained using the polarizer contains more detailed information compared to that obtained by adjusting the exposure time.

4. Discussion

The proposed method captures fringe information in overexposed regions by utilizing a polarizer. This approach differs significantly from the methods of Wang Jianhua [14] and Wang Zhangying [15], whose methods capture fringe information in overexposed regions by adjusting the exposure time; however, this approach often results in the loss of fine fringe details, especially in cases of severe overexposure, which affects the final reconstruction quality. In contrast, a polarizer effectively filters out specular reflection lights, ensuring that fringe information is well preserved even in overexposed regions. This enables our method to outperform others, particularly regarding surface reconstruction and point cloud image details in the fused regions. Additionally, our method offers significant advantages in terms of simplicity and efficiency. Although the methods proposed by Li [13], Wang Jianhua [14], Wang Zhangying [15], and Tang [16] provide solutions for the 3D reconstruction of HDR objects, they often require more complex experimental setups, multiple exposures, or phase-modulated fringe patterns, which increases both the cost and time for data collection and experimentation. In comparison, our method reduces the number of exposures and fringe patterns used, thereby lowering the experimental complexity, cost, and time, making it more feasible for practical applications. Experimental results show that the point cloud image reconstruction performance of our method is comparable to that of Wang Jianhua [14] and Wang Zhangying [15], with some advantages in finer details. Particularly in overexposed areas, we can recover more detailed information, further proving that our method does not sacrifice quality by reducing the number of images. This performance strongly supports the potential of our method as a viable alternative to traditional, complex, and expensive techniques. Despite these advantages, our method does have some limitations. Under extreme lighting conditions or when dealing with highly reflective surfaces (such as mirrors), our method may struggle to accurately reconstruct 3D information. Therefore, although our method demonstrates robustness in HDR object reconstruction, further optimization may be required in these special cases.

In conclusion, the method proposed provides a more efficient, simpler, and cost-effective solution for the 3D reconstruction of HDR objects compared to existing technologies. In the future, we plan to integrate deep learning techniques to optimize the extraction of polarization information and the fringe reconstruction process, aiming to improve processing speed and accuracy. Additionally, we intend to explore the combination of this method with other imaging technologies, such as laser scanning or structured light, to create a multimodal 3D reconstruction system that further expands the applicability of the method proposed.

5. Conclusions

This paper presents a simple and effective 3D reconstruction method combining polarized images with non-highlight features of unpolarized images to improve reconstruction quality under HDR conditions. By analyzing the reflection characteristics of an object's surface at different polarization angles, we extract polarization information that enhances the capture of surface details. The method effectively identifies the optimal fringe regions, ensuring the integrity of fringe information in HDR scenarios. Non-overexposed areas in the set of unpolarized images typically contain complete fringe information and are less affected by highlight interference. By gradually replacing incomplete and unclear fringe information, we achieve effective replacement of the fringes in overexposed areas, resulting in a globally complete set of fringe information.

The experimental results show that the method is capable of reconstructing the 3D information of the overexposed regions in the HDR object's fringe images, providing an effective solution for 3D reconstruction in HDR scenes. The method can perform 3D reconstruction of HDR surfaces without requiring complex algorithms or advanced experimental instruments. However, it is currently limited to the 3D reconstruction of stationary objects and cannot be applied to moving objects.

Author Contributions: Conceptualization, X.S. and Z.L.; methodology, Z.L.; software, Z.L. and D.Z.; validation, Z.L. and J.W.; formal analysis, Z.L.; investigation, S.W.; resources, X.S.; data curation, S.W.; writing—original draft preparation, Z.L.; writing—review and editing, X.S. and Z.L., visualization, Y.Z.; supervision, X.S.; project administration, X.S.; funding acquisition, X.S. All authors have read and agreed to the published version of the manuscript.

Funding: National Natural Science Foundation of China (52405591); Jiangxi Provincial Natural Science Foundation (20224BAB214053, 20242BAB25104).

Informed Consent Statement: Not applicable.

Data Availability Statement: Data underlying the results presented in this paper are not publicly available at this time but may be obtained from the authors upon reasonable request.

Conflicts of Interest: The authors declare no conflicts of interest.

References

1. Geng, J. Structured-light 3D surface imaging: A tutorial. *Adv. Opt. Photonics* **2011**, *3*, 128–160. [CrossRef]
2. Chen, F.; Brown, G.M.; Song, M. Overview of 3-D shape measurement using optical methods. *Opt. Eng.* **2000**, *39*, 10–22.
3. Yang, D.; Qiao, D.; Xia, C.; He, Q. Adaptive horizontal scaling method for speckle-assisted fringe projection profilometry. *Opt. Express* **2022**, *31*, 328–343. [CrossRef] [PubMed]
4. Dong, G.; Sun, X.; Kong, L.; Peng, X. Suppression for Phase Error of Fringe Projection Profilometry Using Outlier-Detection Model: Development of an Easy and Accurate Method for Measurement. *Photonics* **2023**, *10*, 1252. [CrossRef]
5. Song, Z.; Jiang, H.; Lin, H.; Tang, S. A high dynamic range structured light means for the 3D measurement of specular surface. *Opt. Lasers Eng.* **2017**, *95*, 8–16. [CrossRef]
6. Wang, J.; Yang, Y. A new method for high dynamic range 3D measurement combining adaptive fringe projection and original-inverse fringe projection. *Opt. Lasers Eng.* **2023**, *163*, 107490. [CrossRef]
7. Hu, J.; Zhu, J.; Zhou, P. Efficient 3D measurement of a HDR surface based on adaptive fringe projection. *Appl. Opt.* **2022**, *61*, 9028–9036. [CrossRef]
8. Sun, X.; Kong, L.; Wang, X.; Peng, X.; Dong, G. Lights off the Image: Highlight Suppression for Single Texture-Rich Images in Optical Inspection Based on Wavelet Transform and Fusion Strategy. *Photonics* **2024**, *11*, 623. [CrossRef]
9. Zhu, J.; Yang, F.; Hu, J.; Zhou, P. High dynamic reflection surface 3D reconstruction with sharing phase demodulation mechanism and multi-indicators guided phase domain fusion. *Opt. Express* **2023**, *31*, 25318–25338. [CrossRef]
10. Salahieh, B.; Chen, Z.; Rodriguez, J.J.; Liang, R. Multi-polarization fringe projection imaging for high dynamic range objects. *Opt. Express* **2014**, *22*, 10064–10071. [CrossRef]
11. Feng, S.; Chen, Q.; Zuo, C.; Asundi, A. Fast three-dimensional measurements for dynamic scenes with shiny surfaces. *Opt. Commun.* **2017**, *382*, 18–27. [CrossRef]
12. Sun, X.; Kong, L.; Wang, X. Prior-Guided Restoration of Intense Local Specular Highlight in Fringe Projection Profilometry Images. *Appl. Opt.* **2024**, *63*, 8656–8669. [CrossRef]
13. Li, J.; Guan, J.; Chen, X.; Le, X.; Xi, J. Exposure map fusion for precise 3-D reconstruction of high dynamic range surfaces. *IEEE Trans. Instrum. Meas.* **2022**, *71*, 5022911. [CrossRef]
14. Wang, J.; Zhou, Y.; Yang, Y. A novel and fast three-dimensional measurement technology for the objects surface with non-uniform reflection. *Results Phys.* **2020**, *16*, 102878. [CrossRef]
15. Wang, Z.; Li, K.; Gao, N.; Meng, Z.; Zhang, Z. High dynamic range 3D shape measurement based on crosstalk characteristics of a color camera. *Opt. Express* **2023**, *31*, 38318–38333. [CrossRef]
16. Tang, S.; Zhang, X.; Li, C.; Gu, F. High dynamic range three-dimensional shape reconstruction via an auto-exposure-based structured light technique. *Opt. Eng.* **2019**, *58*, 064108. [CrossRef]
17. Zheng, Y.; Wang, Y.; Suresh, V.; Li, B. Real-time high-dynamic-range fringe acquisition for 3D shape measurement with a RGB camera. *Meas. Sci. Technol.* **2019**, *30*, 075202. [CrossRef]
18. Liu, Y.; Fu, Y.; Zhuan, Y.; Zhong, K.; Guan, B. High dynamic range real-time 3D measurement based on Fourier transform profilometry. *Opt. Laser Technol.* **2021**, *138*, 106833. [CrossRef]
19. Lee, S.; Yoon, K.; Kim, J.; Kim, K.G. Reduction of specular reflection based on linear polarization control for fluorescence-induced diagnostic evaluation. *Diagnostics* **2022**, *12*, 1990. [CrossRef]
20. Wu, X.; Zhang, H.; Hu, X.; Shakeri, M.; Fan, C.; Ting, J. HDR reconstruction based on the polarization camera. *IEEE Robot. Autom. Lett.* **2020**, *5*, 5113–5119. [CrossRef]
21. Fatima, T.; Pistellato, M.; Torsello, A.; Bergamasco, F. One-shot hdr imaging via stereo pfa cameras. In Proceedings of the International Conference on Image Analysis and Processing, Lecce, Italy, 23–27 May 2022; pp. 467–478.

22. Pistellato, M.; Fatima, T.; Wimmer, M. Exploiting Light Polarization for Deep HDR Imaging from a Single Exposure. *Sensors* **2023**, *23*, 5370. [CrossRef]
23. Yang, X.; Xu, K.; Song, Y.; Zhang, Q.; Wei, X.; Lau, R.W. Image correction via deep reciprocating HDR transformation. In Proceedings of the IEEE Conference on Computer Vision and Pattern Recognition, Salt Lake City, UT, USA, 18–23 June 2018; pp. 1798–1807.
24. Liu, Y.-L.; Lai, W.-S.; Chen, Y.-S.; Kao, Y.-L.; Yang, M.-H.; Chuang, Y.-Y.; Huang, J.-B. Single-image HDR reconstruction by learning to reverse the camera pipeline. In Proceedings of the IEEE/CVF Conference on Computer Vision and Pattern Recognition, Seattle, WA, USA, 13–19 June 2020; pp. 1651–1660.
25. Endo, Y.; Kanamori, Y.; Mitani, J. Deep reverse tone mapping. *ACM Trans Graph.* **2017**, *36*, 1–10. [CrossRef]
26. Baek, S.-H.; Jeon, D.S.; Tong, X.; Kim, M.H. Simultaneous acquisition of polarimetric SVBRDF and normals. *ACM Trans. Graph.* **2018**, *37*, 268. [CrossRef]
27. Morel, O.; Gorria, P. Polarization imaging for 3D inspection of highly reflective metallic objects. *Opt. Spectrosc.* **2006**, *101*, 11–17. [CrossRef]
28. Huang, X.; Wu, C.; Xu, X.; Wang, B.; Zhang, S.; Shen, C.; Yu, C.; Wang, J.; Chi, N.; Yu, S. Polarization structured light 3D depth image sensor for scenes with reflective surfaces. *Nat. Commun.* **2023**, *14*, 6855. [CrossRef]
29. Huang, X.; Bai, J.; Wang, K.; Liu, Q.; Luo, Y.; Yang, K.; Zhang, X. Target enhanced 3D reconstruction based on polarization-coded structured light. *Opt. Express* **2017**, *25*, 1173–1184. [CrossRef]
30. Wang, X.-L.; Li, Y.; Chen, J.; Guo, C.-S.; Ding, J.; Wang, H.-T. A new type of vector fields with hybrid states of polarization. *Opt. Express* **2010**, *18*, 10786–10795. [CrossRef]
31. Lerman, G.M.; Stern, L.; Levy, U. Generation and tight focusing of hybridly polarized vector beams. *Opt. Express* **2010**, *18*, 27650–27657. [CrossRef]
32. Lin, S.-S.; Yemelyanov, K.M.; Pugh, E.N., Jr.; Engheta, N. Polarization enhanced visual surveillance techniques. In Proceedings of the IEEE International Conference on Networking, Sensing and Control, Taipei, Taiwan, 21–23 March 2004; pp. 216–221.
33. Yemelyanov, K.M.; Lin, S.-S.; Luis, W.Q.; Pugh, E.N., Jr.; Engheta, N. Bio-inspired display of polarization information using selected visual cues. *Polariz. Sci. Remote Sens.* **2003**, *5158*, 71–84.
34. Strutt, H.J. LVIII. On the scattering of light by small particles. *Lond. Edinb. Dublin Philos. Mag. J. Sci.* **1871**, *41*, 447–454. [CrossRef]
35. Fetić, A.; Jurić, D.; Osmanković, D. The procedure of a camera calibration using Camera Calibration Toolbox for MATLAB. In Proceedings of the 35th International Convention MIPRO, Opatija, Croatia, 21–25 May 2012; pp. 1752–1757.
36. Falcao, G.; Hurtos, N.; Massich, J. Plane-based calibration of a projector-camera system. *VIBOT Master* **2008**, *9*, 1–12.
37. Zhang, Z.H.; Towers, C.E.; Towers, D.P. Time efficient color fringe projection system for 3D shape and color using optimum 3-frequency Selection. *Opt. Express* **2006**, *14*, 6444–6455. [CrossRef] [PubMed]

photonics

Article

Double-Sided Metasurfaces for Dual-Band Mid-Wave and Long-Wave Infrared Reflectors

Yeong Hwan Ko [1] and **Robert Magnusson** [2,*]

1 Division of Electrical, Electronics, and Control Engineering, Kongju National University, Cheonan 31080, Republic of Korea; yhk@kongju.ac.kr
2 Electrical Engineering Department, University of Texas at Arlington, Arlington, TX 76016, USA
* Correspondence: magnusson@uta.edu

Abstract: We present an innovative method for dual-band mid-wave infrared (MWIR) and long-wave infrared (LWIR) reflectors. By using double-sided metasurfaces, two high reflection bands can be generated with a single device. As individual guided-mode resonance (GMR) reflectors are combined with interlayer (or substrate) on the top and bottom sides, we achieved high reflection in the MWIR and LWIR bands simultaneously. Each GMR reflector was optimized as a germanium (Ge) grating structure on a potassium bromide (KBr) substrate. In our analysis, it was found that the transparency of the interlayer is critical to produce the dual-band reflection. The simulation results on the Ge/KBr/Ge double-sided metasurfaces demonstrated wideband reflection from ~3.3 to 4.8 μm and ~8.8 to 11 μm. Additionally, the device exhibited favorable angular tolerance. The work contributes to developing capability of metasurface technologies in various application fields.

Keywords: mid-wave infrared; long-wave infrared; guided-mode resonance; dual-band reflector; double-sided metasurfaces

Citation: Ko, Y.H.; Magnusson, R. Double-Sided Metasurfaces for Dual-Band Mid-Wave and Long-Wave Infrared Reflectors. *Photonics* **2024**, *11*, 1132. https://doi.org/10.3390/photonics11121132

Received: 29 October 2024
Revised: 16 November 2024
Accepted: 26 November 2024
Published: 30 November 2024

1. Introduction

Mid-wave infrared (MWIR, $\lambda \approx 3$ to 5 μm) and long-wave infrared (LWIR, $\lambda \approx 8$ to 14 μm) technologies play important common roles in thermal imaging, spectroscopy and remote sensing [1,2]. These wavelengths correspond to transparent atmospheric windows allowing high transmission of thermal radiation. MWIR imaging systems are capable of excellent resolution with enhanced sensitivity and contrast to temperature difference [3], while LWIR systems offer high quality imaging of near-ambient temperature objects and complex environmental conditions [4]. As a synergistic approach to merging these technologies, recent efforts have focused on combining MWIR and LWIR capabilities to yield enhanced performance and broader spectral coverage [5,6].

Meanwhile, optical metasurface technologies have been rapidly advanced through the implementation of resonant photonic lattices (PLs) [7,8]. Contrasting traditional thin-film technology, the PLs enable a single-layer operation for various optical components including band-pass filters [9], wideband reflectors [10] and polarizers [11]. Particularly, for the MWIR and LWIR bands, guided-mode resonance (GMR)-fashioned PLs are alternatives to traditional multi-layer designs based on quarter-wavelength thickness [12]. However, designing optical components that simultaneously operate effectively in both the MWIR and LWIR bands remains a considerable challenging in metasurface technology. In this work, we introduce an approach for dual-band operation via double-sided metasurface structures. As an essential optical component, we focus on design and analysis of MWIR and LWIR reflectors, potentially useful for dual-band operation in various application fields.

2. Double-Side Metasurface Reflector

As shown in Figure 1, double-sided metasurfaces enable high reflection in MWIR and LWIR bands. The dual-band reflection is attributed to cooperative interaction between two distinct resonant reflectors, optimally designed for each MWIR (3–5 μm) and LWIR (8–12 μm) region. When light illuminates the device, the first metasurface reflector generates a high reflection band in the MWIR spectral region while partially reflecting light in LWIR region, as depicted in (i). In an ideal lossless system, the remaining light entirely passes through the substrate and interacts with the second reflector, as shown in (ii). Thereafter, the LWIR light is mostly reflected into air. As a result, seen in (iii), light undergoes high reflection in both MWIR and LWIR regions. Regarding infrared transparent materials such as germanium (Ge), calcium difluoride (CaF_2), potassium bromide (KBr), here, we successfully develop and analyze double-sided metasurfaces for dual-band MWIR and LWIR reflectors.

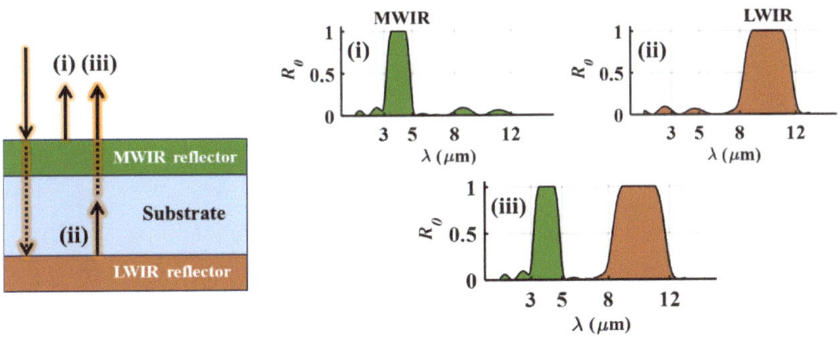

Figure 1. Concept of a double-sided metasurfaces fashioning a dual-band reflector in air. Cooperative interaction between (i) MWIR and (ii) LWIR reflectors enables dual bands of high reflection as presented in (iii).

3. Numerical Modeling

Dual-band reflectors as proposed here can be constructed using two individually optimized resonant wideband reflectors. In this work, we employ particle swarm optimization (PSO) algorithms [13] to achieve a high reflection in each 3–5 μm MWIR and 8–12 μm LWIR band. For calculation of the reflection, a rigorous coupled-wave analysis (RCWA) method [14] is used with a commercial software package (version 2023.12-SP1) [15]. As a simple construction, we treat a one-dimensional (1D) grating array based on high refractive index dielectric material where we keep 19 harmonics for reliable simulations.

Figure 2a presents the schematic of the 1D MWIR grating reflector. In this design, we incorporate a homogeneous sublayer beneath the one-dimensional (1D) grating using the same high refractive index material. This forms a zero-contrast grating (ZCG), which facilitates superior wideband reflection to achieve [16]. The grating parameter set $\{\Lambda_1, F_1, d_{g1}, d_{h1}\}$ determines the geometry of the 1D ZCG, where the Λ_1, F_1, d_{g1} and d_{h1} denote a period, fill factor, grating depth, and sublayer thickness of the first part of the metasurface. By setting the refractive indices (n_h = 4, n_s = 1.4) and using TM polarization (i.e., electric field is perpendicular to the grooves), we determine the optimal parameter set as $\{\Lambda_1$ = 1.73 μm, F_1 = 0.73, d_{g1} = 0.91 μm, d_{h1} = 0.5 μm$\}$ for the MWIR reflector. The simulated zeroth-order reflectance (R_0) spectrum demonstrates a high reflection band spanning from 3.06 to 4.85 μm, with R_0 values exceeding 0.97 remarkably achieved by the single grating layer. The wideband reflection is attributed to the merging of multiple guided-mode resonance (GMR) modes in the subwavelength regime [17]. At longer wavelengths, high reflection does not appear because GMRs are not excited.

For supporting high reflection in the LWIR region, as shown in Figure 2b, the second metasurface is designed by increasing the grating period where the optimal parameter set is $\{\Lambda_2$ = 6 μm, F_2 = 0.38, d_{g2} = 1.7 μm, d_{h2} = 0.33 μm$\}$ under TE polarization (i.e., electric field is parallel to the grooves). In the simulated R_0 spectrum, wideband reflection is observed from 8.54 to 12.22 μm, with high reflectance (R_0 > 0.99). As the second metasurface, it can efficiently reflect LWIR light transmitted from the first metasurface.

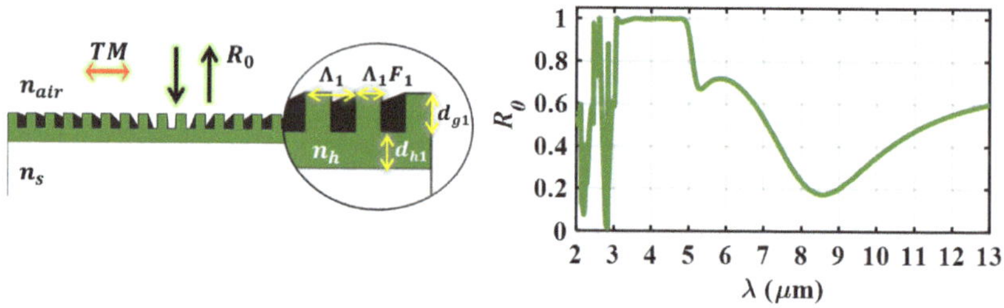

(a)

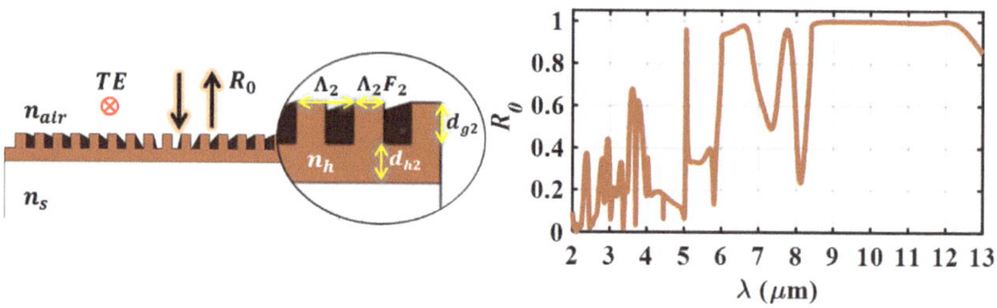

(b)

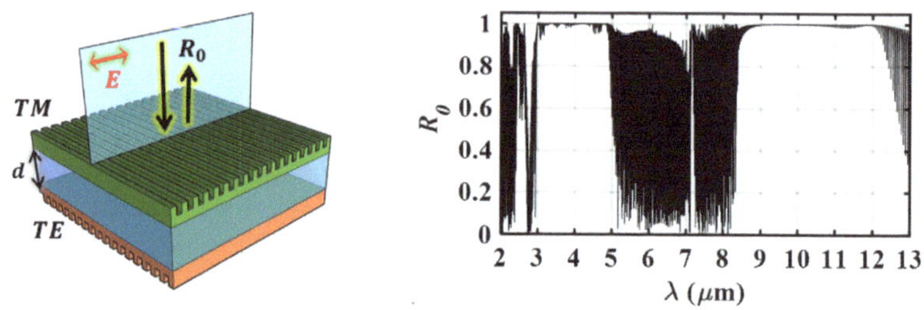

(c)

Figure 2. Modeling of a double-side metasurface for the dual-band reflector. The (**a**) first and (**b**) second metasurface for each MWIR and LWIR reflector designed by a resonant 1D grating structure. The grating parameters includes a period (Λ), fill factor (F), grating depth (d_g) and sublayer thickness (d_h). The n_h and n_s indicate high refractive index of grating layer and low refractive index of substrate. (**c**) The double-sided metasurfaces composed of each MWIR and LWIR reflector with a thick interlayer of substrate (d). In the simulated R_0 spectrum, both MWIR and LWIR reflection bands are clearly observed.

Figure 2c shows a double-sided metasurfaces composed the of two individual reflectors from Figure 2a,b. The substrate thickness (d) is set to 1 mm, which is a commonly used value. To ensure that each reflector operates with the proper polarization, two 1D grating are aligned in orthogonal directions under linearly polarized light illumination. When the incident electric field is perpendicular to the first grating grooves, the first and second metasurface interact with TM and TE polarized electromagnetic waves, respectively. The simulation result of the R_0 spectrum clearly shows dual MWIR and LWIR reflection with high fluctuations due to Fabry-Pérot (FP) resonances.

4. Analysis

To explain a mechanism of the dual-band reflector, we provide an analytical formula based on a simplified optical configuration as illustrated in Figure 3a. The total output R_0

is derived from the sum of the nth reflectance ($R_{(n)}$) between the two metasurface which can be expressed by (1) to (3).

$$R_0 = \sum_{n=0}^{\infty} R_{(n)} \quad (1)$$

$$= R_1 + R_2(1 - R_1)^2 e^{-2\alpha d} \sum_{n=0}^{\infty} \left(R_1 R_2 e^{-2\alpha d}\right)^n \quad (2)$$

$$= R_1 + \frac{R_2(1 - R_1)^2 e^{-2\alpha d}}{1 - R_1 R_2 e^{-2\alpha d}} \quad (3)$$

The R_1 and R_2 represent the reflectance of the first and the second metasurfaces, and α denotes the absorption coefficient of the interlayer (i.e., $\alpha = 4\pi k/\lambda$, k being the extinction coefficient of the complex refractive index). In the equation, a phase interference is ignored to avoid FP resonance effects.

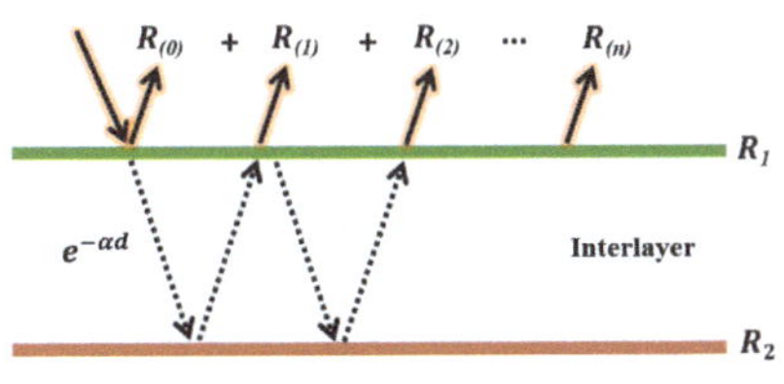

(a)

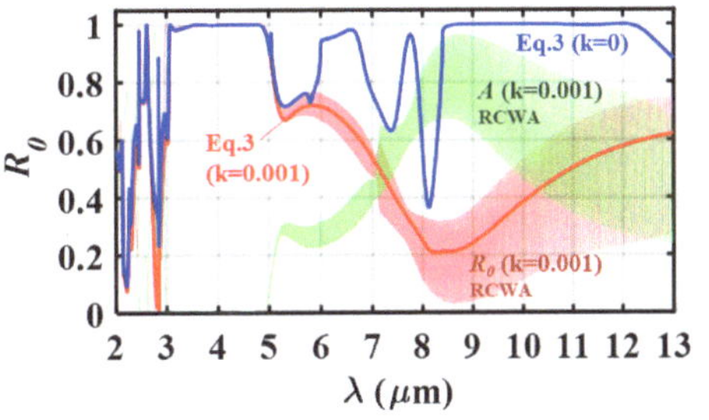

(b)

Figure 3. Analytical modeling for the double-sided metasurfaces in lossless and lossy materials. (**a**) Schematic for total reflection between two interfaces separated by a distance (d) where the R_1 and R_2 are reflectance at the first and second interface, respectively. The absorption coefficient (α) is obtained by $\alpha = 4\pi k/\lambda$ where the k is an extinction coefficient of the interlayer. (**b**) Calculation results from solving (3) for k = 0 (blue bold line) and 0.001 (red bold line) with d = 1 mm. The R_1 and R_2 are the zeroth-order reflectance of MWIR and LWIR reflector in Figure 2a,b. For comparison, the numerical simulation results are also added for k = 0.001. The thin red and green lines represent absorption (A) and R_0 spectrum.

Figure 3b shows calculation results obtained by solving (3) using the R_1 and R_2 reflectance from each R_0 spectrum of Figure 2a,b. Comparing Figure 2c, which is in case of a lossless interlayer (extinction coefficient k = 0, depicted by the blue bold line), the analytical results match to the simulation (RCWA) results, except for the dense oscillation. In realistic fabrication, where mathematically perfect parallelism between two interfaces is unlikely to be formed, FP interference effects will be diminished. Thus, the double-sided metasurfaces can cooperate the two individual reflectors without the rapid fluctuation seen in the RCWA result, producing a response similar to that predicted by the analytical model. However, when the interlayer is lossy material, the dual-band reflection capability is significantly degraded. As shown in red bold line, which indicates analytical result for k = 0.001, R_0 dramatically decreases in the LWIR region while matching the RCWA results (red thin line). This substantial reduction is caused by the high absorption of the thick interlayer (d = 1 mm) as evidenced by the absorption (A) spectrum (i.e., RCWA results, depicted by the green thin line). The LWIR light is substantially absorbed during the long propagation, multi-wavelength path in the interlayer, whereas the MWIR light is predominantly reflected not penetrating the interlayer. Unfortunately, for this reason, it is hard to operate the second reflector when utilizing lossy materials for the interlayer.

5. Design and Results

As investigated in the previous section, transparency of the interlayer is critical to achieve the dual-band reflection. Prior to design, as seen in Figure 4a, we examined the zeroth-order transmittance (T_0) of 1 mm-thickCaF_2 and KBr substrates as the interlayer. The red line represents a measured T_0 spectrum of the CaF_2 substrate obtained by using a Fourier transform IR (FTIR) spectrometer (Thermo Scientific Nicolet iN10, Thermo Fisher Scientific, Waltham, MA, USA). In the MWIR region, it exhibits a high T_0~0.92 indicating that the light remaining after external reflection (R~0.08 at surface and bottom) propagates without internal absorption. However, the T_0 rapidly decreases when the wavelength is longer than λ = 8 μm due to the increased internal loss. The blue line shows a simulated T_0 spectrum where we have adjusted the CaF_2 dispersion to closely match the measured values based on previous experimental results [18]. At long wavelengths λ > 8 μm, the FP resonance is diminished by a fact that the internal absorption prevents the light from making a round trip within the cavity. The T_0 spectrum of a KBr substrate (green line) is also simulated by accounting for material dispersion based on previous experiments [19]. In both MWIR and LWIR regions, KBr exhibits apparently high transparency without absorption.

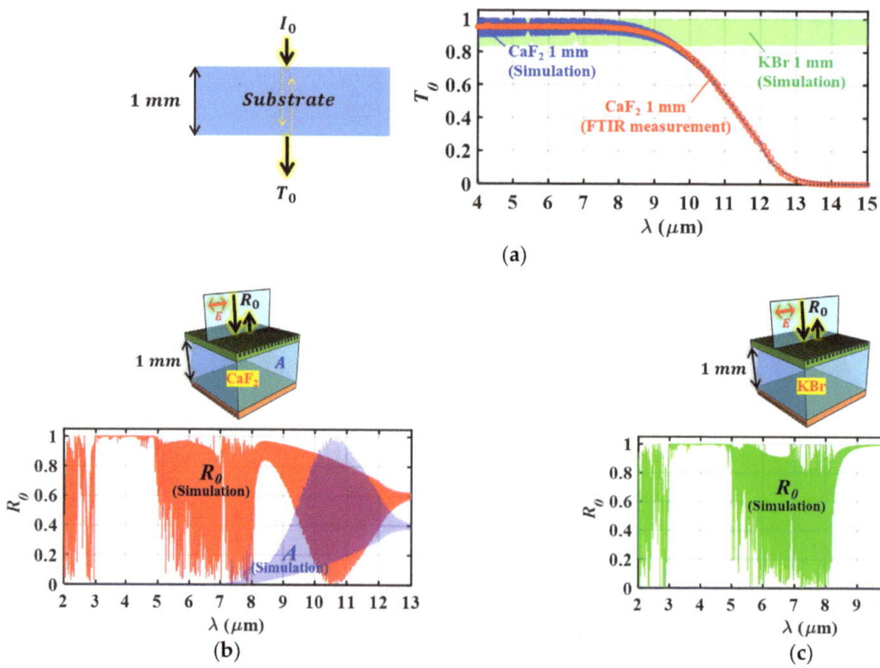

Figure 4. Effects of interlayer on the double-sided metasurfaces. (**a**) Zeroth-order transmittance (T_0) spectra of 1 mm-thick-substrate as the interlayer. The T_0 spectra of CaF_2 (blue line) and KBr (green line) substrates are simulated by RCWA. Additionally, the measured T_0 spectrum of the CaF_2 substrate is also compared to simulation result. (**b**) Design and simulation results of double-sided metasurfaces based on Ge/CaF_2/Ge material system where the top and bottom grating parameter sets are {Λ_1 = 1.725 μm, F_1 = 0.732, d_{g1} = 0.9 μm, d_{h1} = 0.49 μm} and {Λ_2 = 6 μm, F_2 = 0.38, d_{g2} = 1.7 μm, d_{h2} = 0.32 μm}. Due to the absorption at longer wavelengths, the LWIR reflection is degraded. (**c**) Design and simulation results of double-sided metasurfaces based on Ge/KBr/Ge materials. The top and bottom grating parameter sets are {Λ_1 = 1.73 μm, F_1 = 0.73, d_{g1} = 0.92 μm, d_{h1} = 0.5 μm} and {Λ_2 = 5.39 μm, F_2 = 0.38, d_{g2} = 1.75 μm, d_{h2} = 0.34 μm}.

Figure 4b shows the design result of double-sided metasurfaces based on $Ge/CaF_2/Ge$ material system where the first and second optimal grating parameter sets are $\{\Lambda_1 = 1.725$ μm, $F_1 = 0.732$, $d_{g1} = 0.9$ μm, $d_{h1} = 0.49$ μm$\}$ and $\{\Lambda_2 = 6$ μm, $F_2 = 0.38$, $d_{g2} = 1.7$ μm, $d_{h2} = 0.32$ μm$\}$ for the individual reflectors. The dispersion of Ge is accounted for [20]. As anticipated, dual-band reflection is not available due to the internal absorption of the CaF_2 interlayer as noted quantitatively by the significantly high absorbance (A) spectrum displayed in blue. Only the first reflector effectively operates in the MWIR region. In contrast, the Ge/KBr/Ge double-sided metasurfaces achieves the dual-band reflection without the internal absorption as presented in Figure 4c. In each region, high reflection is observed from 3.3 to 4.77 μm and 8.85 to 11 μm where the first and second optimal grating parameter sets are $\{\Lambda_1 = 1.73$ μm, $F_1 = 0.73$, $d_{g1} = 0.92$ μm, $d_{h1} = 0.5$ μm$\}$ and $\{\Lambda_2 = 5.39$ μm, $F_2 = 0.38$, $d_{g2} = 1.75$ μm, $d_{h2} = 0.34$ μm$\}$.

To investigate the angular tolerance of the double-sided metasurfaces, we examine the dual-band reflection while varying two different angles. Figure 5a illustrates angular variation with the plane of incidence (POI) perpendicular to the grating grooves of the first reflector. In this case, the MWIR reflector is subjected to classical mounting while the LWIR reflector experiences fully conical mounting. In general, the classic incidence varies the resonance more since it induces greater deviation of the diffracted waves than conic incidence [21]. Likewise, as shown in the $R_0(\lambda, \theta)$ color map, it is observed that the MWIR reflection band is split under classic mounting, but the LWIR reflection band remains robust under fully conical mounting. When the POI is parallel to the grooves, as seen in Figure 5b, the first reflector undergoes fully conical mounting, thereby enhancing angular tolerance in the MWIR band. However, reflectance is somewhat degraded in the LWIR band due to the classical mounting of the second reflector for this case.

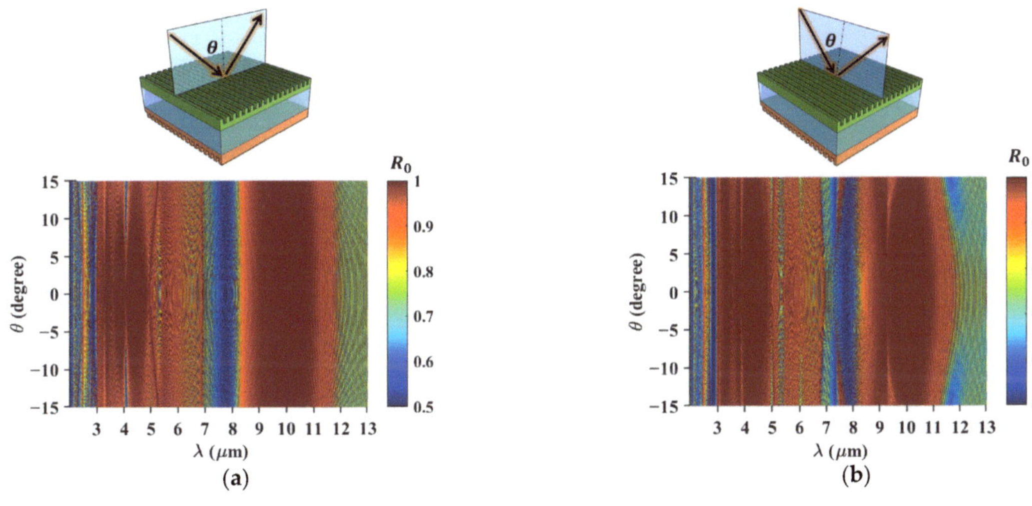

Figure 5. Angular tolerance of the double-sided metasurfaces. Simulated $R_0(\lambda, \theta)$ maps for the different angle of incidence (θ) as the plane of incidence (POI) is (**a**) parallel and (**b**) perpendicular to the grating groove of the first reflector.}.

6. Discussion

We present an analysis and a design of dual-band resonant reflectors operating in the infrared spectral region. We emphasize the fact that the dual-band functionality is realized with a pair of single-layer films appropriately patterned laterally. The reflector is shown to exhibit wideband performance with high efficiency engaging guided-mode resonance, also termed lattice resonance, effects as physical basis. Thereby, efficient and simultaneous reflection of MWIR and LWIR spectra is achieved as presented in detail herein. Challenges due to material loss in available substrate media are explained.

In these bands, there is a dearth of available lossless dielectric materials that limit the supply and functionality of corresponding components for broad spectral handling and manipulation. This contrasts with wide availability of materials in the visible and near-IR domains where component technology is dominated by classic thin-film solutions that are widely offered commercially. We note that thin-film versions of the dual-band reflector presented here will be challenging in fabrication due to substantial thickness of quarter-wave films particularly in the LWIR band. Often demanded specifications on size, weight, and power (SWaP) attributes are very favorable for resonance lattice MWIR/LWIR reflectors relative to the metallic reflectors often deployed in these bands.

Physical construction of the proposed device is challenging due to the necessity of protecting one surface during the pattering of the other. Nevertheless, fabrication is feasible with careful processing and procedures. Alternate implementation involves the serial arrangement of individual resonators on separate substrates; corresponding devices have not yet been designed. The basic concept of dual reflection band operation is extendable to other spectral regions spanning 400–6000 nm wherein the material selection limitation is not as extreme as in the LWIR band. For polarization independence, the GMR resonators may be patterned as 2D metasurfaces.

Few comparable results exist in the literature. A dual-band reflector operating at two precisely specified frequencies was designed and fabricated for 6G applications and operating in the mm-wave region using the principles of diffractive optics [22]. The reflector consists of a 2D array of unit cells containing metallic elements on a dielectric substrate over a metallic ground plane. For a fixed incident direction of plane waves at 220 GHz and 293.3 GHz, the 3rd and 4th diffraction orders propagate colinearly in reflection with ~80% efficiency excluding inherent metal losses [22]. In another work, for lower frequencies, a high-gain low-profile reflector dual-band antenna was reported [23]. The antenna exhibited reasonable gain in bands centered at 31 GHz and 158 GHz. Here, the innovation was a periodic subwavelength array patterned in 2D to achieve a useful performance [23] but the operational principle is totally different from the lattice resonance concept presented herein. Finally, a multi-band reflector using sequential planes of metamaterials was proposed, designed, and fabricated to operate at still lower frequencies [24]. The metasurfaces were composed of wood-pile photonic crystals, a classic element operating principally on Bragg effects and not via GMR. Reflectance data in the 0.6–2.2 GHz band were presented [24].

These very-long-wave microwave reflectors are lossy, bulky, and complex. In contrast, the optical dual-band reflector presented here is ultracompact and simple. We estimate that the substrate could be as thin as ~200 μm while remaining robust and manageable in optical systems. It would appear as a single-layer device to a user, excellent for lightweight systems in space and air.

In closing, we consider the effect of incident illumination deviating from a perfect TM polarization state. As shown in Figure 6a, the double-sided 1D reflector of Figure 4c maintains the dual band of high reflection up to 10° in polarization angle (ϕ). The inset indicates the ϕ between the POI and grating vector of the first reflector. However, the dual band is degraded by further increases in $\phi = 20°$, $30°$, and $90°$ shown in Figure 6b–d. Polarization independence at normal incidence is attained via 2D photonic lattices.

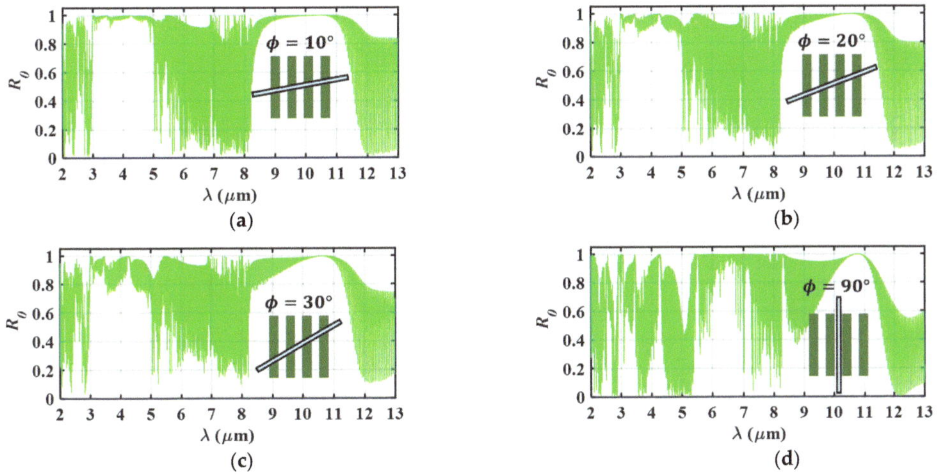

Figure 6. Polarization tolerance of the double-sided metasurfaces. Simulated R_0 spectra for the different polarization angle (ϕ) at (**a**) 10°, (**b**) 20°, (**c**) 30° and (**d**) 90°. The POI relative to the grating grooves of the first reflector are illustrated in the inset.

7. Conclusions

In summary, we introduced a dual-band reflector that incorporates double-sided metasurfaces. This innovative design pertains to the cooperation of two GMR reflectors, where two metasurface are combined with an interlayer positioned between them. The top and bottom metasurfaces can be independently designed to achieve the desired spectral response and both functionalities are integrated into a single device with a transparent interlayer. For the MWIR and LWIR reflection bands, two 1D ZCG structures were optimized by utilizing Ge grating layers on a KBr substrate. The Ge/KBr/Ge design demonstrates wideband reflection from 3.3 to 4.77 μm (MWIR region) and 8.85 to 11 μm (LWIR region). In addition, we provided an analytical formula, which will be useful to estimate output spectral response of double-sided metasurfaces in this device class.

We note that the work presented is completely original with no similar results previously published. The contribution is solid on account of potential applications in imaging, night vision systems, thermal systems, and environmental monitoring. A future challenge is to fabricate examples of this device class while establishing attendant processes and methodology.

Author Contributions: Conceptualization, Y.H.K. and R.M.; methodology, Y.H.K. and R.M.; formal analysis, Y.H.K.; investigation, Y.H.K. and R.M.; resources, Y.H.K. and R.M.; writing—original draft preparation, Y.H.K.; writing—review and editing, R.M.; visualization, Y.H.K.; supervision, R.M.; project administration, R.M.; funding acquisition, R.M. All authors have read and agreed to the published version of the manuscript.

Funding: This research was supported, in part, by the UT System Texas Nanoelectronics Research Superiority Award funded by the State of Texas Emerging Technology Fund and by the Texas Instruments Distinguished University Chair in Nanoelectronics endowment.

Institutional Review Board Statement: Not applicable.

Informed Consent Statement: Not applicable.

Data Availability Statement: The data underlying the results presented in this paper are not publicly available at this time but may be obtained from the authors upon reasonable request.

Acknowledgments: We thank Pawarat Bootpakdeetam for collecting the experimental data in Figure 4a.

Conflicts of Interest: The authors declare no conflicts of interest. The authors declare that they have no known competing financial interest or personal relationships that could have appeared to influence the work reported in this paper.

References

1. Morris, N.J.W.; Avidan, S.; Matusik, W.; Pfister, H. Statistics of Infrared Images. In Proceedings of the 2007 IEEE Conference on Computer Vision and Pattern Recognition, Minneapolis, MN, USA, 17–22 June 2007; TR2007-020.
2. Neinavaz, E.; Schlerf, M.; Darvishzadeh, R.; Gerhards, M.; Skidmore, A.K. Thermal Infrared Remote Sensing of Vegetation: Current Status and Perspectives. *Int. J. Appl. Earth Obs. Geoinf.* **2021**, *102*, 102415. [CrossRef]
3. Siciliani de Cumis, M.; Viciani, S.; Borri, S.; Patimisco, P.; Sampaolo, A.; Scamarcio, G.; De Natale, P.; D'Amato, S.V. Widely-Tunable Mid-Infrared Fiber-Coupled Quartz-Enhanced Photoacoustic Sensor for environmental Monitoring. *Opt. Express* **2014**, *17*, 22. [CrossRef] [PubMed]
4. Huang, L.; Han, Z.; Wirth-Singh, A.; Saragadam, V.; Mukherjee, S.; Fröch, J.E.; Tanguy, Q.A.A.; Rollag, J.; Gibson, R.; Hendrickson, J.R.; et al. Broadband Thermal Imaging Using Meta-Optics. *Nat. Commun.* **2024**, *15*, 1662. [CrossRef] [PubMed]
5. Shi, X.; Hou, E.; Liang, Z.; Zhang, S.; Dai, R.; Xin, W.; Meng, D.; Liu, H.; Xu, H.; Liu, Y. Broadband Metamaterial Absorber Based on Hybrid Multi-Mode Resonance in Mid-Wave and Long-Wave Infrared Region. *Res. Phys.* **2022**, *42*, 105972. [CrossRef]
6. Zhou, X.; Huang, J.; Wang, H.; Kong, J.; Li, J.; Mu, Y.; Zhang, Y.; Ma, W.; Ren, H.; Li, D.; et al. Dual-band MWIR/LWIR Superlattice Infrared Focal Plane Arrays for Simultaneous Detection. *Infrared Phys. Technol.* **2023**, *128*, 104539. [CrossRef]
7. Gambino, F.; Giaquinto, M.; Ricciardi, A.; Cusano, A. A Review on Dielectric Resonant Gratings: Mitigation of Finite Size and Gaussian Beam Size Effects. *Results Opt.* **2022**, *6*, 100210. [CrossRef]
8. Ji, W.; Chang, J.; Xu, H.X.; Gao, J.R.; Gröblacher, S.; Urbach, H.P.; Adam, A.J.L. Recent Advances in Metasurface Design and Quantum Optics Applications with Machine Learning, Physics-Informed Neural Networks, and Topology Optimization Methods. *Light Sci. Appl.* **2023**, *12*, 169. [CrossRef]
9. Ferraro, A.; Zografopoulos, D.C.; Caputo, R. Guided-Mode Resonant Narrowband Terahertz Filtering by Periodic Metallic Stripe and Patch Arrays on Cyclo-olefin Substrates. *Sci. Rep.* **2018**, *8*, 17272. [CrossRef]
10. Ko, Y.H.; Magnusson, R. Wideband Dielectric Metamaterial Reflectors: Mie Scattering or Leaky Bloch Mode Resonance? *Optica* **2018**, *5*, 289. [CrossRef]
11. Hu, Y.; Wang, X.; Luo, X.; Ou, X.; Li, L.; Chen, Y.; Ping, Y.; Wang, S.; Duan, H. All-Dielectric Metasurfaces for Polarization Manipulation: Principles and Emerging Applications. *Nanophotonics* **2020**, *9*, 3755. [CrossRef]
12. Simlan, F.A.; Lee, K.J.; Ko, Y.H.; Gupta, N.; Magnusson, R. Fabrication of Single-Layer Resonant Infrared Filters with High Optical Density. *J. Lightwave Technol.* **2024**, *42*, 2345. [CrossRef]
13. Shokooh-Saremi, M.; Magnusson, R. Particle Swarm Optimization and Its Application to the Design of Diffraction Grating Filters. *Opt. Lett.* **2007**, *32*, 894. [CrossRef] [PubMed]
14. Moharam, M.G.; Grann, E.B.; Pommet, D.A.; Gaylord, T.K. Formulation for Stable and Efficient Implementation of the Rigorous Coupled-Wave Analysis of Binary Gratings. *J. Opt. Soc. Am.* **1995**, *12*, 1068. [CrossRef]
15. *RSoft Products, DiffractMod*; Synopsys, Inc.: Sunnyvale, CA, USA, 2024.
16. Magnusson, R. Wideband Reflectors with Zero-Contrast Gratings. *Opt. Lett.* **2014**, *39*, 4337. [CrossRef]
17. Magnusson, R.; Shokooh-Saremi, M. Physical Basis for Wideband Resonant Reflectors. *Opt. Express* **2008**, *16*, 3456. [CrossRef]
18. Kaiser, W.; Spitzer, W.G.; Kaiser, R.H.; Howarth, L.E. Infrared Properties of CaF_2, SrF_2, and BaF_2. *Phys. Rev.* **1962**, *127*, 1950. [CrossRef]
19. Li, H.H. Refractive Index of Alkali Halides and Its Wavelength and Temperature Derivatives. *J. Phys. Chem. Ref. Data* **1976**, *5*, 329. [CrossRef]
20. Li, H.H. Refractive Index of Silicon and Germanium and Its Wavelength and Temperature Derivatives. *J. Phys. Chem. Ref. Data* **1980**, *9*, 561. [CrossRef]
21. Ko, Y.H.; Niraula, M.; Lee, K.J.; Magnusson, R. Properties of Wideband Resonant Reflectors under Fully Conical Light Incidence. *Opt. Express* **2016**, *24*, 4542. [CrossRef]
22. Kato, Y.; Ameya, M.; Tezuka, M.; Kobuke, H.; Sanada, A. 220/293 GHz Dual-Band Anomalous Reflectors Using Higher-Order Diffraction Modes and Their Precise Characterization Using a Compact Antenna Test Range System. *IEEE Access* **2023**, *11*, 139295–139305. [CrossRef]
23. Lu, S.; Qu, S.-W. Low-Profile Dual-Band Reflector Antenna for High-Frequency Applications. *Sensors* **2023**, *23*, 5781. [CrossRef] [PubMed]
24. So, H.; Maruta, K. Sector Design Using Multiband Antenna with Metamaterial Reflector for Cellular UAV System. *IEEE Access* **2022**, *10*, 4924–4933. [CrossRef]

photonics

Article

An Objective Evaluation Method for Image Sharpness Under Different Illumination Imaging Conditions

Huan He [1], Benchi Jiang [1,2], Chenyang Shi [1,3,*], Yuelin Lu [1] and Yandan Lin [4,*]

1 School of Artificial Intelligence, Anhui Polytechnic University, Wuhu 241000, China; lime_yhhuan@163.com (H.H.); benchi@ahpu.edu.cn (B.J.); luyuelin@ahpu.edu.cn (Y.L.)
2 Industrial Innovation Technology Research Co., Ltd., Anhui Polytechnic University, Wuhu 241000, China
3 Anhui Engineering Research Center of Vehicle Display Integrated Systems, School of Integrated Circuits, Anhui Polytechnic University, Wuhu 241000, China
4 Department of Illuminating Engineering & Light Sources, School of Information Science and Technology, Fudan University, Shanghai 200433, China
* Correspondence: shichenyang@ahpu.edu.cn (C.S.); ydlin@fudan.edu.cn (Y.L.)

Abstract: Blurriness is troublesome in digital images when captured under different illumination imaging conditions. To obtain an accurate blurred image quality assessment (IQA), a machine learning-based objective evaluation method for image sharpness under different illumination imaging conditions is proposed. In this method, the visual saliency, color difference, and gradient information are selected as the image features, and the relevant feature information of these three aspects is extracted from the image as the feature value for the blurred image evaluation under different illumination imaging conditions. Then, a particle swarm optimization-based general regression neural network (PSO-GRNN) is established to train the above extracted feature values, and the final blurred image evaluation result is determined. The proposed method was validated based on three databases, i.e., BID, CID2013, and CLIVE, which contain real blurred images under different illumination imaging conditions. The experimental results showed that the proposed method has good performance in evaluating the quality of images under different imaging conditions.

Keywords: image sharpness evaluation; different illumination imaging conditions; real blur images; PSO-GRNN

Citation: He, H.; Jiang, B.; Shi, C.; Lu, Y.; Lin, Y. An Objective Evaluation Method for Image Sharpness Under Different Illumination Imaging Conditions. *Photonics* **2024**, *11*, 1032. https://doi.org/10.3390/photonics11111032

Received: 18 September 2024
Revised: 15 October 2024
Accepted: 28 October 2024
Published: 1 November 2024

1. Introduction

With the continuous development of imaging science, imaging technology has been widely applied in many fields such as video conferencing, medical imaging, remote sensing, compressive sensing, and social media [1,2]. In machine vision systems, extreme external illumination can further lead to a decrease in the image quality [3]. Illumination plays a critical role in capturing the image procedure, and a change in the illumination is the main factor causing image blurring and distortion [4]. The evaluation of image quality consists of analyzing and quantifying the degree of distortion and developing a quantitative evaluation index. A subjective quality evaluation is relatively reliable. However, it is time-consuming, labor-intensive, and not conducive to application in intelligent evaluation systems [5]. Studies relating to objective image quality assessment (IQA), e.g., image sharpness assessments, are becoming increasingly important in assessing the impact of the variation in an image's appearance on the resulting visual quality and ensuring the reliability of image processing systems [6,7].

According to the availability of a reference image, IQA can usually be divided into three categories: full-reference IQA (FR-IQA) [8,9], reduced-reference IQA (RR-IQA) [10], and no-reference IQA (NR-IQA) [11]. In practical applications, it is usually impossible to obtain undistorted images or their features as references, so NR-IQA has practical significance. In this field, many mature algorithms have achieved good results in image quality evaluation. Gaussian blur is one of the common and dominant types of distortion perceived in images when captured under low-light conditions. Therefore, a suitable and efficient blurriness or sharpness evaluation method should be explored [7].

Usually, statistical data on the structural features of an image are important in NR-IQA research. When the blurred image is captured under nonideal illumination conditions, the structural features will change accordingly. This structural change can be characterized by some specific structural statistics of the image, which can be used to evaluate the image quality [11–13]. Bahrami and Kot [11] proposed a model to measure sharpness based on the maximum local variation (MLV). Li [12] proposed a blind image blur evaluation (BIBLE) algorithm based on discrete orthogonal moments, where gradient images are divided into equally sized blocks and orthogonal moments are calculated to characterize the image sharpness. Gvozden [13] proposed a fast blind image sharpness/ambiguity evaluation model (BISHARP), and the local contrast information of the image was obtained by calculating the root mean square of the image. These methods are all built in the traditional way on the spatial domain/spectral domain of an image.

Unlike traditional methods, learning-based methods can improve the accuracy of the evaluation results [14]. Li [15] proposed a reference-free and robust image sharpness assessment (RISE) method, which evaluates image quality by learning multi-scale features extracted in spatial and spectral domains. A no-reference image sharpness metric based on structural information using sparse representation (SR) [16] was proposed. Yu [17] proposed a blind image sharpness assessment by using a shallow convolutional neural network (CNN). Kim [18] applied a deep CNN to the NR-IQA by separating the training of the NR-IQA into two stages: (1) an objective distortion part and (2) a part related to the human visual system. Liu [19] developed an efficient general-purpose no-reference NR-IQA model that utilizes local spatial and spectral entropy features on distorted images. Li [20] proposed a method based on semantic feature aggregation (SFA) to alleviate the impact of image content variation. Zhang [21] proposed a deep bilinear convolutional neural network (DB-CNN) model for blind image quality assessment that works for both synthetically and authentically distorted images.

These methods can solve simulated blur evaluation problems well, but the majority of them cannot accurately evaluate the realistic blur introduced during image capturing, especially under different illumination imaging conditions. Moreover, evaluating realistic blur is undoubtedly more significant than the evaluation of simulated blur. Therefore, it is necessary to design a sharpness assessment method that is effective for image sharpness under different illumination imaging conditions.

In this research, an NR-IQA method for blurred images under different illumination imaging conditions is proposed to evaluate image sharpness based on a particle swarm optimization-based general regression neural network (PSO-GRNN). Firstly, some basic image feature maps are extracted, i.e., the visual saliency (VS), color difference, and gradient, and the feature values of all maps are obtained by using statistical calculation. Secondly, the feature values are trained and optimized by a PSO-GRNN. Lastly, after the PSO-GRNN is determined, an evaluation result for the real blurry images will be calculated. The experimental results show that the evaluation performance of the proposed method on real blur databases, i.e., BID [22], CID2013 [23], and CLIVE [24], is better than the state-of-the-art and recently published NR methods.

2. Feature Extraction

2.1. Visual Saliency (VS) Index

The vs. of an image can reflect how "salient" a local region is [25,26]. The relationship between the vs. and image quality has been integrated into IQA studies [27]. For a blurred image that is captured under non-ideal illumination conditions, the important areas of the scene will decrease, and, consequently, the vs. map of the image will also change. The extraction method for the vs. in this study is based on the SDSP method [27]. In Figure 1, the blurred and distorted images with the same content under different lighting conditions and their vs. maps (pseudo-color maps) are presented. From Figure 1, it can be seen that the vs. maps can accurately extract important regions. Figure 1c,d shows the vs. maps of Figure 1a,b, which show the blurred level of an image under specific lighting conditions, in line with the human visual perception characteristics.

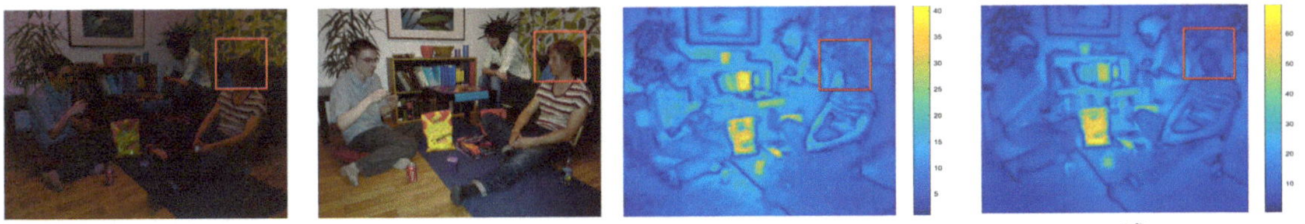

(a) (b) (c) (d)

Figure 1. Images of the same content under different lighting conditions and the corresponding vs. maps. (**a**,**b**) are images of the same content under different lighting conditions [23], while (**c**,**d**) are the corresponding vs. maps.

2.2. Color Difference (CD) Index

The previous section introduces the overall structural information of images, i.e., the vs. index. For a color image, color information is also important for image quality. The CD index [28] can reflect the color distortion by different illumination imaging conditions. Therefore, the CD is used to evaluate the image quality for the color information. For an RGB image, mapping the image to a specific color space where each pixel contains three color components (brightness $S_1{}^*$, color channel $S_2{}^*$, and $S_3{}^*$) allows the CD index between adjacent pixels to be calculated using Equation (1):

$$\Delta E = \sqrt{(S_{11}{}^* - S_{12}{}^*) + (S_{21}{}^* - S_{22}{}^*) + (S_{31}{}^* - S_{32}{}^*)}, \tag{1}$$

where $S_{11}{}^*$ and $S_{12}{}^*$ are the lightness channels of two neighbor pixels; $S_{21}{}^*$ and $S_{22}{}^*$ are the color channels of two neighbor pixels; $S_{31}{}^*$ and $S_{32}{}^*$ are the other color channels of two neighbor pixels. The CD operators in the horizontal and vertical directions of the k channel are defined as

$$\Delta E_h(i,j) = \sqrt{\sum_{n=1}^{k} (S_n{}^*(i,j) - S_n{}^*(i,j+1))^2}, \tag{2}$$

$$\Delta E_v(i,j) = \sqrt{\sum_{n=1}^{k} (S_n{}^*(i,j) - S_n{}^*(i+1,j))^2}, \tag{3}$$

where $S_n{}^*(i,j)$ represent the intensity at the pixel location for each color channel. The color channel number is represented by k.

By combining the local CD operators in the above two directions, the local CD (ΔE_L) for pixel (i, j) is obtained, which is given by Formula (4):

$$\Delta E_L(i,j) = (\Delta E_h(i,j) + \Delta E_v(i,j))/2 \tag{4}$$

Figure 2 shows the CD pseudo-color maps corresponding to the images with the same content under different lighting conditions. After comparing, it can be seen that there are obvious differences in the CD maps of images under different lighting environments. Therefore, the CD index can be used to evaluate the quality of real blurred images.

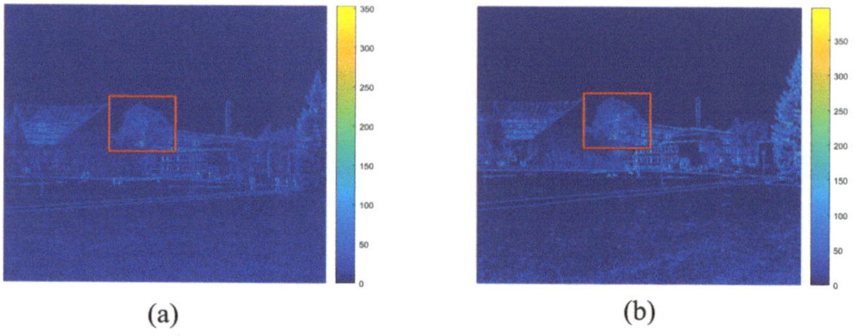

Figure 2. (**a**) and (**b**) are two CD pseudo-color maps of different images in CID2013.

2.3. Gradient Index

In IQA studies, a grayscale image gradient is also a commonly used feature [29]. The calculation of an image gradient refers to the magnitude of image changes. For the edge parts of an image, the grayscale values change significantly, and the gradient values also change significantly. On the contrary, for the smoother parts of an image, the grayscale values change less, and the corresponding gradient values also change less. As shown in Figure 3, the red rectangular area is more prominent. The degree of blur in an image is positively correlated with the change in edge location. By calculating the image gradient, the corresponding change in edge location can be determined. In this study, the Roberts operator is utilized to calculate the image gradient. The gradient value of the pixel point at (i, j) is defined as $G(i, j)$, and the gradient calculation formula is

$$G(i,j) = \sqrt{[I(i,j) - I(i,j+1)]^2 + [I(i,j) - I(i+1,j)]^2} \tag{5}$$

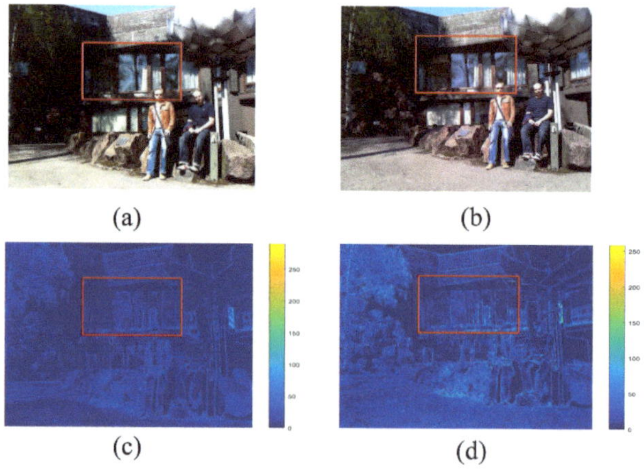

Figure 3. Blurred images of the same content under different lighting conditions and corresponding gradient maps: (**a**,**b**) are blurry images of the same content under different lighting conditions [23], while (**c**,**d**) are corresponding gradient maps.

2.4. Image Feature Value

Based on the above analysis, the VS, CD, and gradient information are selected to extract image features in this study. As shown in Figure 4, for a blurred image, the vs. map, CD map, and gradient map are processed at first. Then, these three feature maps are subjected to maximum (Max), relative change (RC), and variance (Var) calculations. After that, nine feature values can be obtained. These values are used to construct the feature value of an image, which is then input into the following parts of the proposed method.

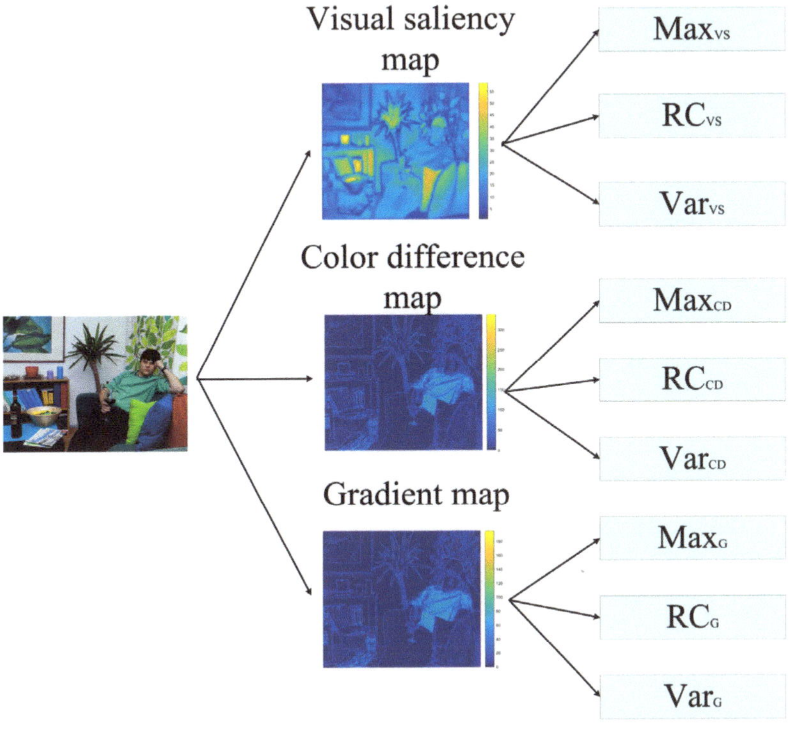

Figure 4. Flowchart for extracting image feature values.

The calculation process for the Max, RC, and Var eigenvalues of the obtained feature map M, e.g., the VS, CD, and gradient maps, is as follows:

$$\text{Max} = \max(\text{M}(i,j)) \tag{6}$$

$$\text{RC} = \frac{\max(\text{M}(i,j)) - \min(\text{M}(i,j))}{mean(\text{M}(i,j))} \tag{7}$$

$$\text{Var} = \frac{\left((x_1 - \overline{x})^2 + (x_2 - \overline{x})^2 + \ldots + (x_n - \overline{x})^2\right)}{n} \tag{8}$$

where $x_1, x_1, \ldots, x_n$ represents the pixel value of each feature map, $\overline{x}$ represents the average pixel value of a feature map, and n represents the total number of pixels.

3. Algorithm Framework

3.1. Generalized Regression Neural Network (GRNN)

GRNN is a powerful regression tool with a dynamic network structure [30]. It is a radial basis function neural network based on non-parametric estimation for nonlinear regression. The network structure of a GRNN is shown in Figure 5, which includes an input layer for conditional samples, a corresponding network pattern layer, a summation layer, and an output layer for the final network training results. The number of neurons in the input layer is equal to the dimension of the input vector in the learning sample. In this paper, the number of neurons in the input layer is n, which equals the number of feature values in an image. Each neuron is a simple distribution unit that directly passes input variables to the pattern layer. The number of neurons in the pattern layer is equal to the number n in the input layer, and each neuron corresponds to a different sample. Each neuron in the pattern layer is connected to two neurons in the summation layer, and the output layer calculates the quotient of the two outputs of the summation layer to generate feature-based predictions. The corresponding network input and network output are

$$Y(X) = \frac{\sum_{i=1}^{n} Y_i \exp(-D_i^2/2\sigma^2)}{\sum_{i=1}^{n} \exp(-D_i^2/2\sigma^2)} \tag{9}$$

where n is the number of observed values in the sample. X_i and Y_i are the values of the sample. $p_i = \exp\left[(X - X_i)^T (X - X_i)/2\sigma^2\right] D_i^2 = (X - X_i)^T (X - X_i)$ is the transfer function of the pattern layer neuron, and σ is the transfer parameter. The larger the value of σ, the smoother the function approximation.

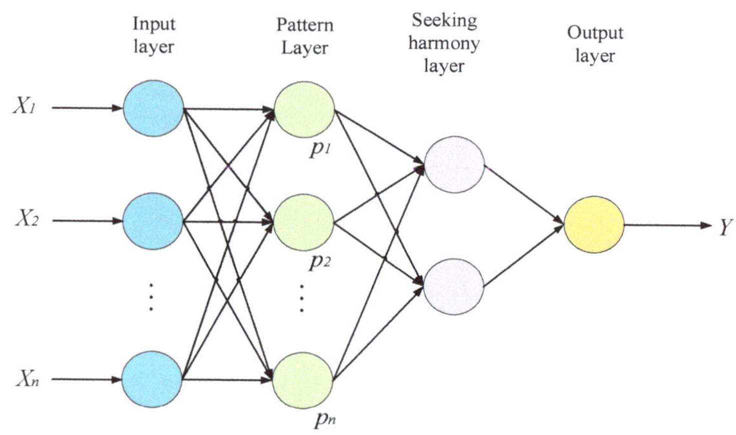

Figure 5. The network structure of GRNN.

To achieve the optimal performance of a GRNN, it is necessary to determine the ideal variable value σ. In IQA studies, the method of controlling variables is commonly used to calculate the variable values. However, adaptive optimization methods are selected in the proposed method to obtain better performance. In Table 1, three adaptive optimization methods, i.e., fruit fly optimization algorithm (FOA), firefly algorithm (FA), and particle swarm optimization (PSO), are tested on the BID database. In addition, a GRNN without an adaptive optimization method is also tested. The best results are highlighted in boldface in Table 1. Based on the corresponding evaluation standard value (SROCC and PLCC), PSO performs better than other methods. Therefore, PSO-GRNN is selected as the main framework of the proposed IQA method.

Table 1. Performance on different adaptive optimization methods.

Database	Criteria	GRNN	FOA-GRNN	FA-GRNN	PSO-GRNN
BID	SROCC	0.880	0.880	0.876	**0.885**
	PLCC	0.885	0.887	0.883	**0.890**

3.2. Particle Swarm Optimization (PSO) Algorithm

The calculation steps of PSO are shown in Figure 6. The specific implementation steps are as follows:

(1) Initializing the population: Randomly initialize the position (P_i) and velocity of each particle in the population (v_i), the maximum number of iterations of the algorithm, etc.
(2) Calculate the fitness value of each particle based on the fitness function, save the optimal position of each particle (i), and save the individual best fitness value ($pbest_i$) and the global best position of the population ($gbest_i$).
(3) Update the velocity and position based on the velocity and position the update formula according to the following equations:

$$v_i^{t+1} = \omega \cdot v_i^t + c_1 r_1 \left(pbest_i - P_i^t\right) + c_2 r_2 \left(gbest_i - P_i^t\right) \tag{10}$$

$$P_i^{t+1} = P_i^t + v_i^t \tag{11}$$

where c_1, c_2 are the learning factors, also known as the acceleration constant. r_1, r_2 are the uniform random numbers within the range of [0, 1]. ω is the inertia weight. t is the iteration number. Calculate the fitness value of each particle after updating and compare the best fitness value of each particle with its historical best position fitness value. If it is good, use its current position as the optimal position for that particle. For each particle, compare the fitness value corresponding to its optimal position with the population's optimal fitness value. If it is better, update the population's optimal position and fitness value.
(4) Determine whether the search results meet the stopping conditions (reach the maximum number of iterations or meet the accuracy requirements). If the stopping conditions are met, output the optimal value. Otherwise, proceed to the second step and continue running until the stopping conditions are met.

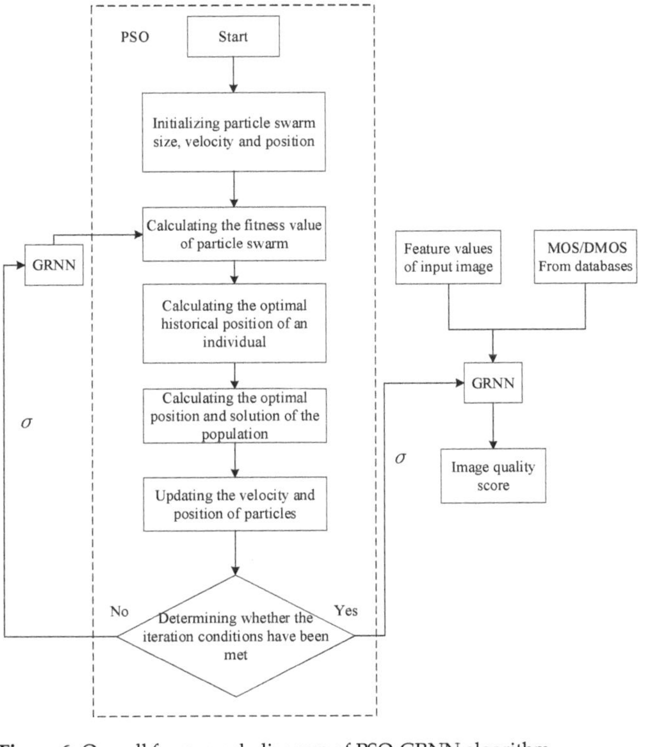

Figure 6. Overall framework diagram of PSO-GRNN algorithm.

3.3. PSO-GRNN Image Quality Evaluation Model

A GRNN has a strong nonlinear mapping ability and a flexible network structure. The PSO-GRNN prediction model was introduced in reference [31]. It can solve the problems of easy convergence to local minima, slow the convergence speed during network training, and improve the generalization ability of the neural network by optimizing the expansion speed of GRNN functions.

This article extends the PSO-GRNN model to blurred images quality evaluation under different illumination imaging conditions. The design of the PSO-GRNN-based IQA method is proposed in Figure 6. Firstly, the VS, CD, and gradient processing is performed on all color blurred images to obtain feature values by Equations (6)–(8). Then, 80% of the images in the database are randomly selected, and the feature values and benchmark values, i.e., MOS or DOMS, of these images are input in the GRNN for training. Later, the variable value of the GRNN is optimized by PSO. Finally, the trained PSO-GRNN is used to evaluate the quality of the other 20% of images in the database and obtain the predicted value for image quality.

4. Experiments and Discussion

4.1. Database and Evaluation Indicators

In our study, experiments are performed on the BID [22], CID2013 [23], and CLIVE [24] databases. These three databases are all public realistic blur image databases. The information of each database is introduced in Table 2. Eight different scenes are included in CID2013. Scenes 1, 2, 3, 6 include six datasets (I–VI), Scene 4 includes four datasets (I–IV), Scene 5 includes five datasets (I–V), Scene 7 includes one dataset (VI), and Scene 8 includes two datasets (V, IV). Table 3 shows the descriptions and example images for each scene. The BID database contains 586 images, while the CLIVE database contains 1162 images, all of which are real blurry images in real-world environments. These three databases are commonly utilized collections in real IQA studies, covering a wide range of ordinarily authentic distortions in real-world applications.

Table 2. Database information description.

Database	Blur Images	Subjective Scores	Typical Size	Score Range
BID	586	MOS	1280 × 960	[0, 5]
CID2013	474	MOS	1600 × 1200	[0, 100]
CLIVE	1162	MOS	500 × 500	[0, 100]

Table 3. Introduction to CID2013.

Cluster	Subject luminance (lux)	Subject-Camera Distance (m)	Scene Description	Example Images	Image Set	Motivation
1	2	0.5	Close-up in dark lighting conditions		I–VI	Bar and restaurant setting
2	100	1.5	Close-up in typical indoor lighting conditions		I–VI	Living room environment, Indoor portrait

Table 3. *Cont.*

Cluster	Subject luminance (lux)	Subject-Camera Distance (m)	Scene Description	Example Images	Image Set	Motivation
3	10	4.0	Small group in dim lighting conditions		I–VI	Living room environment, group picture
4	1000	1.5	Studio image		I–IV	Studio image generally used in image quality testing
5	>3400	3.0	Small group in cloudy bright to sunny lighting conditions		I–V	Typical tourist image
6	>3400	>50	Close-up in high dynamic range lighting conditions		I–VI	Landscape image
7	>3400	3.0	Small group in cloudy bright to sunny lighting conditions (~3× optical or digital zoom)		VI	General zooming
8	>3400 (outdoors) and <100 (indoors)	1.5	Close-up in high dynamic range lighting conditions		V, VI	High dynamic range scene

The most commonly used sharpness performance evaluation indicators are the Pearson linear correlation coefficient (PLCC) and Spearman's rank-ordered correlation coefficient (SROCC). The PLCC and SROCC reaching unity 1 means that the prediction performance of an objective method is performing better.

1. Prediction Accuracy

The PLCC is used to measure the prediction accuracy of an IQA method. To compute the PLCC, a logistic regression with the five parameters is used to obtain the same scale values with subjective ratings [31]:

$$p(x) = \beta_1 \left[\frac{1}{2} - \frac{1}{1 + \exp(\beta_2 (x - \beta_3))} \right] + \beta_4 x + \beta_5, \tag{12}$$

where x denotes the objective quality scores directly from an IQA method, $p(x)$ denotes the IQA scores after the regression step, and $\beta_1, \ldots, \beta_5$ are model parameters that are found numerically to maximize the correlation between subjective and objective scores. The PLCC value of an IQA method is then calculated as

$$\mathrm{PLCC} = \frac{1}{n-1}\sum\nolimits_{i=1}^{n}\left(\frac{x_i - \overline{x}}{\sigma_x}\right)\left(\frac{y_i - \overline{y}}{\sigma_y}\right), \tag{13}$$

where $\overline{x}$ and $\overline{y}$ are the mean values of x_i and y_i, respectively, and σ_x and σ_y are the corresponding standard deviations.

2. Prediction Monotonicity

SROCC is used to predict the monotonicity of an IQA method. The SROCC value of an IQA method on a database with n images is calculated as [32]

$$\mathrm{SROCC} = 1 - \left[6\sum_{i=1}^{n}\frac{(r_{xi} - r_{yi})^2}{n(n^2-1)}\right] \tag{14}$$

where r_{xi} and r_{yi} represent the ranks of the prediction score and the subjective score, respectively.

4.2. Performance Comparison

In this section, real blurry images from the BID, CID2013, and CLIVE databases are tested to obtain the prediction scores for each image. The predicted scores are then linearly fitted with the subjective scores of the images to obtain the corresponding PLCC and SROCC values. Each database was tested 20 times by the proposed method, and the average value of 20 tests was taken as the final fitting result for the entire database. The PLCC and SROCC results of the proposed method and comparison methods are shown in Table 4. The comparison methods are related to the spatial domain, the frequency domain, machine learning, and deep learning. The best results are highlighted in boldface for the two indices in Table 4. RISE [15] (2017), SR [16] (2016), Yu's CNN [17] (2017), DIQA [18] (2019), SSEQ [19] (2014), SFA [20] (2019), DB-CNN [21] (2020), DIVINE [33], and NIQE [34] are learning-based algorithms, while MLV [11] (2014), BIBLE [12] (2017), BISHARP [13] (2018), and GCDV [28] (2024) are the methods related to the spatial and frequency domains. Moreover, DIVINE [33] and NIQE [34] are general-purpose NR-IQA methods, and the other compared methods are all image sharpness assessment methods.

Table 4. Comparison between the proposed method and others.

Databases	BID [22]		CID2013 [23]		CLIVE [24]	
Criteria	PLCC	SROCC	PLCC	SROCC	PLCC	SROCC
BISHARP [13]	0.356	0.307	0.678	0.681	-	-
BIBLE [12]	0.392	0.361	0.698	0.687	0.515	0.427
MLV [11]	0.375	0.317	0.689	0.621	0.400	0.339
GCDV [28]	0.338	0.294	0.681	0.596	0.405	0.334
RISE [15]	0.602	0.584	0.793	0.769	0.555	0.515
SR [16]	0.415	0.467	0.621	0.634	-	-
Yu's CNN [17]	0.560	0.557	0.715	0.704	0.501	0.502
DIQA [18]	0.506	0.492	0.720	0.708	0.704	0.703
SSEQ [19]	0.604	0.581	0.689	0.676	-	-
SFA [20]	0.840	0.826	-	-	0.833	0.812
DB-CNN [21]	0.471	0.464	0.686	0.672	0.869	0.851
DIVINE [33]	0.506	0.489	0.499	0.477	0.558	0.509
NIQE [34]	0.471	0.469	0.693	0.633	0.478	0.421
Proposed	**0.890**	**0.885**	**0.924**	**0.913**	**0.873**	**0.867**

It can be seen that the results of the proposed method on these three databases are all above 0.85. Overall, learning-based algorithms perform better than the methods related to the spatial and frequency domains in evaluating the quality of real blurry and distorted images. The PSO-GRNN-based proposed method yields better performance than other advanced network structure-based methods, i.e., Yu's CNN [17] (2017), DIQA [18] (2019), SFA [20] (2019), and DB-CNN [21] (2020). Therefore, a fully connected GRNN is suitable for dealing with IQA problems.

4.3. Performance of Image Feature Selection

In this section, the impact of feature selection on the performance of the proposed method is verified, and the features are the VS, CD, and gradient. Three feature value calculation methods, i.e., Max, RC, and Var, are selected for these features. In Table 5, seven different feature value combinations are tested and the best results are highlighted in boldface for the two indices.

Table 5. The performance of different feature selections.

Databases	Feature Value	PLCC	SROCC
BID	Max	0.828	0.825
	RC	0.802	0.788
	Var	0.852	0.850
	Max + RC	0.846	0.844
	Max + Var	0.886	0.877
	RC + Var	0.877	0.869
	Max + RC + Var	**0.890**	**0.885**
CID2013	Max	0.902	0.892
	RC	0.843	0.828
	Var	0.917	0.909
	Max + RC	0.902	0.888
	Max + Var	**0.925**	0.915
	RC + Var	0.922	**0.916**
	Max + RC + Var	0.924	0.913
CLIVE	Max	0.834	0.821
	RC	0.785	0.769
	Var	0.851	0.858
	Max + RC	0.866	0.857
	Max + Var	0.871	0.863
	RC + Var	0.861	0.855
	Max + RC + Var	**0.873**	**0.867**

From Table 5, it can be seen that the results of all combinations on these two databases are almost all above 0.8. Thus, the proposed method can yield better performance by all combinations. Especially, the combination of all three feature value calculation methods yields the best results. Furthermore, analyzing and comparing the data reveals that the Var plays a slightly greater role in feature combination than the Max and RC.

4.4. Performance on Different Scenarios on CID2013

The CID2013 database consists of six datasets (I–VI) and each dataset contains six different scenes [23]. This section focuses on testing images from different groups in the CID2013 database to verify the evaluation effect of the proposed method on images with the same content under different lighting conditions. The test results are shown in Table 6. In Table 6, the results of the same scenarios with different subjects are set in the same background color.

Table 6. Test results in different scenarios on CID2013.

Scenarios	SROCC	PLCC	Scenarios	SROCC	PLCC	Scenarios	SROCC	PLCC
IS_I_C01	0.789	0.801	IS_I_C02	0.878	0.934	IS_I_C03	0.989	0.991
IS_II_C01	0.709	0.860	IS_II_C02	0.781	0.745	IS_II_C03	0.841	0.934
IS_III_C01	0.979	0.991	IS_III_C02	0.772	0.741	IS_III_C03	0.985	0.990
IS_IV_C01	0.908	0.890	IS_IV_C02	0.955	0.997	IS_IV_C03	0.960	0.960
IS_V_C01	0.902	0.890	IS_V_C02	0.968	0.990	IS_V_C03	0.993	0.999
IS_VI_C01	0.866	0.938	IS_VI_C02	0.928	0.937	IS_VI_C03	0.966	0.991
IS_I_C04	0.964	0.956	IS_I_C05	0.884	0.930	IS_I_C06	0.964	0.984
IS_II_C04	0.968	0.948	IS_II_C05	0.877	0.836	IS_II_C06	0.775	0.733
IS_III_C04	0.972	0.971	IS_III_C05	0.844	0.871	IS_III_C06	0.791	0.807
IS_IV_C04	0.908	0.968	IS_IV_C05	0.977	0.996	IS_IV_C06	0.952	0.969
IS_VI_C07	0.849	0.967	IS_V_C05	0.975	0.961	IS_V_C06	0.765	0.833
IS_V_C08	0.965	0.984	IS_VI_C08	0.804	0.968	IS_VI_C06	0.743	0.983

A total of 36 groups were tested, and after analysis, more than 50% (20 groups) of the fitting results were above 0.90, and more than 75% (28 groups) of the fitting results were above 0.80. The worst SROCC and PLCC results from the above test were 0.7090 and 0.7326, respectively. In this part, the content of the test images was the same in each group, but the lighting conditions were different. From the test data, it can be concluded that the proposed method yields good and stable performance on IQA under different lighting conditions.

Based on the results from Table 6, two box charts of SROCC and the PLCC on different scenarios are shown in Figure 7. From Figure 7, the proposed method shows better performance in Scenes 3 and 4, which are all indoor scenarios with subject illuminance between 10 and 1000 lux. In addition, the proposed method also shows stable performance in Scenes 3 and 4. For the other scenarios, the proposed method yields similar performance.

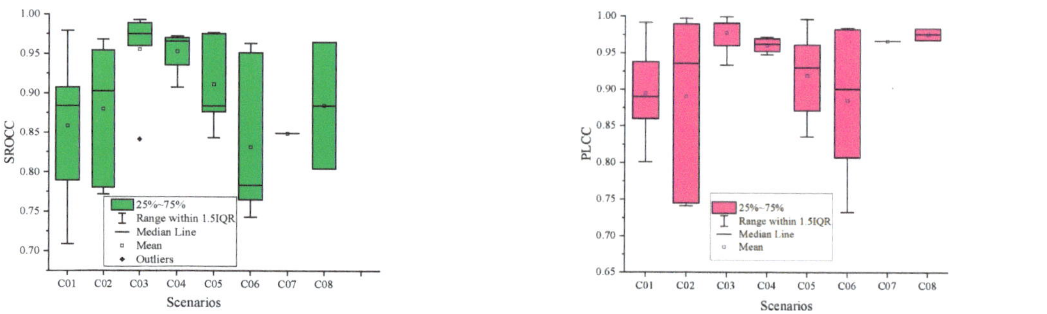

Figure 7. Box chart results of different scenarios on CID2013.

4.5. Scatter Plot and Fitting Curve

A total of 20 experiments were conducted on the BID, CID2013, and CLIVE databases to obtain 20 PLCC and SROCC values. The average value of these 20 values was selected as the final PLCC and SROCC values of the proposed method, and the results are presented in Table 4. Here, a random test was conducted on these databases, and the scatter plot of the proposed method's prediction results and subjective scores from the databases are shown in Figure 8. From Figure 8, it can be seen that the proposed methods perform well in evaluating the quality of real blurry images. The regression curve of the proposed method has a better correlation with the subjective observation values.

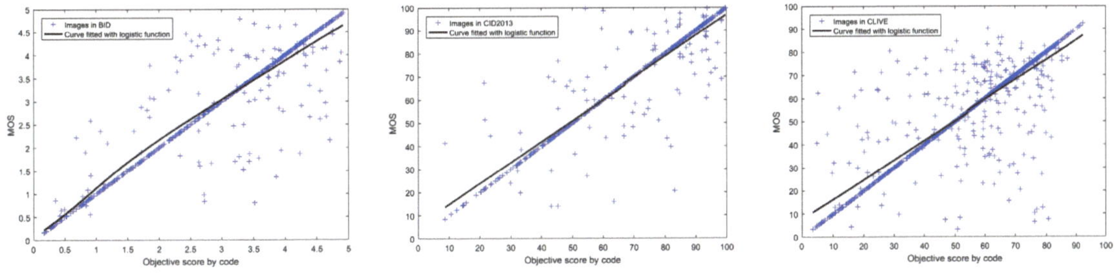

BID CID2013 CLIVE

Figure 8. Scatter plot and fitting curve of BID, CID2013, and CLIVE databases.

5. Conclusions

An NR image sharpness evaluation method for images under different lighting imaging conditions is proposed in this article. The proposed method consists of two parts, namely the feature values extraction part and the machine learning part using a PSO-GRNN. Firstly, the VS, CD, and gradient feature information are extracted from the test image and the related feature maps are obtained. Then, the Max, RC, and Var calculations are conducted on these feature maps to obtain the feature values. Lastly, the PSO algorithm is used to optimize the GRNN, and the image feature values are input into the PSO-GRNN to predict the image sharpness. Some tests are conducted on real databases, i.e., the BID, CID2013, and CLIVE databases, and some other state-of-the-art or widely cited leaning-based IQA methods are selected for comparison with the proposed method. The results indicate that the proposed method produces better prediction accuracy than all other competing methods. In the future, further studies can be conducted on the evaluation of the different specific illumination parameters.

Author Contributions: Conceptualization, H.H. and C.S.; methodology, H.H.; software, H.H.; validation, H.H., B.J., and C.S.; formal analysis, C.S.; investigation, Y.L. (Yandan Lin); resources, B.J.; data curation, H.H.; writing—original draft preparation, H.H.; writing—review and editing, C.S.; visualization, C.S.; supervision, Y.L. (Yuelin Lu); project administration, C.S.; funding acquisition, B.J. and C.S. All authors have read and agreed to the published version of the manuscript.

Funding: This research was funded by Key Project of Scientific Research in universities of Anhui Province, grant number 2022AH050983 and 2024AH050117, Anhui Future Technology Research Institute enterprise cooperation project, grant number 2023qyhz02, Anhui Engineering Research Center of Vehicle Display Integrated Systems Open Fund, grant number VDIS2023C01, Research Start-up Foundation for Introduction of Talents of AHPU, grant number 2021YQQ027, and Scientific Research Fund of AHPU, grant number Xjky20220003.

Institutional Review Board Statement: Not applicable.

Informed Consent Statement: Not applicable.

Data Availability Statement: Data are contained within the article.

Conflicts of Interest: The authors declare no conflicts of interest.

References

1. Sachin; Kumar, R.; Sakshi; Yadav, R.; Reddy, S.G.; Yadav, A.K.; Singh, P. Advances in Optical Visual Information Security: A Comprehensive Review. *Photonics* **2024**, *11*, 99. [CrossRef]
2. Xu, W.; Wei, L.; Yi, X.; Lin, Y. Spectral Image Reconstruction Using Recovered Basis Vector Coefficients. *Photonics* **2023**, *10*, 1018. [CrossRef]
3. Sun, X.; Kong, L.; Wang, X.; Peng, X.; Dong, G. Lights off the Image: Highlight Suppression for Single Texture-Rich Images in Optical Inspection Based on Wavelet Transform and Fusion Strategy. *Photonics* **2024**, *11*, 623. [CrossRef]

4. Qiu, J.; Xu, H.; Ye, Z.; Diao, C. Image quality degradation of object-color metamer mismatching in digital camera color reproduction. *Appl. Opt.* **2018**, *57*, 2851–2860. [CrossRef]
5. Liu, C.; Zou, Z.; Miao, Y.; Qiu, J. Light field quality assessment based on aggregation learning of multiple visual features. *Opt. Express* **2022**, *30*, 38298–38318. [CrossRef]
6. Kim, B.; Heo, D.; Moon, W.; Hahh, J. Absolute Depth Estimation Based on a Sharpness-assessment Algorithm for a Camera with an Asymmetric Aperture. *Curr. Opt. Photonics* **2021**, *5*, 514–523.
7. Baig, M.A.; Moinuddin, A.A.; Khan, E. A simple spatial domain method for quality evaluation of blurred images. *Multimed. Syst.* **2024**, *30*, 28. [CrossRef]
8. Wang, Z.; Bovik, A.; Sheikh, H. Image Quality Assessment: From error visibility to structural similarity. *IEEE Trans. Image Process.* **2004**, *13*, 600–612. [CrossRef]
9. Shi, C.; Lin, Y. Full reference image quality assessment based on visual salience with color appearance and gradient similarity. *IEEE Access* **2020**, *8*, 97310–97320. [CrossRef]
10. Dost, S.; Saud, F.; Shabbir, M.; Khan, M.G.; Shahid, M.; Lovstrom, B. Reduced reference image and video quality assessments: Review of methods. *EURASIP J. Image Video Process.* **2022**, *2022*, 1–31. [CrossRef]
11. Bahrami, K.; Kot, A.C. A fast approach for no-reference image sharpness assessment based on maximum local variation. *IEEE Signal Process. Lett.* **2014**, *21*, 751–755. [CrossRef]
12. Li, L.; Lin, W.; Wang, X.; Yang, G.; Bahrami, K.; Kot, A.C. No-reference image blur assessment based on discrete orthogonal moments. *IEEE Trans. Cybern.* **2017**, *46*, 39–50. [CrossRef] [PubMed]
13. Gvozden, G.; Grgic, S.; Grgic, M. Blind image sharpness assessment based on local contrast map statistics. *J. Vis. Commun. Image Represent.* **2018**, *50*, 145–158. [CrossRef]
14. Zhu, M.; Yu, L.; Wang, Z.; Ke, Z.; Zhi, C. Review: A Survey on Objective Evaluation of Image Sharpness. *Appl. Sci.* **2023**, *13*, 2652. [CrossRef]
15. Li, L.; Xia, W.; Lin, W.; Fang, Y.; Wang, S. No-Reference and Robust Image Sharpness Evaluation Based on Multiscale Spatial and Spectral Features. *IEEE Trans. Multimed.* **2017**, *19*, 1030–1040. [CrossRef]
16. Lu, Q.; Zhou, W.; Li, H. A no-reference image sharpness metric based on structural information using sparse representation. *Inf. Sci.* **2016**, *369*, 334–346. [CrossRef]
17. Yu, S.; Wu, S.; Wang, L.; Jiang, F.; Xie, Y.; Li, L. A shallow convolutional neural network for blind image sharpness assessment. *PLoS ONE* **2017**, *12*, e0176632. [CrossRef]
18. Kim, J.; Nguyen, A.D.; Lee, S. Deep CNN-based blind image quality predictor. *IEEE Trans. Neural Netw. Learn. Syst.* **2019**, *30*, 11–24. [CrossRef]
19. Liu, L.; Liu, B.; Huang, H.; Bovik, A.C. No-reference image quality assessment based on spatial and spectral entropies. *Signal Process. Image Commun.* **2014**, *29*, 856–863. [CrossRef]
20. Li, D.; Jiang, T.; Lin, W.; Jiang, M. Which has better visual quality: The clear blue sky or a blurry animal? *IEEE Trans. Multimed.* **2019**, *21*, 1221–1234. [CrossRef]
21. Zhang, W.X.; Ma, K.D.; Yan, J.; Deng, D.; Wang, Z. Blind image quality assessment using a deep bilinear convolutional neural network. *IEEE Trans. Circuits Syst. Video Technol.* **2020**, *30*, 36–47. [CrossRef]
22. Ciancio, A.; da Costa, A.L.N.T.; Silva, E.A.B.D.; Said, A.; Samadani, R.; Obrador, P. No-reference blur assessment of digital pictures based on multifeature classifiers. *IEEE Trans. Image Process.* **2011**, *20*, 64–75. [CrossRef] [PubMed]
23. Toni, V.; Mikko, N.; Mikko, V.; Pirkko, O.; Jukka, H. CID2013: A database for evaluating no-reference image quality assessment algorithms. *IEEE Trans. Image Process.* **2015**, *24*, 390–402.
24. Deepti, G.; Alan, C.B. Massive online crowd sourced study of subjective and objective picture quality. *IEEE Trans. Image Process.* **2016**, *25*, 372–387.
25. Kim, W.; Kim, C. Saliency detection via textural contrast. *Opt. Lett.* **2012**, *37*, 1550–1552. [CrossRef]
26. Zahra, S.S.; Karim, F. Visual saliency detection via integrating bottom-up and top-down information. *Optik* **2019**, *178*, 1195–1207.
27. Zhang, L.; Shen, Y.; Li, H. VSI: A visual saliency-induced index for perceptual image quality assessment. *IEEE Trans. Image Process.* **2014**, *23*, 4270–4281. [CrossRef]
28. Shi, C.; Lin, Y. No reference image sharpness assessment based on global color difference variation. *Chin. J. Electron.* **2024**, *33*, 293–302. [CrossRef]
29. Varga, D. Full-Reference Image Quality Assessment Based on Grünwald–Letnikov Derivative, Image Gradients, and Visual Saliency. *Electronics* **2022**, *11*, 559. [CrossRef]
30. Li, C.; Bovik, A.C.; Wu, X. Blind Image Quality Assessment Using a General Regression Neural Network. *IEEE Trans. Neural Netw.* **2011**, *22*, 793–799.
31. Zhao, M.; Ji, S.; Wei, Z. Risk prediction and risk factor analysis of urban logistics to public security based on PSO-GRNN algorithm. *PLoS ONE* **2020**, *15*, e0238443. [CrossRef] [PubMed]
32. Rahim, H.S.; Farooq, M.S.; Conrad, A.B. A statistical evaluation of recent full reference image quality assessment algorithms. *IEEE Trans. Image Process.* **2006**, *15*, 3440–3451.

33. Moorthy, A.K.; Bovik, A.C. Blind image quality assessment: From natural scene statistics to perceptual quality. *IEEE Trans. Image Process.* **2011**, *20*, 3350–3364. [CrossRef]
34. Mittal, A.; Soundararajan, R.; Bovik, A.C. Making a 'completely blind' image quality analyzer. *IEEE Signal Process. Lett.* **2013**, *20*, 209–212. [CrossRef]

photonics

Article

Accurate Inspection and Super-Resolution Reconstruction for Additive Manufactured Defects Based on Stokes Vector Method and Deep Learning

Shangrongxi Sun [1], Xing Peng [1,2,*] and Hongbing Cao [1,2]

1 College of Intelligent Science and Technology, National University of Defense Technology, Changsha 410073, China; 18786111718@163.com (S.S.); hb_cao@163.com (H.C.)
2 Hunan Provincial Key Laboratory of Ultra-Precision Machining Technology, Changsha 410073, China
* Correspondence: pengxing22@nudt.edu.cn

Abstract: Defects in additive manufacturing processes are closely related to the mechanical and physical properties of the components. However, the extreme conditions of high temperatures, intense light, and powder during the manufacturing process present significant challenges for defect detection. Additionally, the high reflectivity of metallic components can cause pixels in image sensors to become overexposed, resulting in the loss of many defect signals. Thus, this paper mainly focuses on proposing an accurate inspection and super-resolution reconstruction method for additive manufactured defects based on Stokes vector and deep learning, where the Stokes vectors, polarization degree, and polarization angles of the inspected defects are effectively utilized to suppress the high reflectivity of metallic surfaces, enhance the contrast of defect regions, and highlight the boundaries of defects. Furthermore, a modified SRGAN model designated SRGAN-H is presented by employing an additional convolutional layer and activation functions, including Harswish and Tanh, to accelerate the convergence of the SRGAN-H network and improve the reconstruction of the additive manufactured defect region. The experiment results demonstrated that the SRGAN-H model outperformed SRGAN and traditional SR reconstruction algorithms in terms of the images of Stokes vectors, polarization degree, and polarization angles. For the scratch and hole test sets, the *PSNR* values were 33.405 and 31.159, respectively, and the *SSIM* values were 0.890 and 0.896, respectively. These results reflect the effectiveness of the SRGAN-H model in super-resolution reconstruction of scratch and hole images. For the scratch and hole images chosen in this study, the *PSNR* values of SRGAN-H for single image super-resolution reconstruction ranged from 31.86786 to 43.82374, higher than the results obtained by the pre-improvement SRGAN algorithm.

Keywords: polarization technology; super resolution; additive manufacturing; SRGAN

Citation: Sun, S.; Peng, X.; Cao, H. Accurate Inspection and Super-Resolution Reconstruction for Additive Manufactured Defects Based on Stokes Vector Method and Deep Learning. *Photonics* **2024**, *11*, 874. https://doi.org/10.3390/photonics11090874

Received: 15 August 2024
Revised: 12 September 2024
Accepted: 14 September 2024
Published: 18 September 2024

1. Introduction

Additive manufacturing technology is a highly competitive, cost-effective, and highly flexible manufacturing technology widely used in fields such as aerospace, military, medical equipment, energy, and automotive manufacturing. However, there are still significant limitations in the production process of additive manufacturing, mainly due to two reasons: quality and repeatability, which may be severely affected by certain defects (such as cracks, and spheroidization) in the additive manufacturing process [1]. Defects in the SLM process are closely related to the mechanical and physical properties of the parts, and extreme environments such as high temperature, intense light, and powder splashing during processing pose great challenges to defect detection. In this paper, in response to the above problems and detection requirements, visual inspection is adopted to enhance defect detection capabilities and improve the quality of target images. The focus is on defect image detection and defect image super-resolution reconstruction. This provides solutions and references for defect detection and optimization of processing parameters for laser additive manufacturing workpieces, which are of great significance for the theory and application of defect detection.

So far, SLM technology has successfully processed various alloys and metals, including aluminum alloys [2], stainless steel [3], nickel-based superalloys [4], and titanium alloys [5]. In the SLM process, many factors that affect part quality, including powder size, laser power, scanning speed, etc., and improper parameter control can lead to defect formation and serious degradation of the physical and mechanical properties of parts [6–9]. The generation of defects will seriously degrade the physical and mechanical properties of the workpiece, so defect detection technology is of great significance for improving the quality and repeatability of processed parts. Many defect detection methods used in the SLM process have been applied in fields such as the nuclear industry, aerospace, mechanical manufacturing, and petrochemical industry [10].

Depending on the sensor type, additive manufacturing defect detection technologies include the use of high-speed cameras, infrared thermal imagers, photodiodes and thermocouples, X-ray microimaging, and acoustic techniques. Valeev et al. found that any changes in input thermal parameters during the SLM process would lead to direct changes in thermal radiation [11]. Therefore, photodiodes and thermocouples are used to measure radiation conditions. Zhang et al. employed a high-speed camera to monitor the dynamic changes in the melt pool and proposed a novel approach based on the use of a wavelength-selective filter (350–800 nm) to enhance the contrast between the melt pool and the surrounding fluid, which was further developed into a new image processing method and extracted features from the melt pool, the fluid, and the particles ejected from the melt pool [12]. This method was designed to provide a comprehensive understanding of the SLM process. Bisht et al. [13] introduced a novel SLM process melt pool monitoring tool, designated as DMP (developed melt pool), and employed a Ge photodiode to monitor the melt pool and a manual data analysis method to assess the quality of the manufactured parts, demonstrating that the DMP melt pool tool is capable of detecting and marking changes in the signal during the manufacturing process, thereby enabling effective prediction of the quality of the manufactured parts.

Yamamoto et al. [14] investigated the melting and solidification processes, employing a dual-color thermometer to assess the energy equilibrium and the quality of the solidifying material during SLM. Their findings indicated that the temperature distribution in the heat-affected zone was asymmetrical, with a more gradual temperature gradient observed in the direction of material solidification than in the direction of powder solidification.

Krauss et al. [15] investigated the impact of insufficient heat dissipation on the precision of manufacturing processes, employing thermal distribution analysis to identify porosity and irregularities. Bartlett [16] developed a novel and full-field monitoring system based on infrared thermography, which was utilized to assess the quality of AlSi10Mg components during SLM processing, demonstrating that infrared thermography was capable of detecting 82% of the identified defects in the SLM-fabricated parts.

However, the aforementioned detection methods employ single-signal measurement techniques, which are limited in terms of precision, scope, and information content. This renders them inadequate for the detection of surface defects in complex environments. In response to these challenges, a detection system based on polarization imaging has been proposed. By optimizing the optical configuration of the imaging system, clear images of the defects can be obtained.

In comparison with traditional optical detection techniques, polarization imaging technology offers a distinctive advantage, as it enables the acquisition of information on the spectral characteristics, polarization, and spatial configuration of the target. By leveraging polarization technology for defect detection, it is possible to extract valuable insights, such as structural patterns, surface materials, and surface roughness from the polarization information of the target. This approach can effectively enhance the precision and reliability of the detection process [17,18]. Ref. [19] demonstrated that, in the context of laser-based additive manufacturing, the use of metallic components with high reflectivity can result in the saturation of image sensors, leading to the obfuscation of a significant proportion of the defect information. Consequently, the incorporation of polarization technology within the framework of defect detection systems has been proposed as a means of mitigating the adverse effects of metallic surfaces with high reflectivity, enhancing the contrast between defect regions and facilitating the delineation of defect boundaries.

The representation of defects on the surface of additive manufactured components is a challenging process, requiring not only improvements to hardware but also significant time and resources. Consequently, enhancing image quality through the application of super-resolution software algorithms is represented as a crucial area of research in this field. The accurate detection and extraction of defect regions, along with their geometric characteristics, are essential for the effective and efficient characterization of manufactured components.

Many scholars have made significant contributions to the field of super-resolution reconstruction algorithms. Existing super-resolution reconstruction algorithms can be broadly classified into three categories: those based on interpolation, those based on reconstruction, and those based on learning. Algorithms based on interpolation primarily utilize the surrounding pixel values of existing images to insert pixels at the center point, resulting in a reconstructed image. Commonly used interpolation algorithms include nearest neighbor interpolation, bilinear interpolation, and trilinear interpolation. However, these algorithms often result in reconstructed images with blurred edges and certain geometric distortions, as discussed in [20–22]. In the domain of learning-based approaches, Dong et al. [23] pioneered the integration of convolutional neural networks (CNN) into super-resolution reconstruction, proposing a convolutional neural network-based super-resolution reconstruction algorithm, namely SRCNN, which learns the mapping between low-resolution (LR) and high-resolution (HR) images through a convolutional neural network, enabling the reconstruction of HR images from LR inputs. In contrast to traditional algorithms and machine learning methods, a mere three layers of convolution are employed in SRCNN. This superior performance paves the way for subsequent research. RCAN [24] marked the first introduction of channel attention, which effectively leverages channel-specific characteristics, significantly enhancing reconstruction quality over previous methods, and enabling the optimal utilization of channel features, resulting in a significant enhancement in reconstruction quality compared with previous algorithms. The large-scale convolution kernel is capable of extracting large-scale characteristics, while the small-scale convolution kernel is adept at extracting small-scale characteristics.

Li et al. [25] proposed a method for image super-resolution reconstruction based on a multi-scale residual network (MSRN), employing a multi-scale residual block (MSRB) as its fundamental nonlinear mapping unit, utilizing a dual-path structure with a 3×3 and 5×5 convolution to extract multi-scale features. Ledig et al. [26] proposed a GAN-based image super-resolution reconstruction network (GAN for SR, SRGAN), whose object was to recover high-frequency details in images that have been subjected to a loss of perceptual fidelity. It was capable of enlarging low-resolution natural images by a factor of four and enhancing the visual perception quality of the resulting SR images. Then, a ranking-based generative adversarial network (Rank-SRGAN) was proposed by Zhang et al. [27], employing a learning-based approach to simulate perceptual metrics, thereby enhancing the visual quality of generated images without modifying the underlying generative network. A physical GAN framework [28] was proposed to enhance the fidelity of generated images by refining the discriminator network, achieved by introducing a new criterion for evaluating the consistency between HR and SR images.

However, the majority of super-resolution reconstruction algorithms are designed for the recovery and reconstruction of natural images. Consequently, there is a significant gap in the research literature on super-resolution reconstruction of SLM defect images. This paper proposes a super-resolution reconstruction algorithm based on an improved SRGAN model for the reconstruction of defect images on the surface of laser-based additive manufacturing parts. As a continuation of previous work on SRGAN [26], we present a defect detection and deep learning system, called SRGAN-H. The objective is to enhance the quality of real SLM defect images reconstructed as SR images. This model is based on the SRGAN framework and comprises a generator and a discriminator. In the generator model, the Hardswish activation function [29] is introduced into the depth residual module to enhance the accuracy of the neural network. As with the existing SRGAN model, SRGAN-H employs multiple depthwise separable convolutions for feature extraction, and recurrent connections are utilized to link the input to the network output, thereby ensuring network stability [26]. At the output layer, a convolutional layer is added and the Tanh activation function is introduced to normalize the output to the range of $[-1, 1]$, shown to alleviate the issues of gradient vanishing and gradient explosion that are commonly encountered in deep learning networks, with the result that the networks in question can converge more rapidly.

In this paper, the detection and representation of defects in SLM processes are discussed. By introducing polarization techniques into defect detection systems, the detection capability of SLM defects is enhanced. Four sets of defect detection images were collected at polarization angles of 0°, 45°, 90°, and 135°. These images were converted into Stokes vectors, polarization degree, and polarization angles using mathematical formulas. Subsequently, SRGAN-H was applied to the aforementioned four groups of images. The SR images exhibited enhanced PNSR, *SSIM*, and *SD* parameters, providing a theoretical and practical foundation for further improvements to the processing parameters and the development of defect classification and prediction capabilities. The following is the organizational structure of this paper: Sections 2 and 3 introduce the defect detection and characterization in the SLM process, respectively, providing a methodology for the research. Section 4 presents experimental findings, discussing the image results of defect detection and the SR images processed by SRGAN-H, thereby verifying the effectiveness of the aforementioned methods. Finally, Section 5 summarizes the research and plan future work.

2. Defect Detection System Based on Stokes Properties

Common metal materials have strong reflective properties and the high reflectivity of metal workpiece surfaces can cause pixels of the image sensor to be overexposed. To address this issue, a polarization technology and a defect detection system based on Stokes properties have been introduced. By changing the polarization angle to obtain multiple sets of polarized images, the Stokes vector images, polarization degree images, and polarization angle images were solved. Analysis was conducted on the contrast of defect regions in the target image, defect contour information, and the effects of high reflectivity suppression, demonstrating the importance of Stokes properties in defect detection.

2.1. Polarization of Light and Stokes Vector Method

In the defect detection process based on polarization technology, the Stokes vector method was used to represent polarized light to extract the polarization state information of the defect detection images. The Stokes vector method can be used to represent the polarization state information and intensity of polarized light [30], indicating that any polarization state of light can be represented by four Stokes vectors, defined as:

$$S(x,y) = \begin{bmatrix} S_0(x,y) \\ S_1(x,y) \\ S_2(x,y) \\ S_3(x,y) \end{bmatrix} = \begin{bmatrix} I_0(x,y) + I_{90}(x,y) \\ I_0(x,y) - I_{90}(x,y) \\ I_{45}(x,y) - I_{135}(x,y) \\ I_{right}(x,y) - I_{left}(x,y) \end{bmatrix} \quad (1)$$

where $S(x,y)$ represents the Stokes vector. According to the Stokes vector method, any polarized light can be represented by the angle of polarization (*Aop*), the degree of polarization (*Dop*), and the ellipticity of polarization ϖ:

$$\begin{cases} Aop = \frac{1}{2}\arctan\frac{S_2}{S_1} \\ Dop = \frac{\sqrt{S_1{}^2+S_2{}^2+S_3{}^2}}{S_0} \\ \varpi = \frac{1}{2}\arcsin\frac{S_3}{S_0} \end{cases} \tag{2}$$

When the incident light is imaged by the polarization detection system, the Stokes vector of the incident light is denoted by $S(x,y)$, while the Stokes vector of the outgoing light is denoted as follows:

$$S'(x,y) = M_u \cdot S(x,y) = \begin{bmatrix} M_{11} & M_{12} & M_{13} & M_{14} \\ M_{21} & M_{22} & M_{23} & M_{24} \\ M_{31} & M_{32} & M_{33} & M_{34} \\ M_{41} & M_{42} & M_{43} & M_{44} \end{bmatrix} \cdot \begin{bmatrix} S_0(x,y) \\ S_1(x,y) \\ S_2(x,y) \\ S_3(x,y) \end{bmatrix} \tag{3}$$

2.2. Detection System Based on Polarization Technology

The subject of this study was an SLM workpiece, as illustrated in Figure 1. The instrumentation employed in this investigation included the subject workpiece, a light-emitting diode (LED) light source, a linear polarizer, an electric rotating platform, a charge-coupled device (CCD) camera, and a computer. The linear polarizer utilized in this experiment was the THORLABS LPVISE100-A linear polarizer, with a dimension of 25. The CMOS image sensor had a resolution of 2448 × 2448 pixels, with a single pixel size of 3.45 μm. The lens focal length was 50 mm. The motorized rotating platform was a Standa FPSTA-8MPR16-1 (Standa, Wrocław, Poland), which enabled 360° rotation and 0.75 arcmin step resolution for the polarization filter rotation control, and the controller was an 8SMC4-USB.

Figure 1. Polarization-based visual detection system.

The computer utilized in the experiment for image acquisition and analysis was a ThinkPad S2, equipped with an Intel Core i5 8250U processor, operating at a maximum frequency of 1.8 GHz. During the experiment, the test specimen was illuminated by an LED light source and the reflected light was captured by a CMOS image sensor mounted on a rotating platform. The metallic test specimen exhibited a high degree of reflection. By controlling the rotation of the platform and adjusting the angle of the polarizing filter, the computer was used to capture and analyze four sets of images—S_1, S_2, *Aop*, and *Dop*—with different polarization angles, using the Stokes vector method. The Ti6Al4V powder was used on a commercial PBF system Arcam A2X, specifically designed to withstand extremely high process temperatures over 1100 °C. The power of the electron beam could be adjusted between 50 and 3000 W. A layer thickness of 50 μm was chosen for the process, with a spot size of approximately 250 μm in the focal position. The scanning speed was 4530 mm/s and the vacuum degree was 0.2 Pa. The observed specimen was manufactured on a 150 mm × 150 mm stainless steel baseplate mounted on a 200 mm × 200 mm plate. Thus, the preliminary detection of workpiece defects was completed.

3. Fundamentals for the SRGAN-H Defect Reconstruction Model

After the detection of defects in the S_1, S_2, Aop, and Dop, the objective was to enhance the quality of the images obtained through the use of super-resolution software algorithms. This would facilitate the precise extraction of the boundaries and fundamental characteristics of the defects. An improved SRGAN [26] model was presented in this section, which proposed the SRGAN-H model by optimizing the generator structure to effectively reconstruct SLM process defect images. This method employed a generator-discriminator network to achieve super-resolution reconstruction, enabling the generation and training of a generator to deceive the discriminator. The discriminator was trained to distinguish between true images and super-resolution images. In a self-generating and self-discriminating process, a high-quality SR (super-resolution reconstruction) image was generated.

3.1. Generator

During the adversarial process between the generator and discriminator, the goal of the generator is to generate SR images that can confuse the discriminator as much as possible, i.e., to minimize the probability that the discriminator classifies its generated images as fake; while the goal of the discriminator is to distinguish between real and fake images as accurately as possible, i.e., to maximize the probability of correct classification. Ledig [26] solved the minimax problem by alternately optimizing the discriminator and generator, which can be represented by Formula (4).

$$\underset{\theta_G}{min}\underset{\theta_D}{max} \quad \mathbb{E}_{I^{HR}\sim p_{train}(I^{HR})}\left[\log D_{\theta_D}\left(I^{HR}\right)\right] + \mathbb{E}_{I^{LR}\sim p_G(I^{LR})}\left[\log\left(1 - D_{\theta_D}\left(G_{\theta_G}\left(I^{LR}\right)\right)\right)\right] \tag{4}$$

The formula is used to train a generator G, which is then used to deceive the discriminator D. The discriminator is designed to distinguish between SR images and real images. This approach enables the generator to be trained to produce images that are highly similar to genuine images.

Regarding the structure of the generator network, SRGAN-H is based on the SRResNet model, comprising three convolutional layers with 3 × 3 kernels and 64 features, and incorporates Parametric ReLU [31] and Hardswish [29] as activation functions to overcome the issues of blurred edges and gradient disappearance caused by the convolutional layers. Then, a batch normalization layer [32] is incorporated. Prajit et al. [33] have demonstrated that Swish is more effective than ReLU in deeper models, proving that a mere replacement of the ReLU unit with the Swish unit is sufficient to achieve an increase in the classification accuracy of 0.9% on the ImageNet dataset. In contrast, Andrew et al. [29] demonstrated that Hardswish is superior to Swish in numerical stability and computation speed. Consequently, this section introduces Hardswish activation functions to enhance the accuracy and speed of the model, which can be expressed as follows:

$$h - swish(x) = x\frac{\mathrm{ReLU6}(x+3)}{6} \tag{5}$$

After the completion of the residual module, the image is processed through convolutional layers and batch normalization layers [32], and the sub-pixel convolution module [34] is used to achieve a ×4 magnification of the image. At the image output stage, a convolutional layer and Tanh function are introduced in the generator network to normalize the output to the [−1, 1] range. This not only reduces the differentiation between pixels during upscaling but also prevents gradient explosion during neural network training. This establishes the generator network, with its specific structure shown in Figure 2.

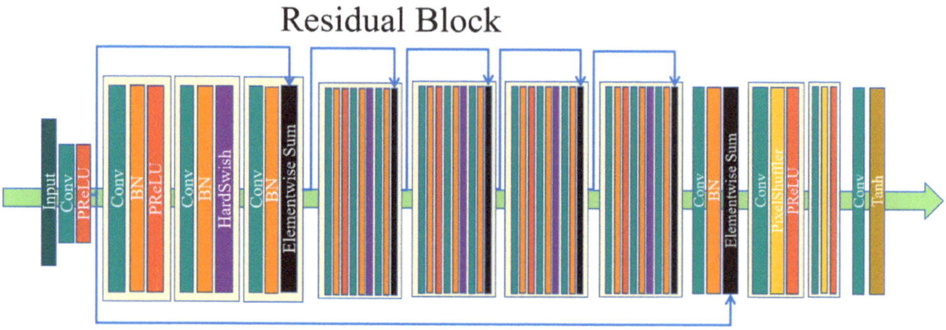

Figure 2. The network architecture of the generator.

3.2. Discriminator

According to Radford [35], Ledig [26] trained a discriminator network with 3 × 3 filter kernels, increasing the number of filter kernels from 64 to 512 and using Leaky ReLU activation (α = 0.2), avoiding the entire max-pooling network to solve the maximization problem in Formula (4). The resulting 512 feature maps are passed through two dense layers with Leaky ReLU and sigmoid activations, ultimately resulting in a possible classification of real or fake images. The network structure is illustrated in Figure 3. In the output, Dense(1024) and Dense(1) are used, where 'Dense(1024)' represents a fully connected layer with 1024 neurons, and 'Dense(1)' represents a fully connected layer with one neuron.

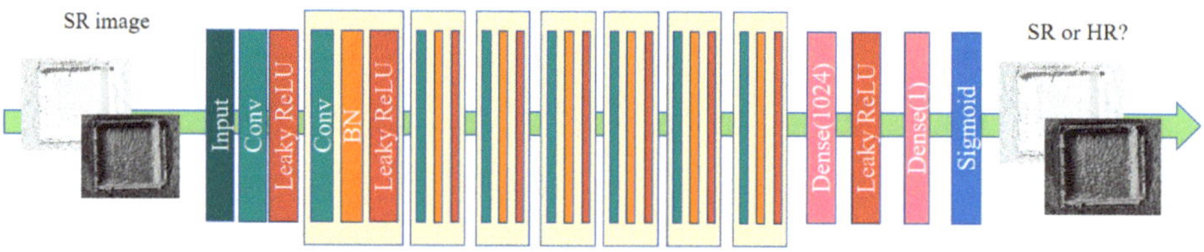

Figure 3. The network architecture of the discriminator.

3.3. Perceptual Loss Function

The definition of the loss function is crucial to the quality of the reconstructed images. SRGAN-H continues Ledig et al.'s design of the perceptual loss function, which represents the perceptual loss as a combination of content loss and adversarial loss [26], with the specific formulas as follows:

$$l^{SR} = l_X^{SR} + 10^{-3} l_{Gen}^{SR} \tag{6}$$

where l_X^{SR} represents content loss and l_{Gen}^{SR} represents an adversarial loss. For the training of SRResNet, Ledig et al. utilized the MSE loss function to improve the evaluation of *PSNR*, which can be expressed as:

$$l_X^{SR} = l_{MSE}^{SR} = \frac{1}{r^2 \cdot W \cdot H} \sum_{x=1}^{rW} \sum_{y=1}^{rH} \left(I_{x,y}^{HR} - G_{\theta_G}\left(I^{LR}\right)_{x,y} \right)^2 \tag{7}$$

For the training of SRGAN, Ledig et al. defined the VGG loss based on the ReLU activation layers of a 19-layer VGG network pre-trained by Simonyan and Zisserman [36], where $\phi_{i,j}$ indicates the feature maps obtained by activating the j-th convolutional layer before the i-th max pooling layer within the VGG19 network. Finally, the VGG loss function is defined as the Euclidean distance between the reconstructed image $G_{\theta_G}(I^{LR})$ and the reference image I^{HR}, which can be expressed as:

$$l^{SR}_{\frac{VGG}{i}.j} = \frac{1}{W_{i,j}H_{i,j}} \sum_{x=1}^{W_{i,j}} \sum_{y=1}^{H_{i,j}} \left(\phi_{i,j}(I^{HR})_{x,y} - \phi_{i,j}(G_{\theta_G}(I^{LR}))_{x,y}\right)^2 \tag{8}$$

For the adversarial loss function, Ledig et al. defined it as shown in Formula (9), where l^{SR}_{Gen} represents the generative loss, which is defined based on the discriminator's probabilities over all training samples, and $D_{\theta_D}(G_{\theta_G}(I^{LR}))$ indicates the probability of the reconstructed image $G_{\theta_G}(I^{LR})$ being classified as a real image [26]. By minimizing this loss, it aims to deceive the discriminator by generating outputs with higher discriminative values:

$$l^{SR}_{Gen} = \sum_{n=1}^{N} -logD_{\theta_D}(G_{\theta_G}(I^{LR})) \tag{9}$$

3.4. Representation of Defective Images of SLM

For scratch-type (#1) and hole-type (#2) defect images, four original images are obtained by varying the polarizer angles (0°, 45°, 90°, and 135°). However, these four images exhibit poor visual quality and cannot display the defect contours and information. Therefore, using Formulas (2) and (3), two sets of transformed images (*Aop*, *Dop*, S_1, S_2) are calculated. The transformed images demonstrate significant improvements in contrast and brightness, though the defect contours remain unclear. Consequently, the SRGAN-H model is employed to perform super-resolution reconstruction (SR) on the images (*Aop*, *Dop*, S_1, S_2). This approach ultimately facilitates the detection and processing of all defect images. The detailed process is illustrated in Figure 4.

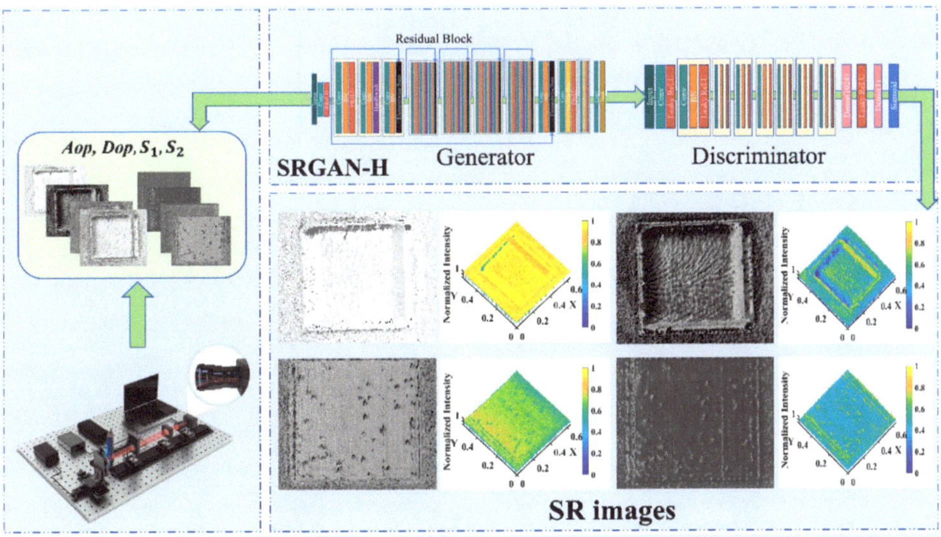

Figure 4. Defect detection and characterization system based on Stokes property and SRGAN-H.

4. Experiments

4.1. Quantitative Evaluation Metrics

To quantitatively evaluate the quality of the detected and characterized images, the experimental process involves using the V-channel value to assess the quality of defect detection images in the SLM manufacturing process. Then, *PSNR*, *SSIM*, and *SD* metrics are employed to evaluate the quality of defect image super-resolution reconstruction.

4.1.1. Peak Signal-to-Noise Ratio (*PSNR*)

The *PSNR* stands for peak signal-to-noise ratio, which is a metric used to measure image quality, typically employed to assess the similarity between reconstructed images and the original images [37]. It can be represented by Formula (10):

$$PSNR = 10 \cdot \log_{10}\left(\frac{MAX_I^2}{MSE}\right) = 20 \cdot \log_{10}\left(\frac{MAX_I}{\sqrt{MSE}}\right) \tag{10}$$

where *MSE* represents mean squared error, and a higher *PSNR* value indicates a higher similarity between the reconstructed image and the original image, resulting in lower image distortion.

4.1.2. Structural Similarity Degree (*SSIM*)

SSIM stands for the structural similarity index, which is a metric used to measure image quality. It takes into account factors such as luminance, contrast, and structure to evaluate the similarity between two images [38], which can be represented by formula (11):

$$SSIM(x,y) = \frac{(2\mu_x\mu_y + C_1)(2\sigma_{xy} + C_2)}{(\mu_x^2 + \mu_y^2 + C_1)(\sigma_x^2 + \sigma_y^2 + C_2)} \tag{11}$$

where μ_x, μ_y represent the means of x, y. σ_x^2, σ_y^2 represent the variances of x, y. σ_{xy} represents the covariance. C_1, C_2 are constants. The value of *SSIM* closer to 1 indicates a higher similarity in structure, luminance, and contrast between the reconstructed image and the original image, indicating better image quality.

4.1.3. Standard Deviation (*SD*)

Standard deviation (*SD*) typically refers to the standard deviation of the brightness or color distribution of an image, serving as a statistical measure of the differences between image pixels. It can be represented by Formula (12), where a larger *SD* value indicates bigger differences between pixels.

$$\mathrm{SD} = \sqrt{\frac{1}{N}\sum_{i=1}^{N}(x_i - \overline{x})^2} \tag{12}$$

4.2. *Generating Defect Detection Images of SLM*

Four sets of defect detection images for scratch-type (#1) and hole-type (#2) were acquired at polarization angles of 0°, 45°, 90°, and 135°, as shown in Figure 5.

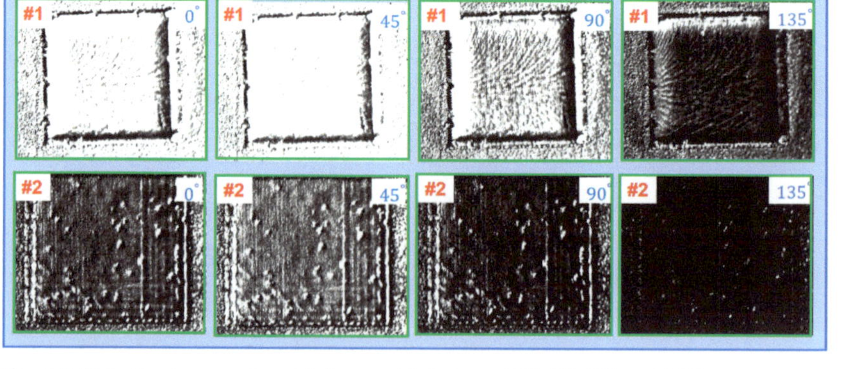

Figure 5. Detection images of scratch-type and hole-type at 0°, 45°, 90°, and 135°.

According to Formulas (2) and (3), the polarization angle images *Aop*, the degree of polarization images *Dop*, and the Stokes vector images S_1 and S_2 were calculated, as shown in Figures 6 and 7, respectively.

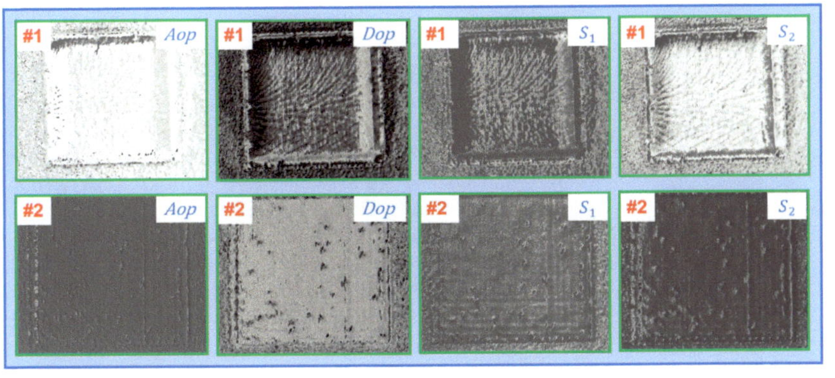

Figure 6. Transformed polarization angle image *Aop*, degree of polarization image *Dop*, and Stokes vector S_1 and S_2 images.

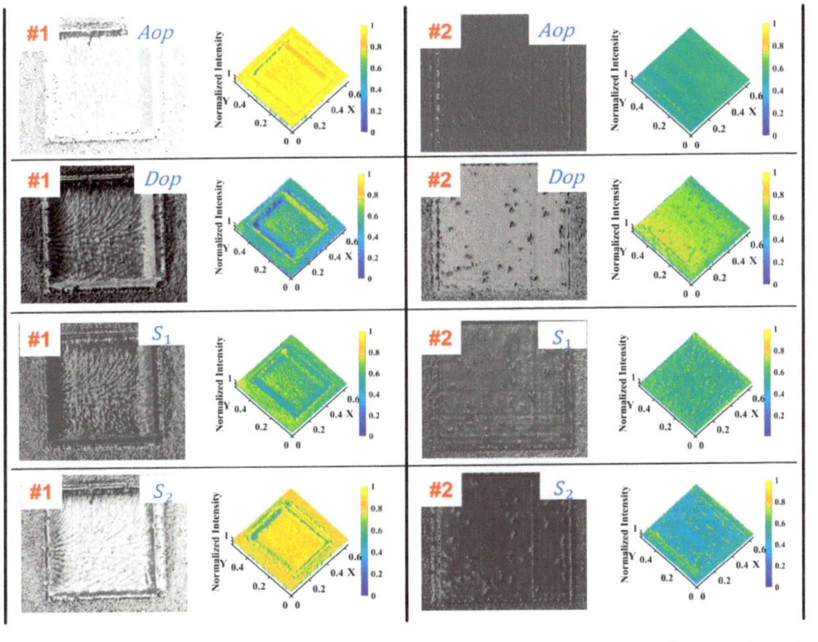

Figure 7. The detection results of scratch-type and hole-type defects in the *Aop*, *Dop*, S_1, and S_2 images.

Furthermore, the V-channel value of *Aop*, *Dop*, S_1, and S_2 were evaluated. As shown in Figure 7, the Stokes vector images S_1 and S_2 represented the linear polarization information of the defect image, which had a more uniform grayscale distribution compared with the original image. The edge contours of some key defect areas in the image were highlighted, but some details were lost.

In contrast, the degree of polarization images *Dop* and the polarization angle images *Aop* provided clearer details of the defect areas compared with the Stokes vector images S_1

and S_2. It can be seen that the defect detection method based on Stokes properties effectively suppressed high radiation metal, enhanced contrast, and highlighted defect contours.

Next, applying the SRGAN-H technology, the *Aop*, *Dop*, S_1, and S_2 images were reconstructed through super resolution (SR) to further enhance their detail and edge information.

4.3. Super-Resolution (SR) Image Reconstruction

4.3.1. Dataset and Experiment Details

We randomly selected 25,372 images from the COCO2014 dataset as the training set. The test set included commonly used datasets such as BSD100 [39], Set5 [40], and Set14 [41], along with defect categories of scratch-type (#1) and hole-type (#2) detected by the above detection system. All networks were trained using NVIDIA RTX 3090 24G GPU (NVIDIA Corporation, Santa Clara, CA, USA).

All experiments were conducted using a $\times 4$ scaling factor between low-resolution and high-resolution images, which was equivalent to reducing the image pixels by $\times 16$. Following the training conditions of Ledig et al. [26], for each mini-batch, we randomly cropped 16 of 96 $\times$ 96 HR sub-images from different training images. For the training of SRGAN-H, which is based on the training of SRResNet using MSE as a premise to accelerate the convergence of SRGAN-H, the learning rate was set, and the residual modules were set to 16.

The iteration rounds of SRResNet were set to 130, while the iteration rounds of SRGAN-H were set to 50. In SRGAN-H, the generator and discriminator were alternately updated to train the model.

4.3.2. The Evaluation of Testing Set

In the experiment, widely recognized datasets including BSD100, Set5, and Set14, along with datasets of scratch-type (#1) and hole-type (#2) images, were selected to evaluate the reconstruction performance of SRGAN-H. The testing metrics were the average *PSNR* and *SSIM* of the entire testing set, as shown in Table 1. It can be observed that the SRGAN-H model demonstrated good reconstruction performance for SLM defect images, making it a suitable improvement for SR reconstruction of SLM defect images. Specifically, the *PSNR* of scratch-type images was higher than that of hole-type images, indicating lower distortion in the SR reconstruction of scratch-type images. On the other hand, the *SSIM* of hole-type images was higher than that of scratch-class images, suggesting a higher degree of similarity in terms of brightness, contrast, and structure in the SR reconstruction of hole-type images compared with scratch-type images. For the test sets BSD100, Set5, and Set14, the *PSNR* values were 24.488, 26.788, and 25.390, respectively, with *SSIM* values all below 0.850. In contrast, for scratch-type and hole-type images, the *PSNR* values exceeded 30, and the *SSIM* values were above 0.850. This indicates that SRGAN-H demonstrates better reconstruction performance for scratch-type and hole-type images, depending on the type of image.

Table 1. The evaluation of the testing set.

Indicator	BSD100	Set5	Set14	#1	#2
PSNR	24.488	26.788	25.390	33.405	31.159
SSIM	0.641	0.803	0.693	0.890	0.896

4.3.3. Single Image SR Reconstruction

To validate the single image super-resolution (SR) reconstruction performance of the SRGAN-H model, the experiment randomly cropped portions of the scratch-type and hole-type *Aop*, *Dop*, S_1, and S_2 images obtained from the previous section as samples for SR reconstruction. Additionally, the experiment used the improved SRGAN-H model, the original SRGAN model, and nearest and bicubic interpolation methods as comparisons to demonstrate the significant role of the SRGAN-H model in enhancing image quality.

As shown in Figure 8, for scratch-type images, the image quality is improved in terms of *PSNR*, *SSIM*, and *SD* metrics after processing with the SRGAN-H model, with these metrics outperforming other super-resolution processing methods.

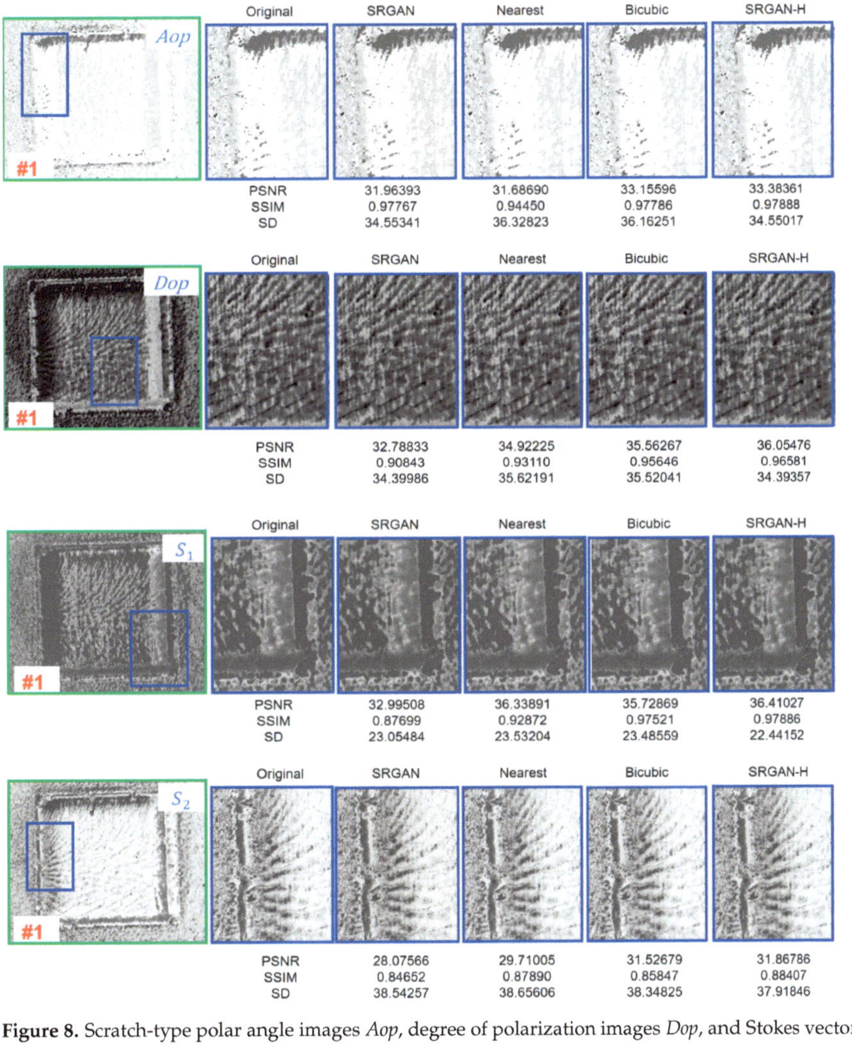

Figure 8. Scratch-type polar angle images *Aop*, degree of polarization images *Dop*, and Stokes vector images S_1 and S_2 after processing with SRGAN-H.

Comparing the results of the reconstruction of the four groups of images, it can be observed from Figure 8 that, in the case of scratch-type images, the Stokes vector images demonstrated the best reconstruction performance, with a higher *PSNR* value, *SSIM* closer to 1, and smaller *SD* value. This confirmed that after SR processing, the image was closer to the original, with lower distortion, closer brightness, contrast, and other aspects resembling the original image, thereby indicating higher image quality. As shown in Figure 8, the *PSNR* values of the reconstructed images by the SRGAN-H model exceeded 30 for the selected regions, with *SSIM* values also above 0.88, demonstrating the effectiveness of the SRGAN-H model.

For the hole-type images, the image quality was also improved in terms of *PSNR*, *SSIM*, and *SD* metrics after processing with the SRGAN-H model. The results are shown in Figure 9. It can be observed that, for the hole-type images, the reconstruction performance of the polar angle images *Aop* was the best, with a higher *PSNR* value, *SSIM* closer to 1, and a smaller *SD* value, indicating lower distortion. The SR image was more in line with the original image in terms of contrast, brightness, and similarity. For super-resolution reconstruction of selected regions, the *PSNR* values for SRGAN-H were all above 36, and the *SSIM* values were all above 0.97. This indicates the effectiveness of the SRGAN-H model in handling defect images of partial regions.

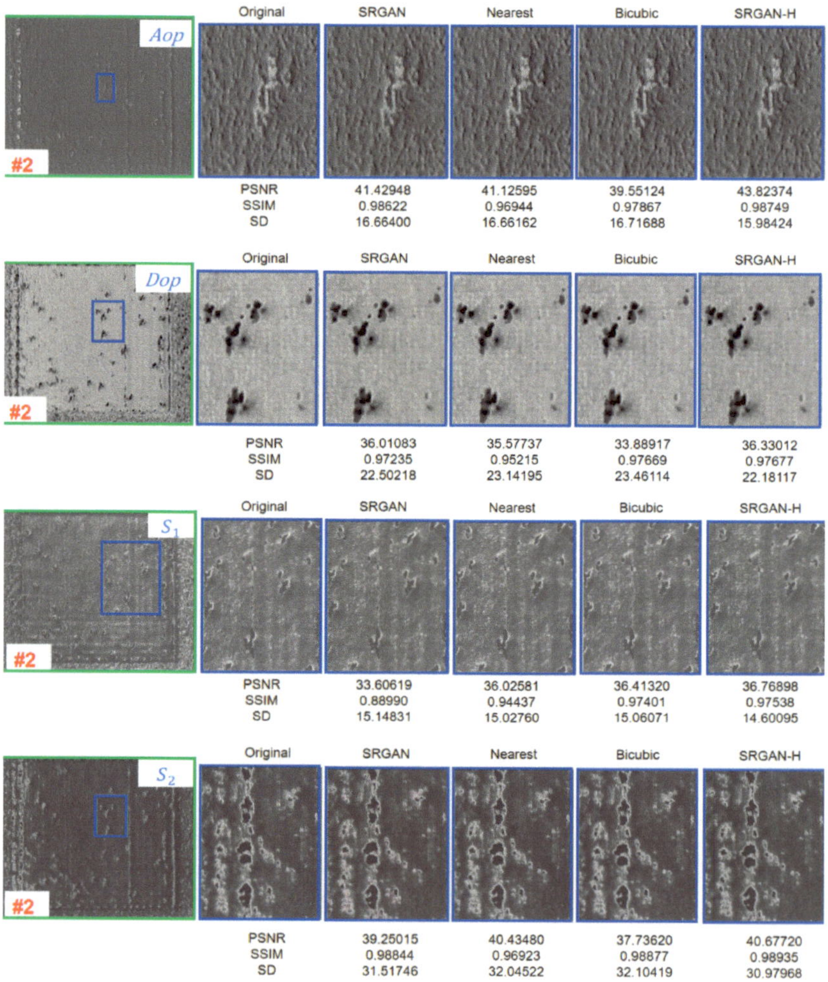

Figure 9. Hole-type polar angle images *Aop*, degree of polarization images *Dop*, and Stokes vector images S_1 and S_2 after processing with SRGAN-H.

From Figure 10, it can be seen that SRGAN-H outperformed SRGAN, nearest, and bicubic methods in terms of *PSNR*, *SSIM*, and *SD* evaluations. Whether for scratch-type (#1) or hole-type images (#2), the image reconstruction quality was the best, and, compared with the other three algorithms, SRGAN-H showed superiority in *PSNR* and *SSIM* evaluations.

The *SD* values of SRGAN-H were smaller than the other three methods, indicating that the images reconstructed by SRGAN-H were more similar to the original images.

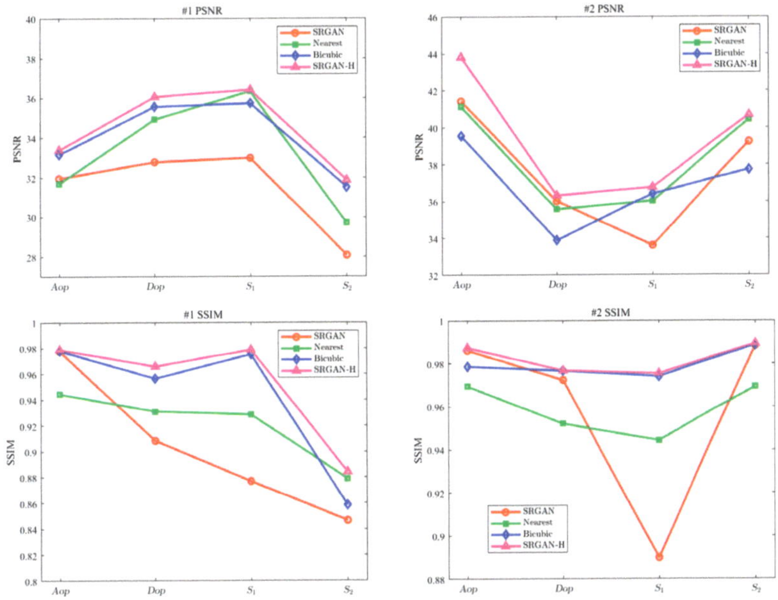

Figure 10. Evaluation results for scratch-type and hole-type images after processing with SRGAN-H.

This indicated that the polarization angle images *Aop*, degree of polarization images *Dop*, and Stokes vector images S_1 and S_2 obtained through polarization technology, after undergoing SR reconstruction with SRGAN-H, highlighted the edge information more prominently and enriched the detail information. This compensated for the limitations of polarization detection technology, making the characterization of SLM process defects more comprehensive.

However, performing super-resolution reconstruction on only a limited region does not result in a significant improvement in visual effect. Therefore, we conducted super-resolution reconstruction on the entire defect area and examined the enlarged reconstructed images, as shown in Figures 11 and 12. It was evident that, compared with the CNN algorithm, the SRGAN-H algorithm showed notable improvements in edge extraction, image contrast, and brightness. However, as indicated in Tables 2 and 3, when performing super-resolution reconstruction on the entire defect area, SRGAN-H's *PSNR* and *SSIM* values were significantly lower than those of CNN, and the overall image evaluation was poorer. Specifically, the *PSNR* values for SRGAN-H in reconstructing the entire scratch defect ranged from 23.62629 to 32.55771, which was lower than the CNN range of 25.89608 to 32.88573. The *SSIM* values for SRGAN-H were also considerably lower than those for CNN. Similar observations can be seen in Tables 4 and 5. This was because SRGAN-H was not designed according to the *PSNR* and *SSIM* evaluation metrics but aimed to enhance visual quality. Thus, when performing super-resolution reconstruction on the entire defect image, SRGAN-H processed images showed a better visual effect compared with CNN, as illustrated in Figures 11 and 12.

However, for the reconstruction of the entire Aop images with holes, the *SSIM* value of SRGAN-H dropped as low as 0.16376, and for the entire Dop images, the *SSIM* value also dropped to 0.48044. This indicates that the SRGAN-H algorithm still has room for improvement in handling images with holes.

Table 2. Reconstruction of scratch-type (#1) images using the CNN algorithm.

Indicator	*Aop*	*Dop*	S_1	S_2
PSNR	25.89608	31.38389	32.07515	32.88573
SSIM	0.86139	0.92353	0.90521	0.93046

Table 3. Reconstruction of scratch-type (#1) images using the SRGAN-H algorithm.

Indicator	*Aop*	*Dop*	S_1	S_2
PSNR	23.62629	27.15678	28.97817	32.55771
SSIM	0.492767	0.85298	0.84497	0.92549

Table 4. Reconstruction of hole-type (#2) images using the CNN algorithm.

Indicator	*Aop*	*Dop*	S_1	S_2
PSNR	24.90034	31.04520	32.13501	31.49505
SSIM	0.86139	0.82485	0.87432	0.83833

Table 5. Reconstruction of hole-type (#2) images using the SRGAN-H algorithm.

Indicator	*Aop*	*Dop*	S_1	S_2
PSNR	15.81080	23.02324	28.35896	26.49653
SSIM	0.16376	0.48044	0.81744	0.68581

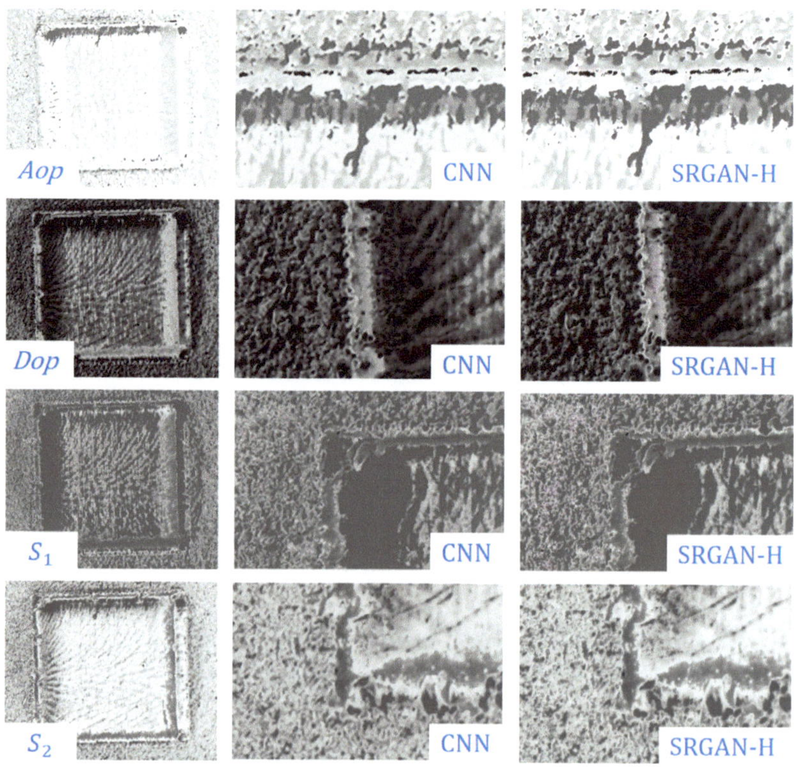

Figure 11. Comparison of the scratch-type (#1) images after super-resolution reconstruction using SRGAN-H and CNN, observed under magnification.

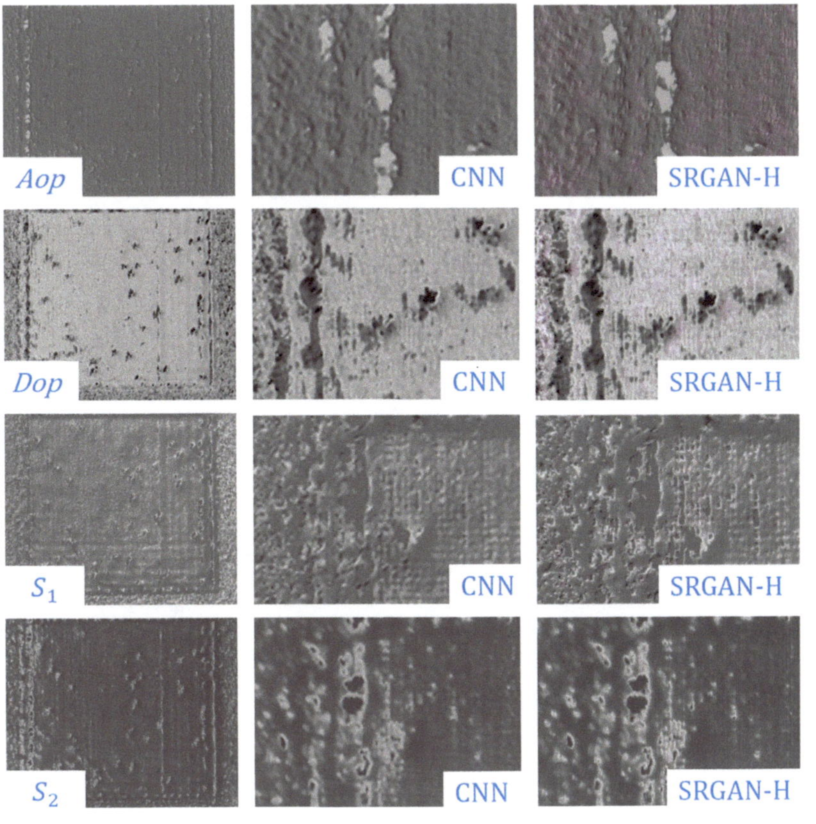

Figure 12. Comparison of the hole-type (#2) images after super-resolution reconstruction using SRGAN-H and CNN, observed under magnification.

5. Conclusions

This study introduced Stokes characteristics into the defect detection system and specifically built a polarization channel imaging system. By changing the polarization angle to obtain multiple sets of polarization images, the Stokes vector, degree of polarization, and polarization angle images were solved. The evaluation of these images using the v-channel value showed a significant improvement in the contrast of the defect areas, defect contour information, and high reflection suppression effect, demonstrating the importance of incorporating polarization technology into defect detection systems. Experimental results indicated that this method effectively extracted and characterized surface defects such as cracks, scratches, and pores, providing solutions and references for defect detection and processing parameter optimization of laser additive manufacturing components.

Subsequently, this study applied the SRGAN-H model to perform super-resolution (SR) reconstruction on the target images. The results indicated that, for SR reconstruction of defect images captured from specific regions, the *PSNR*, *SSIM*, and *SD* values showed good performance. Depending on the selected regions, the *PSNR* value for scratch-type (#1) images reached 36.05476, and the *SSIM* value reached 0.97888; while for hole-type (#2) images, the *PSNR* value reached 43.82374, and the *SSIM* value reached 0.98935. However, the visual improvement compared with traditional methods was not significant. Conversely, when performing SR reconstruction on entire defect images, SRGAN-H yielded poorer *PSNR* and *SSIM* values but demonstrated a notable enhancement in visual quality compared with CNN-based methods. The images processed by SRGAN-H exhibited clearer contrast and edge contours upon magnification, with richer image details, making them more suitable for visual inspection. The lower *PSNR* and *SSIM* values may be attributed to the fact that the SRGAN model was not specifically designed to optimize these evaluation metrics but rather to enhance visual perception for the human eye. Consequently, this resulted in poorer *PSNR* and *SSIM* values but better visual reconstruction quality.

However, the above method still has certain limitations, which can be summarized as follows.

While this study achieved SR reconstruction of defect images in the SLM process using the SRGAN-H model, there was limited research on network adjustment of the SRGAN-H model, and the network architecture was not applied to other algorithms, indicating that its portability requires further investigation.

Training SRGAN-H required high hardware requirements and longer training times. In situations with small datasets, the SR reconstruction effect might even be inferior to traditional interpolation methods. In evaluations on the BSD100, Set5, and Set14 test sets, the *PSNR* and *SSIM* values of SRGAN-H were relatively low, since *PSNR* and *SSIM* metrics might overly smooth the algorithm, leading to less-than-ideal objective evaluation results.

The instability of SRGAN training and the occurrence of artifacts were not effectively addressed, and for SRGAN-H training, the stability and convergence speed of training need to be improved. While the introduction of the perceptual loss function enhanced the representation of more texture details, it also increased the occurrence of artifacts.

Author Contributions: Conceptualization, S.S.; methodology, S.S.; validation, H.C.; writing—original draft preparation, S.S.; writing—review and editing, X.P.; supervision, X.P.; funding acquisition, X.P. All authors have read and agreed to the published version of the manuscript.

Funding: This work is supported by the National Natural Science Foundation of China (Grant No. 52305594); China Postdoctoral Science Foundation Grant (2024M754299); Natural Science Foundation of Hunan Province (Grant Nos. 2024JJ6460, 2023JJ30079); Strategic Priority Research Program of the Chinese Academy of Sciences (Grant No. XDA25020317).

Data Availability Statement: Data underlying the results presented in this paper are not publicly available at this time but may be obtained from the authors upon reasonable request.

Conflicts of Interest: The authors declare no conflicts of interest.

References

1. Yang, X.; Ren, Y.-J.; Liu, S.-F.; Wang, Q.-J.; Shi, M.-J. Microstructure and tensile property of SLM 316L stainless steel manufactured with fine and coarse powder mixtures. *J. Cent. South Univ.* **2020**, *27*, 334–343. [CrossRef]
2. Aboulkhair, N.T.; Simonelli, M.; Parry, L.; Ashcroft, I.; Tuck, C.; Hague, R. 3D printing of Aluminium alloys: Additive manufacturing of aluminium alloys using selective laser melting. *Prog. Mater. Sci.* **2019**, *106*, 100578. [CrossRef]
3. Yan, X.; Chen, C.; Chang, C.; Dong, D.; Zhao, R.; Jenkins, R.; Wang, J.; Ren, Z.; Liu, M.; Liao, H.; et al. Study of the microstructure and mechanical performance of C-X stainless steel processed by selective laser melting (SLM). *Mater. Sci. Eng.* **2020**, *781*, 139227. [CrossRef]
4. Al-Rubaie, K.S.; Melotti, S.; Rabelo, A.; Paiva, J.M.; Elbestawi, M.A.; Veldhuis, S.C. Machinability of SLM-produced Ti6Al4V titanium alloy parts. *J. Manuf. Process.* **2020**, *57*, 768–786. [CrossRef]
5. Chan, Y.F.; Chen, C.J.; Zhang, M. Review of on-line monitoring research on metal additive manufacturing process. *Mater. Rep.* **2019**, *33*, 2839–2867.
6. Tapia, G.; Elwany, A.A. Review on process monitoring and control in metal-based additive manufacturing. *J. Manuf. Sci. Eng.* **2014**, *13*, 060801. [CrossRef]
7. Marco, G.; Bianca, M.C. Process defects and in situ monitoring methods in metal powder bed fusion: A review. *Meas. Sci. Technol.* **2017**, *28*, 044005.
8. Charalampous, P.; Kostavelis, I.; Tzovaras, D. Non-destructive quality control methods in additive manufacturing: A survey. *Rapid Prototyp. J.* **2020**, *26*, 777–790. [CrossRef]
9. Chua, Z.Y.; Ahn, I.H.; Moon, S.K. Process monitoring and inspection systems in metal additive manufacturing: Status and applications. *Int. J. Precis. Eng. Manuf.-Green Technol.* **2017**, *4*, 235–245. [CrossRef]
10. Agyapong, J.; Mateos, D.; Czekanski, A.; Boakye-Yiadom, S. Investigation of effects of process parameters on microstructure and fracture toughness of SLM CoCrFeMnNi. *J. Alloys Compd.* **2024**, *987*, 173998. [CrossRef]
11. Valeev, S.I.; Kharlamov, I.E. Determination of powerful active zones of petrochemical equipment. *IOP Conf. Ser. Mater. Sci. Eng.* **2019**, *537*, 032059. [CrossRef]

12. Zhang, Y.; Fuh, J.Y.; Ye, D.; Hong, G.S. In-situ monitoring of laser-based PBF via off-axis vision and image processing approaches. *Addit. Manuf.* **2019**, *25*, 263–274. [CrossRef]
13. Bisht, M.; Ray, N.; Verbist, F.; Coeck, S. Correlation of selective laser melting-melt pool events with the tensile properties of Ti-6Al-4V ELI processed by laser powder bed fusion. *Addit. Manuf.* **2018**, *22*, 302–306. [CrossRef]
14. Rezaeifar, H.; Elbestawi, M.A. On-line melt pool temperature control in L-PBF additive manufacturing. *Int. J. Adv. Manuf.* **2021**, *112*, 2789–2804. [CrossRef]
15. Krauss, H.; Eschey, C.; Zaeh, M.F. Thermography for monitoring the selective laser melting process. In Proceedings of the 23rd International Solid Freeform Fabrication Symposium, Austin, TX, USA, 6–8 August 2012.
16. Bartlett, J.L.; Heim, F.M.; Murty, Y.V.; Li, X. In situ defect detection in selective laser melting via full-field infrared thermography. *Addit. Manuf.* **2018**, *24*, 595–605. [CrossRef]
17. Lu, R.S.; Wu, A.; Zhang, T.D.; Wang, Y.H. Overview of automatic optical (visual) inspection technology and its application in defect detection. *Acta Opt. Sin.* **2018**, *38*, 23–58.
18. Li, X.; Liu, F.; Shao, X.P. Principles and research progress of polarized three-dimensional imaging technology. *J. Infrared Millim. Waves* **2021**, *40*, 248–262.
19. Peng, X.; Kong, L. Development of a multi-sensor system for defects detection in additive manufacturing. *Opt. Express* **2022**, *30*, 30640–30665. [CrossRef]
20. Wang, L.; Kim, T.K.; Yoon, K.J. EventSR: From asynchronous events to image reconstruction, restoration, and super-resolution via end-to-end adversarial learning. In Proceedings of the IEEE Conference on Computer Vision and Partten Recognition(CVPR), Seattle, WA, USA, 13–19 June 2020; IEEE: Piscataway, NJ, USA; pp. 8315–8325.
21. Kong, X.; Zhao, H.; Qiao, Y.; Dong, C. ClassSR: A general framework to accelerate super-resolution networks by data characteristic. *arXiv* **2021**, arXiv:2103.04039v1.
22. Liang, J.; Cao, J.; Sun, G.; Zhang, K.; Van Gool, L.; Timofte, R. SwinIR: Image restoration using swin transformer. In Proceedings of the IEEE International Conference on Computer Vision (ICCV), Montreal, BC, Canada, 11–17 October 2021; IEEE: Piscataway, NJ, USA, 2021; pp. 1833–1844.
23. Dong, C.; Loy, C.C.; He, K.; Tang, X. Learning a deep convolutional network for image super resolution. In *Proceeding of the European Conference on Computer Vision (ECCV)*; Springer: Cham, Switzerland, 2014; pp. 184–199.
24. Zhang, Y.; Li, K.; Li, K.; Wang, L.; Zhong, B.; Fu, Y. Image super-resolution using very deep residual channel attention networks. In *Proceedings of the European Conference on Computer Vision (ECCV)*; Springer: Cham, Switzerland, 2018; pp. 294–310.
25. Li, J.; Fang, F.; Mei, K.; Zhang, G. Multi-scale residual network for image super-resolution. In *Proceedings of the European Conference on Computer Vision (ECCV)*; Springer: Cham, Switzerland, 2018; pp. 527–542.
26. Ledig, C.; Theis, L.; Huszar, F.; Caballero, J.; Cunningham, A.; Acosta, A. Photo-realistic single image super-resolution using a generative adversaral network. In Proceedings of the IEEE Conference on Computer Vision and Pattern Recognition (CVPR), Honolulu, HI, USA, 21–26 July 2017; IEEE: Piscataway, NJ, USA, 2017; pp. 105–114.
27. Zhang, W.; Liu, Y.; Dong, C.; Qiao, Y. RankSRGAN: Generative adversarial networks with ranker for image super-resolution. In Proceedings of the IEEE International Conference on Computer Vision (ICCV), Seoul, Republic of Korea, 27 October–2 November 2019; IEEE: Piscataway, NJ, USA, 2019; pp. 3096–3105.
28. Pan, J.; Dong, J.; Liu, Y.; Zhang, J.; Ren, J.; Tang, J.; Tai, Y.W.; Yang, M.H. Physics-based generative adversarial models for image restoration and beyond. *IEEE Trans. Pattern Anal. Mach. Intell.* **2020**, *43*, 2449–2462. [CrossRef]
29. Howard, A.; Sandler, M.; Chu, G.; Chen, L.C.; Chen, B.; Tan, M.; Wang, W.; Zhu, Y.; Pang, R.; Vasudevan, V.; et al. Searching for MobileNetV3. In Proceedings of the 2019 IEEE/CVF International Conference on Computer Vision (ICCV), Seoul, Republic of Korea, 27 October–2 November 2019; pp. 1314–1324. [CrossRef]
30. Xue, Q.S.; Chen, W. Optical system design of spaceborne ultraviolet panoramic imager. *Infrared Laser Eng.* **2014**, *43*, 517–522.
31. He, K.; Zhang, X.; Ren, S.; Sun, J. Delving deep into rectifiers: Surpassing human-level performance on imagenet classification. In Proceedings of the IEEE International Conference on Computer Vision (ICCV), Santiago, Chile, 7–13 December 2015; pp. 1026–1034.
32. Ioffe, S.; Szegedy, C. Batch normalization: Accelerating deep network training by reducing internal covariate shift. In Proceedings of the 32nd International Conference on Machine Learning (ICML), Lille, France, 6–11 July 2015; pp. 448–456.
33. Ramachandran, P.; Zoph, B.; Le, Q.V. Searching for Activation Functions. *arXiv* **2018**, arXiv:1710.05941.
34. Shi, W.; Caballero, J.; Huszár, F.; Totz, J.; Aitken, A.P.; Bishop, R.; Rueckert, D.; Wang, Z. Real-Time Single Image and Video Super-Resolution Using an Efficient Sub-Pixel Convolutional Neural Network. In Proceedings of the 2016 IEEE Conference on Computer Vision and Pattern Recognition (CVPR), Las Vegas, NV, USA, 27–30 June 2016; pp. 1874–1883. [CrossRef]
35. Radford, A.; Metz, L.; Chintala, S. Unsupervised representation learning with deep convolutional generative adversarial networks. In Proceedings of the International Conference on Learning Representations (ICLR), San Juan, Puerto Rico, 2–4 May 2016.
36. Simonyan, K.; Zisserman, A. Very deep convolutional networks for large-scale image recognition. In Proceedings of the International Conference on Learning Representations (ICLR), San Diego, CA, USA, 7–9 May 2015.
37. Poobathy, D.; Chezian, R.M. Edge detection operators: Peak signal to noise ratio based comparison. *Int. J. Image Graph. Signal Process.* **2014**, *10*, 55–61. [CrossRef]
38. Hore, A.; Ziou, D. Image quality metrics: PSNR vs. SSIM. In Proceedings of the 2010 20th International Conference on Pattern Recognition, Istanbul, Turkey, 23–26 August 2010; IEEE: Piscataway, NJ, USA, 2010; pp. 2366–2369.

39. Martin, D.; Fowlkes, C.; Tal, D.; Malik, J. A database of human segmented natural images and its application to evaluating segmentation algorithms and measuring ecological statistics. In Proceedings of the IEEE International Conference on Computer Vision (ICCV), Vancouver, BC, Canada, 7–14 July 2001; IEEE: Piscataway, NJ, USA, 2001; Volume 2, pp. 416–423.
40. Bevilacqua, M.; Roumy, A.; Guillemot, C.; Alberi-Morel, M.L. Low-complexity single-image super-resolution based on nonnegative neighbor embedding. In Proceedings of the British Machine Vision Conference (BMVC), Surrey, UK, 3–7 September 2012; BMVA Press: Surrey, UK, 2012; pp. 1–10.
41. Zeyde, R.; Elad, M.; Protter, M. On single image scale-up using sparse-representations. In *Curves and Surfaces*; Springer: Berlin/Heidelberg, Germany, 2012; pp. 711–730.

photonics

Article

Research on Distortion Control in Off-Axis Three-Mirror Astronomical Telescope Systems

En Liu [1,2], Yuquan Zheng [1,*], Chao Lin [1], Jialun Zhang [1], Yanlin Niu [1,2] and Lei Song [1,2]

[1] Changchun Institute of Optics, Fine Mechanics and Physics, Chinese Academy of Sciences, Changchun 130033, China
[2] University of Chinese Academy of Sciences, Beijing 100049, China
[*] Correspondence: zhengyq@sklao.ac.cn

Abstract: With off-axis reflection systems with specific distortion values serving as objectives or collimators, it is possible to compensate and correct for spectral line bending in spectroscopic instruments. However, there is limited research on the precise control of distortion, which poses particular challenges in large field-of-view optical systems. This paper presents a method for controlling distortion in off-axis reflection systems. Based on Seidel aberration theory and the relationship between distortion wavefront error and primary ray error, we construct objective functions with structural constraints and aberration constraints. The initial structure with specific distortion values is then solved using a differential evolution algorithm. The effectiveness and reliability of this method are verified through the design of an off-axis three-reflection system. The method provided in this study facilitates the design of remote sensing instruments.

Keywords: optical design; distortion control; off-axis three-mirror system; differential evolution algorithm; aberration theory

Citation: Liu, E.; Zheng, Y.; Lin, C.; Zhang, J.; Niu, Y.; Song, L. Research on Distortion Control in Off-Axis Three-Mirror Astronomical Telescope Systems. *Photonics* **2024**, *11*, 686. https://doi.org/10.3390/photonics11080686

Received: 19 June 2024
Revised: 20 July 2024
Accepted: 22 July 2024
Published: 23 July 2024

1. Introduction

The imaging spectrometer is based on the theory of spectral analysis of substances. While obtaining the two-dimensional spatial image of the target, it acquires the spectral information of the ground elements corresponding to the pixels. The combination of two-dimensional spatial information with one-dimensional spectral information yields an image cube, which has irreplaceable advantages in monitoring dynamic changes in the Earth's environment [1,2]. The imaging spectrometer consists of a front-end telescope system and a spectral imaging system. The spectral imaging system includes spectral elements, slits, collimators, and focusing mirrors. Off-axis reflective optical systems can simultaneously achieve long focal lengths, wide fields of view, and high imaging quality. Compared to refractive optical systems, they have advantages such as achromatic aberration-free performance, flexible layout, and low sensitivity to temperature and pressure changes. Therefore, they are widely used in spectrometer instruments [3–8]. Due to the non-principal section dispersion caused by the spectral elements, the monochromatic images formed exhibit spectral line curvature, which is particularly pronounced in wide-field spectrometer instruments. This can significantly affect the subsequent data application.

Distortion, as a special type of aberration, does not affect the clarity of the image but causes deformation [9]. The spectral line curvature in spectrometer instruments resembles the radial distortion of imaging systems. Therefore, introducing specific distortion values when designing the front-end telescope system and collimating (converging) mirror can balance the spectral line curvature caused by dispersion elements, thereby achieving images with constant resolution [10]. Distortion control is particularly challenging in the design of wide-field optical systems. In the field of off-axis reflective system design, Meng's group [11] achieved an ultra-wide field of view of 80° × 4° with distortion as high as 28.77% by leveraging the TMA (Three Mirrors Anastigmatic) symmetric property and introducing free-form surfaces. On the other hand, Cao's group [12] utilized the theory of nodal aberration and genetic algorithms to determine the initial configuration of the off-axis system. They designed an off-axis three-mirror optical system with a field of view of $3^\circ \times 4^\circ$, achieving a distortion reduction to 1.84%. Currently, there is relatively little research on precisely controlling distortion, often relying on image processing or complex surface design for distortion control in later stages.

Currently, mainstream optical design software utilizes the Damped Least Squares (DLS) method as the optimization algorithm [13,14]. However, this algorithm is prone to getting stuck in local minima, resulting in design outcomes deviating from the global optimum. Additionally, optical design software often relaxes constraints on distortion during aberration balancing to enhance image quality. Ignoring distortion control in the initial design phase may lead to unacceptable levels of distortion after achieving imaging quality standards [15]. This could necessitate correcting distortion using complex surface shapes, thereby reducing optimization efficiency. Therefore, selecting an initial structure that meets the requirements is crucial for optical system design.

This paper derives the relationship between distortion wavefront error and primary ray error under coaxial conditions and studies the third-order distortion characteristics of off-axis pupils. Based on the required distortion size at the marginal field and solving for the wavefront coefficient W_{311}, Seidel aberrations are calculated by analyzing the paraxial ray information on each surface. An error evaluation function is established, composed of weighted aberrations and defined constraints, to meet configuration requirements. A smaller error function indicates a structure closer to the target solution. The differential evolution algorithm is employed to minimize the error function, thus finding an initial structure with specific distortion values. Distortion is then constrained and further optimized using commercial optical design software CODE V [16].

2. Methods

2.1. Analysis of Third-Order Distortion Characteristics

Primary Wave Aberration describes the optical path difference between the actual wavefront and the ideal wavefront [17]. In traditional rotationally symmetric optical systems, wave aberrations are typically represented by the basic parameters H, ρ, and $H\rho\cos\theta$, as shown in Figure 1a. The distortion being one of the five major aberrations, denoted by W_{311}. Errors generated by primary rays can also be used to represent distortion. As shown in Equation (3), distortion only affects errors in primary rays parallel to the field of view and is independent of the pupil position. Therefore, for a given field angle, all rays within the pupil converge to the same 'incorrect' position, yet still produce a clear image.

$$W = w_{jnm} H_j \rho_n \cos^m \theta, \tag{1}$$

$$W_{311} = w_{311} H^3 \rho \cos\theta, \tag{2}$$

$$\varepsilon = -2F w_{311} H^3, \tag{3}$$

$$\varepsilon = y - y'. \tag{4}$$

H and ρ represent normalized field coordinates and pupil coordinates, respectively, θ is the angular difference between H and ρ, j, n, and m are non-negative integers, ε denotes the error of the primary ray, F is an approximation of the F-number, y is the ideal position of the primary ray on the image plane, and y' is the actual position of the primary ray on the image plane. Distortion as an aberration is a wavefront error and an associated ray error at the image. The effect can be illustrated with just the chief ray as shown below in Figure 1b.

From coaxial reflective systems to off-axis unobstructed reflective systems, there are generally two approaches: aperture off-axis and field offset. Aperture off-axis is commonly used in TMA optical systems where the primary mirror serves as the aperture. Due to the absence of a common rotationally symmetric axis, the characterization of aberrations becomes complex. To address this challenge, the normalized pupil eccentricity vector $\vec{s}$ is introduced for description. The principle of aperture off-axis is illustrated in Figure 2.

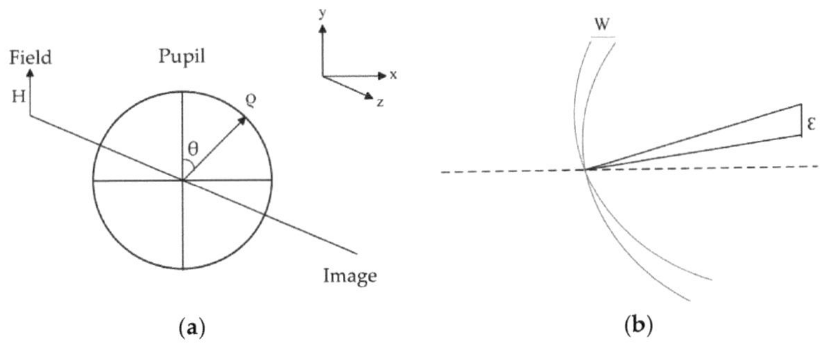

Figure 1. (**a**) Schematic to describe primary aberrations; (**b**) Distortion wavefront error and primary ray error.

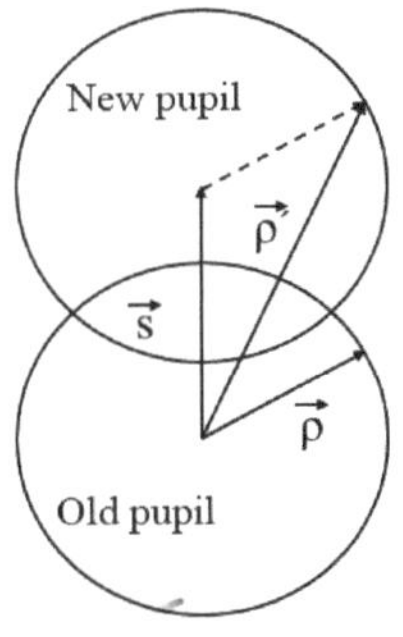

Figure 2. The schematic diagram of the pupil change after the aperture is moved.

For off-axis systems with aperture displacement, when the aperture shifts, the optical axis (OAR) also shifts accordingly. However, the aberrations on the system's image plane are still the sum of aberrations caused by each surface. There are no new types of aberrations introduced, and the wave aberration coefficients remain the same as those for coaxial systems. The only difference is that different types of aberrations are coupled together. The representation of third-order distortion is as follows [18]:

$$
\begin{aligned}
W = 4W_{040}\vec{s}^{\,2}(\vec{s}\cdot\vec{\rho}) + 2W_{131}\vec{s}^{\,2}(\vec{H}\cdot\vec{\rho}) + W_{131}\vec{s}^{\,2}\vec{H}^{*}\cdot\rho + W_{222}\vec{H}^{2}\vec{s}^{\,*}\cdot\vec{\rho} \\
+2W_{220M}(\vec{H}\cdot\vec{H})(\vec{s}\cdot\vec{\rho}) + W_{311}(\vec{H}\cdot\vec{H})(\vec{H}\cdot\vec{\rho}),
\end{aligned}
\tag{5}
$$

where "*" denotes correlation with the Optical Axis Reference (OAR) [11]. From Equation (5), it can be observed that, during the coaxial phase, after correcting the first four monochromatic aberrations, when the aperture is off-axis, the third-order distortion will degrade to coaxial form. Therefore, in designing coaxial reflection systems, we set the coefficients of the first four monochromatic aberrations to zero and solve for the corresponding distortion coefficients based on the required distortion magnitude. The next section will analyze Seidel aberrations and establish their relationship with system structural parameters.

2.2. Calculating Seidel Aberrations

We commence the initial design by constructing a coaxial structure consisting of three quadratic surfaces. The layout diagram of the three-mirror system is depicted in Figure 3, with the aperture stop positioned on the primary mirror (M1). To calculate the overall Seidel aberrations of the system, it is necessary to determine the ray heights, invariant refractive indices, and incident and refracted angles before and after refraction for each paraxial surface. Figure 3a,b, respectively, provide schematic diagrams illustrating the ray tracing of the primary and marginal rays in the coaxial three-mirror structure. For the primary rays, $\overline{y}_i$ denotes the height relative to the optical axis on the i surface, while $\overline{u}_i$ represents the angle relative to the optical axis on the cth surface. Similarly, for the marginal rays, y_i and u_i have the same significance. The variable t_i denotes the distance between mirrors.

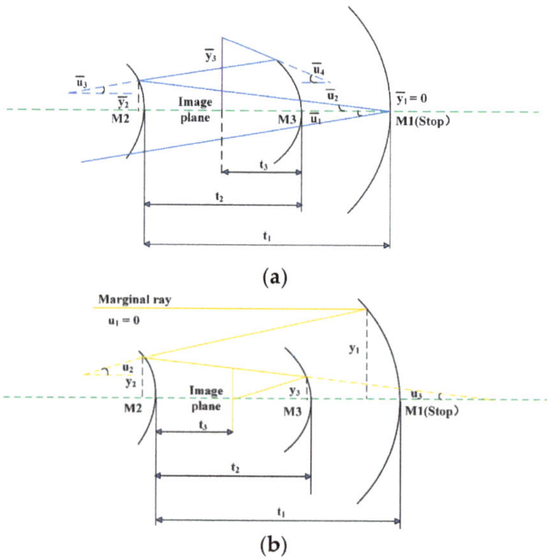

Figure 3. (**a**) Principal ray tracing diagram; (**b**) Edge ray tracing diagram.

In paraxial ray tracing, light rays can be transmitted and transformed through a series of optical components such as lenses, mirrors, etc. The Ray Transfer Matrix (RTM) represents the effects of optical elements as matrix multiplication, thereby simplifying the calculation of ray transfer. It can be algebraically expressed as follows:

$$\begin{aligned} y' &= y + t'u', \\ n'u' &= nu - y\phi, \end{aligned} \tag{6}$$

In the context of the equation, y and y' denote the heights of the incident and reflected rays on the mirror surface, while u and u' represent the angles of incidence and reflection. n and n' denote the refractive indices of the incident and reflected rays upon entry and reflection from the mirror, respectively. ϕ signifies the optical power of the mirror surface.

For a given paraxial ray, we can represent its state using a column vector, which includes information about the ray's height and angle of incidence at each refractive surface. The principal ray can be represented as follows:

$$P_i = \begin{bmatrix} \overline{y_i} \\ n_i \overline{u_i} \end{bmatrix}, \tag{7}$$

When conducting paraxial ray tracing within an optical system, we employ transfer matrices and refraction matrices to compute the ray's state subsequent to traversing the succeeding surface. The transfer matrix establishes the connection between the incident ray's vector at point 1 and the refracted vector at point 2. The transfer matrix can be expressed as follows:

$$T_i = \begin{bmatrix} 1 & \frac{t_i'}{n_i'} \\ 0 & 1 \end{bmatrix}, \tag{8}$$

Similarly, refraction matrix R_i describes the refractive relationship of a ray as it traverses an optical surface. It relates the vector information of the incident ray at point M_i to the vector information of the refracted ray on the same surface. The refraction matrix can be expressed as follows:

$$R_i = \begin{bmatrix} 1 & 0 \\ -\phi_i & 1 \end{bmatrix}, \tag{9}$$

By combining the defined paraxial ray vector with transfer matrices and refraction matrices, we can calculate the height and angle information of the paraxial ray on the next surface:

$$P_{j+1} = T_j R_j P_j. \tag{10}$$

For marginal rays, the computation in paraxial ray tracing is similar to that of principal rays. Since the paraxial angle u_1 and the height $\overline{y}_1$ of principal rays are both zero, the ray heights and paraxial angles of the two characteristic rays on each surface can be calculated as follows:

$$\begin{aligned}
&\overline{y_1} = 0, \\
&\overline{y_2} = -t_1\overline{u_1}, \\
&\overline{y_3} = \overline{u_1}(t_2 + t_1(-1 + (2t_2)/r_2)), \\
&\overline{y_4} = \overline{u_1}(((2t_1t_2 + r_2(-t_1 + t_2))(r_3 - 2t_3))/(r_2r_3) - ((r_2 + 2t_1)t_3)/r_2), \\
&\overline{u_1} = \overline{u_1}, \\
&\overline{u_2} = -\overline{u_1}, \\
&\overline{u_3} = \overline{u_1}(1 + (2t_1)/r_2), \\
&\overline{u_4} = -((\overline{u_1}(r_2r_3 + 2r_2(-t_1 + t_2) + 2t_1(r_3 + 2t_2)))/(r_2r_3)).
\end{aligned} \tag{11}$$

$$\begin{aligned}
&y_1 = y_1, \\
&y_2 = (1 - (2t_1)/r_1)y_1, \\
&y_3 = ((r_1r_2 - 2r_2t_1 + 2(-r_1 + r_2 + 2t_1)t_2)y_1)/(r_1r_2), \\
&y_4 = ((((r_1 - 2t_1)(r_2 - 2t_2) + 2r_2t_2)(r_3 - 2t_3) + 2r_3(r_1 - r_2 - 2t_1)t_3)y_1)/(r_1r_2r_3), \\
&u_1 = 0, \\
&u_2 = -((2y_1)/r_1), \\
&u_3 = (2(-r_1 + r_2 + 2t_1)y_1)/(r_1r_2), \\
&u_4 = -((2(r_1(r_2 - r_3 - 2t_2) + 2t_1(r_3 + 2t_2) + r_2(r_3 - 2t_1 + 2t_2))y_1)/(r_1r_2r_3)).
\end{aligned} \tag{12}$$

According to Seidel aberration theory, the results of paraxial ray tracing in rotationally symmetric optical systems can be described by five monochromatic aberrations. These include spherical aberration (S_I), coma (S_{II}), astigmatism (S_{III}), Petzval field curvature (S_{IV}), and distortion (S_V). In the case of non-spherical surfaces, Seidel aberrations are the sum of spherical and non-spherical components, expressed as follows:

$$\begin{aligned}
&S_I = -\sum A^2 y\Delta(\frac{u}{n}) + a, \\
&S_{II} = -\sum \overline{A}Ay\Delta(\frac{u}{n}) + (\frac{\overline{y}}{y})a, \\
&S_{III} = -\sum \overline{A}^2 y\Delta(\frac{u}{n}) + (\frac{\overline{y}}{y})^2 a, \\
&S_{IV} = -\sum L^2 c\Delta(\frac{1}{n}), \\
&S_V = -\frac{\overline{A}^3}{A} y\Delta(\frac{u}{n}) + \frac{\overline{A}}{A} L^2 c\Delta(\frac{1}{n}) + +(\frac{\overline{y}}{y})^3 a.
\end{aligned} \tag{13}$$

In this context, a serves as a constant coefficient, while A and $\overline{A}$, respectively, denote the refractive invariants of the marginal and principal rays. The term k indicates the conic coefficient, c signifies the curvature of the mirror, and L represents the Lagrangian invariant of the system. Their calculation formulas are as follows:

$$\begin{aligned} & a=-kc^3y^4\Delta(n), \\ & A=n(yc+u), \\ & \overline{A}=n(\overline{y}c+\overline{u}), \\ & \Delta(\tfrac{u}{n})=\tfrac{u'}{n'}-\tfrac{u}{n}, \\ & \Delta(\tfrac{1}{n})=\tfrac{1}{n'}-\tfrac{1}{n}, \\ & L=n\overline{u}y-nu\overline{y}. \end{aligned} \tag{14}$$

When the marginal ray height y_1 and the angle of incidence $\overline{u}_1$ are determined, the five aberration coefficients can be expressed as complex functions of the curvature radius $r_i(i = 1, 2, 3)$, conic coefficient $k_i(i = 1, 2, 3)$, and mirror-to-surface distance $t_i(i = 1, 2)$. Their implicit expressions are as follows:

$$\begin{aligned} & S_I=S_I(r_i,k_i,t_i), \\ & S_{II}=S_{II}(r_i,k_i,t_i), \\ & S_{\ III}=S_{\ III}(r_i,k_i,t_i), \\ & S_{IV}=S_{IV}(r_i,k_i,t_i), \\ & S_V=S_V(r_i,k_i,t_i). \end{aligned} \tag{15}$$

Based on the aforementioned aberration analysis, an error function F is established to evaluate the initial performance of the optical system's structure. This error function sets $S_I = S_{II} = S_{\ III} = S_{IV} = 0$, where S_V' indicates the required distortion inversion value, and S_V represents the true value calculated according to the system's structural parameters. Additionally, constraints are added based on the first-order parameters of the system. The error function is expressed as follows:

$$\begin{aligned} F\ &= f(r_1,r_2,r_3,t_1,t_2,k_1,k_2,k_3) \\ &= w_1|S_I|+w_2|S_{II}|+w_3|S_{\ III}|+w_4|S_{IV}|+w_5|S_V{}'-S_V|+w_6|constraits|, \end{aligned} \tag{16}$$

|| represents the absolute value operator, while w_i (i = 1,2, . . . , 6) denotes the corresponding weight values for each item, which are set according to system configuration requirements. Spherical aberration and comatic aberration are the two most significant aberrations in off-axis systems; thus, their weights can be set to higher values. A smaller value of error function F indicates that the initial structure is closer to the desired one. The next task is to minimize the 8-dimensional error function F through algorithms to find suitable initial structural parameters.

2.3. Using the Differential Evolution Algorithm to Search for the Initial Structure

The inspiration for the differential evolution (DE) algorithm is derived from the evolutionary processes observed in nature, emulating mechanisms such as natural selection, crossover, and mutation. These mechanisms have been proven highly effective in nature for searching and adapting to the environment. By applying these mechanisms to mathematical models, a population-based optimization algorithm is formed. DE is an efficient and robust stochastic optimization algorithm capable of effective global search. When dealing with high-dimensional nonlinear optimization problems, it can explore multiple directions simultaneously, maintaining its search capability without sharp declines as the dimensionality of the problem increases. Moreover, DE demonstrates good robustness in the selection of initial conditions and parameters, making it one of the most commonly used algorithms for solving complex optimization problems [19,20]. The main algorithmic design process is depicted by the green portion inside the dashed line in Figure 4 and summarized as follows:

Step 1: Define the upper and lower bounds of the solution parameters based on the rationality requirements of the system, that is, determine the parameter space.

Step 2: Randomly generate a certain number of individuals from the defined parameter space, where each individual represents a potential solution to the problem. These individuals constitute the initial population. These individuals represent the system's structural parameters $(r_1, r_2, r_3, t_1, t_2, k_1, k_2, k_3)$. The size of the initial population (*NP*) is determined based on the complexity of the problem.

Step 3: Evaluate the fitness of individuals in the initial population. In this study, individuals in the population are evaluated based on the value of the error function *F*. A comparison is made between the fitness value of each individual and the fitness value of the target individual, thereby determining the individuals to be selected. If the fitness value of a trial individual surpasses that of the target individual, the trial individual is chosen as a member of the next generation population; otherwise, the target individual remains unchanged. This can be easily achieved through a greedy selection mechanism. By introducing fitness value comparisons, the algorithm effectively explores the search space while maintaining progress toward regions of higher fitness values.

Step 4: The purpose of the mutation operation is to explore a larger solution space and escape from local optima. Three different individuals (typically referred to as base vectors) are randomly selected from the population, and a mutated individual is generated using the differential operation. The formula for the differential operation is as follows:

$$v_i = x_{r1} + p \cdot (x_{r2} - x_{r3}) \tag{17}$$

In this context, v_i represents the mutated individual of the ith entity, x_{r1}, x_{r2}, x_{r3} denotes three distinct individuals randomly selected from the population, and p stands for the scaling factor (usually taken within the range of [0, 2]).

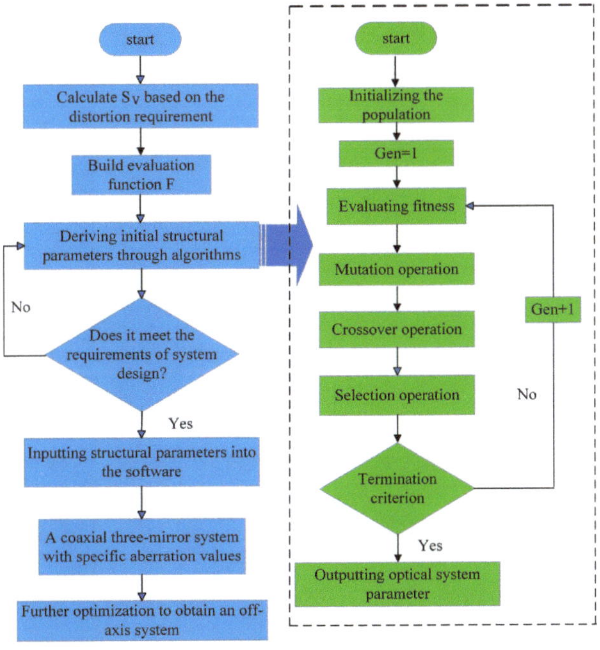

Figure 4. Off-axis three-mirror system distortion control flowchart.

Step 5: The purpose of the crossover operation is to introduce new solutions while maintaining diversity within the population, thereby enhancing the global search capability of the algorithm. During the crossover process, for each dimension, based on a certain probability CR (crossover probability), the corresponding dimension of the mutated individual is replaced with the corresponding dimension of the target individual. For each dimension J, the dimension value of the mutated individual is selected with probability CR; otherwise, the dimension value of the target individual is chosen.

Step 6: Repeat steps 2 through 5 until the termination condition is met. The termination condition is typically reaching the maximum number of iterations or the convergence of the objective function to a certain threshold.

Step 7: Based on the algorithm's output, obtain an initial structure with ultra-low aberration and specific distortion values.

3. Design Example

In this section, two off-axis three-mirror systems were designed to validate the effectiveness of the aforementioned method. To establish an initial structure meeting the requirements, the mathematical model developed in Section 2 was employed. The objective function is defined as the minimization of the absolute difference between the weighted sum of the first four terms of Seidel aberrations and the predefined values of distortion, while assigning a weight of 2 to spherical aberration and coma, which have a significant impact on the imaging quality of the off-axis system, and a weight of 1 to the remaining aberrations. During the system design process, considerations for miniaturization and rationality of mirror structures necessitated constraints on the mirror spacing, curvature radii of the mirrors, and conic constants to ensure that the designed system meets practical application requirements. In algorithm design, the introduction of equality constraints restricts the degrees of freedom in the solution space, thereby reducing the solution space and increasing the difficulty of the optimization problem. Therefore, we constrained the focal length to a specified range.

Using the differential evolution algorithm to solve for the initial structure with specific distortion, the algorithm's key parameters are set empirically as follows: initializing population size NP = 100, scaling factor $F = 5$ in mutation operation, crossover probability CR = 0.9, and maximum iteration count maxFE = 1000. The mathematical model is as follows:

$$
\begin{aligned}
&F = w_1|S_I| + w_2|S_{II}| + w_3|S_{\ III}| + w_4|S_{IV}| + w_5|S_V{}' - S_V|,\\
&890 < f_{EFL}\left(\tfrac{y_1}{u_4}\right) < 910,\\
&-600 < t_1 < -300,\\
&300 < t_2 < 600,\\
&-600 < t_3 < -300,\\
&|r_i| < 3000,\\
&|k_i| < 10.
\end{aligned}
\tag{18}
$$

Based on the constructed model, an off-axis three-mirror optical system with specific aberration values is designed, where the principal ray error $\varepsilon = 180$ μm. The detailed design specifications are outlined in Table 1.

Table 1. Off-axis three-mirror optical system design parameters.

Name	Technical Target
Focal length/mm	900
Entrance pupil diameter/mm	100
Waveband/nm	486–656
Field of view/°	5×1
Modulation transfer function	>0.5@50 lp/mm
Distortion/μm	180/0.46%

Based on the aforementioned error function, the differential evolution (DE) algorithm is utilized to search for the initial structural parameters that meet the requirements. Figure 5 depicts the variation in the error function with the number of iterations. Each blue marker on the curve represents the global optimal value of the error function at the specified iteration point. After 1000 iterations, the error function sharply decreases and then stabilizes, indicating convergence to a local minimum. Beyond 8000 iterations, it exits this local minimum, resulting in further reduction in the error function. By 10,000 iterations, the error function converges close to zero. This demonstrates the effectiveness of DE in navigating high-dimensional nonlinear optimization problems to find global optima.

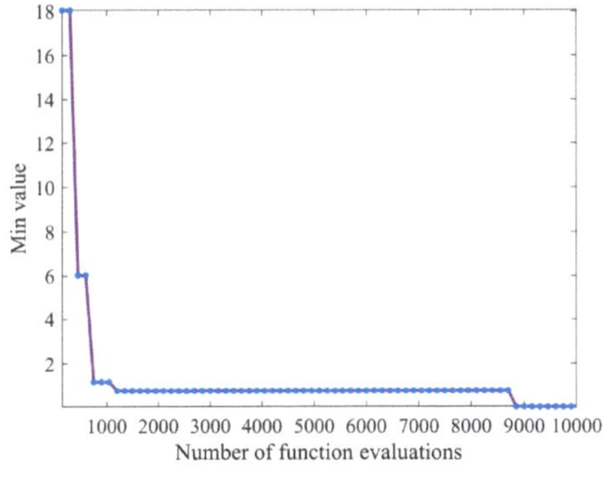

Figure 5. Error function iteration curve.

Table 2 presents the initial structural parameters obtained through the differential evolution algorithm along with their corresponding distortion coefficient S_V. The coefficients calculated align well with those provided in optical design software.

Table 2. Off-axis three-mirror optical system initial parameters.

r_1	r_2	r_3	t_1	t_2	k_1	k_2	k_3	S_V	f/mm
−1827.28	−1590.73	−1120.78	−409.14	387.45	−1.91	−7.16	−9.41	−0.02	901.80
−2490.75	1827.29	615.03	−490.01	339.24	−8.48	−9.87	5.62	−0.02	898.80
−2902.28	1762.55	619.74	−459.35	390.46	2.64	−7.89	5.93	−0.02	905.00

We selected the first configuration among the three initial setups because it exhibits lower spherical aberration and coma coefficients, making it easier to achieve good imaging quality off-axis. The initial configuration and distortion curve are depicted in Figure 6. The first surface corresponds to the object plane at infinity, the aperture stop is located on the primary mirror, which is the second surface, the third surface represents the secondary mirror, the fourth surface is the tertiary mirror, and the fifth surface corresponds to the image plane. The distortion curve illustrates that the maximum distortion at the edge of the field is 0.4524%.

After obtaining a coaxial reflection system with specific distortion, it is imported into optical design software for further optimization. During the optimization process, eccentricity and tilt are often introduced to increase the degrees of freedom for correcting aberrations. Utilizing the built-in module JMRCC within the software, macro functions are scripted to calculate the distances between points and lines, which are then incorporated into the macro language as constraints to eliminate ray occlusion and overlap. Following optimization, the final system structural parameters are presented in Table 3, while the system's two-dimensional layout and three-dimensional layout are illustrated in Figure 7.

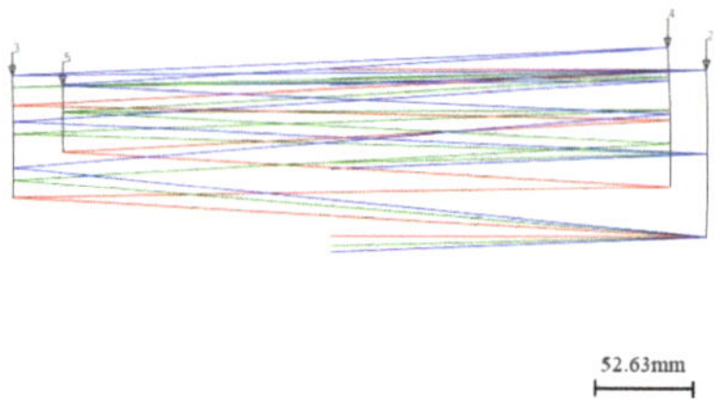

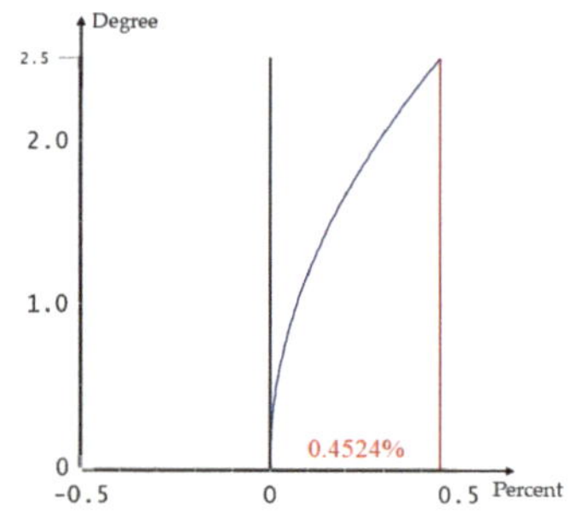

(a) (b)

Figure 6. (a) Initial configuration layout diagram; (b) F—tan θ distortion Diagram.

Table 3. Structural parameters of the off-axis triaxial optical system.

Mirror	r (mm)	d (mm)	Conic
Primary	−1275.00	−397.00	−1.17
Secondary	−456.58	395.00	−1.12
Tertiary	−676.73	−359.98	−0.005

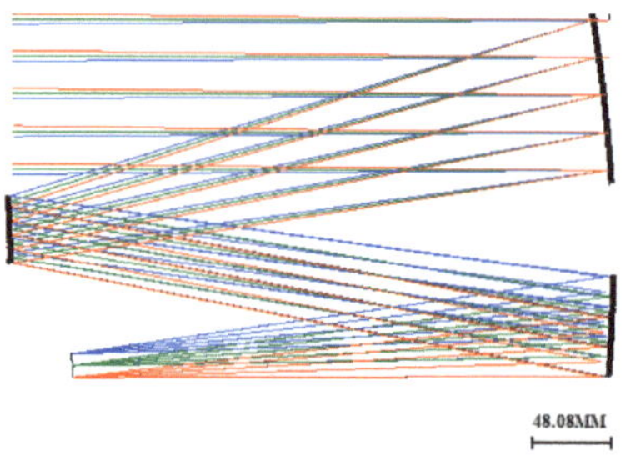

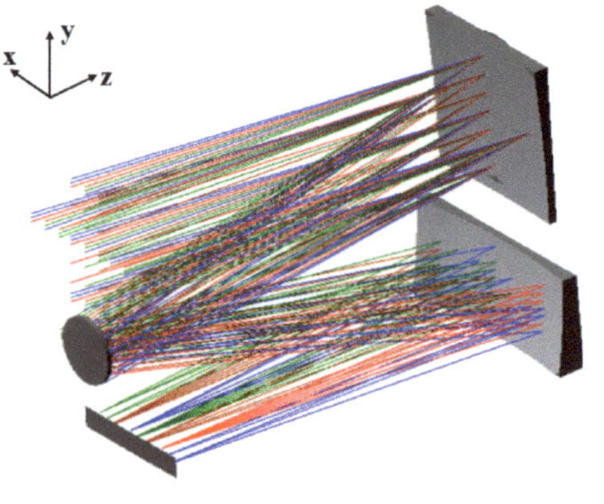

(a) (b)

Figure 7. (a) The optical system's two-dimensional layout diagram; (b) The three-dimensional layout diagram of the optical system.

The modulation transfer function (MTF) is a crucial indicator for evaluating the imaging quality of optical systems. As shown in Figure 8, the MTF of the system is illustrated, and across the entire field of view, it exceeds 0.5 lp/mm, approaching the diffraction limit, indicating excellent imaging quality. The spot diagram effectively reflects the concentration of energy in the spot. The spot diagram for this system is depicted in Figure 9, wherein the black circles represent Airy discs, with a diameter of 6.45 μm calculated using the formula $r = 1.22\lambda \cdot F$. The spot radius of all fields of view is smaller than that of the Airy disc.

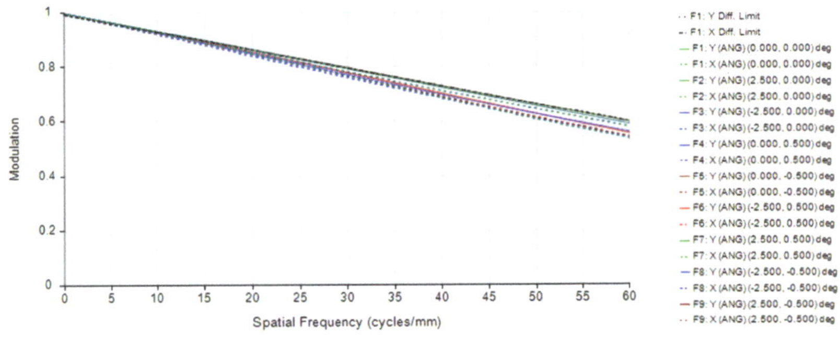

Figure 8. The modulation transfer function (MTF) curves.

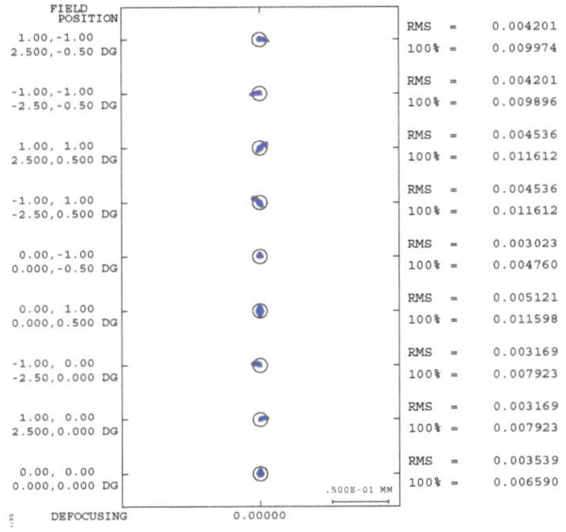

Figure 9. System spot diagram.

The distortion grid can effectively characterize the deviation between ideal and actual imaging. The distortion grid for off-axis three-mirror optical systems is shown in Figure 10, where the red lines represent actual imaging, and the black lines represent ideal imaging. It can be observed from the figure that the two are essentially overlapping, indicating minimal distortion across the full field of view. The specific distortion magnitudes are shown in Table 4.

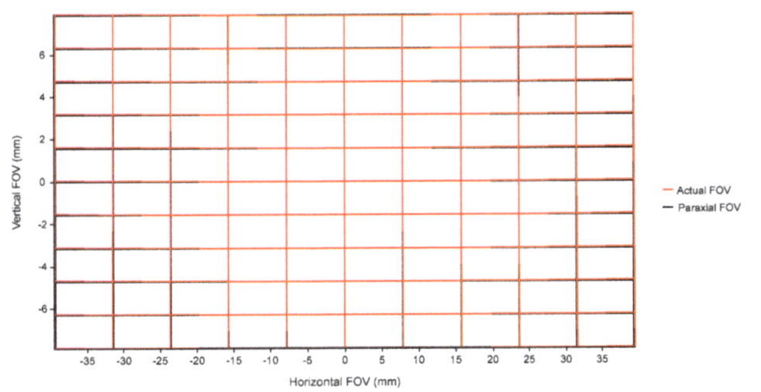

Figure 10. System distortion grid.

Table 4. Distortion magnitude across different fields of view.

FOV (Normalization)	Rad Dist%
(−1, −1)	0.46
(−0.8, −1)	0.38
(−0.6, −1)	0.32
(−0.4, −1)	0.28
(−0.2, −1)	0.25
(0, −1)	0.22
(0.2, −1)	0.25
(0.4, −1)	0.28
(0.6, −1)	0.32
(0.8, −1)	0.38
(1, −1)	0.46

"Rad Dist" represents radial distortion, including barrel distortion and pincushion distortion. During the optical system design phase, radial distortion can be utilized to compensate for the spectral line curvature generated by dispersive elements. As shown in Table 4, it can be seen that the distortion reaches a maximum value of 0.46% at the edge of the field of view, which is close to the initial structure's 0.4524%. This indicates the effectiveness and reliability of the design method in controlling distortion.

Based on the aforementioned system, following the principle of "gradual progression", we further expand the field of view and optimize to obtain a 10° × 1° off-axis three-mirror optical system. The optimized structural parameters of the system are presented in Table 5.

Table 5. Off-axis three-mirror system structural parameters.

Mirror	R (mm)	D (mm)	Conic
Primary	−1357.63	−414.10	−1.34
Secondary	−461.41	337.75	−1.21
Tertiary	−687.63	−402.00	−0.08

The two-dimensional layout of the system is illustrated in Figure 11, showing a compact system structure after further expanding the field of view. The modulation transfer function (MTF) is depicted in Figure 12, indicating that the image quality is good with values exceeding 0.5 across the entire field of view at 60 lp/mm.

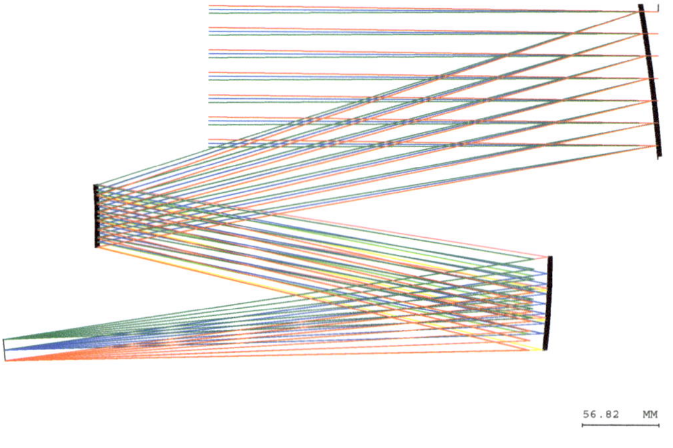

Figure 11. The optical system's two-dimensional layout diagram.

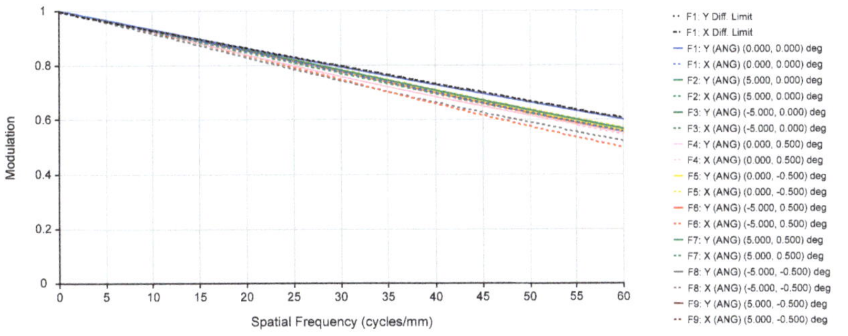

Figure 12. The modulation transfer function (MTF) curves.

After further expanding the field of view, the system's total field distortion is less than 0.76%, and there is no significant fluctuation in distortion between ($-2.5°$, $-1°$) and ($2.5°$, $-1°$), further indicating the effectiveness of this method in distortion control. The system's grid distortion is illustrated in Figure 13.

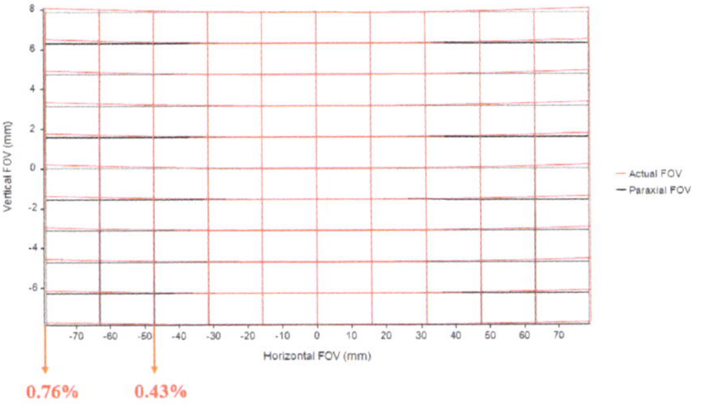

Figure 13. System distortion grid.

4. Conclusions

During the optical system design phase, compensating for spectral line curvature caused by radial distortion generated by the front telecentric system and collimating objective is feasible. However, designing optical systems with specific distortion values presents challenges. This paper analyzes the tertiary aberration characteristics and derives the aberration formula for coaxial triple-mirror structures based on Seidel aberration theory, combined with ray tracing methods. The relationship between the distortion wavefront error and chief ray error is derived. This transforms the problem of solving for initial structural parameters with specific distortion values into an optimization problem of solving a nonlinear single-objective function, which is tackled using a differential evolution algorithm. The unique advantage of the differential evolution algorithm lies in its effectiveness in searching for global minimum values in high-dimensional, highly nonlinear parameter spaces, enabling efficient determination of initial structural parameters with specific distortion values. Taking a triple-mirror optical system as an example, optimization is performed based on a coaxial system with specific distortion values, significantly simplifying subsequent optimization processes and quickly obtaining an off-axis triple-mirror optical system that meets distortion requirements. The effectiveness and reliability of this method are verified, which is equally applicable to the design of off-axis two-mirror and multi-mirror systems.

Author Contributions: Conceptualization, E.L. and Y.Z.; methodology, E.L.; software, E.L.; validation, E.L., J.Z., C.L. and L.S.; formal analysis, E.L. and Y.N.; investigation, E.L.; resources, Y.Z.; data curation, E.L.; writing—original draft, E.L.; writing—review and editing, L.S.; visualization, C.L. and J.Z.; supervision, Y.Z.; project administration, Y.Z.; funding acquisition, Y.Z. All authors have read and agreed to the published version of the manuscript.

Funding: This research was funded by the National Key Research and Development Program of China (No. 2022YFB3904800, No. 2022YFB3904804), and the National Natural Science Foundation of China (No. 62334010).

Institutional Review Board Statement: Not applicable.

Informed Consent Statement: Not applicable.

Data Availability Statement: The original contributions presented in the study are included in the article, further inquiries can be directed to the corresponding author.

Conflicts of Interest: The authors declare no conflicts of interest.

References

1. Liu, Y.; Bauer, A.; Viard, T.; Rolland, J.P. Freeform hyperspectral imager design in a CubeSat format. *Opt. Express* **2021**, *29*, 35915–35928. [CrossRef] [PubMed]
2. Stuart, M.B.; McGonigle, A.J.S.; Willmott, J.R. Hyperspectral Imaging in Environmental Monitoring: A Review of Recent Developments and Technological Advances in Compact Field Deployable Systems. *Sensors* **2019**, *19*, 3071. [CrossRef] [PubMed]
3. Meng, Q.; Wang, W.; Ma, H.; Dong, J. Easy-aligned off-axis three-mirror system with wide field of view using freeform surface based on integration of primary and tertiary mirror. *Appl. Opt.* **2014**, *53*, 3028–3034. [CrossRef] [PubMed]
4. Qu, Z.; Zhong, X.; Zhang, K.; Wang, Y.; Li, L.; Liu, J.; Zeng, C. Automatic compact-volume design strategy for unobscured reflective optical systems based on conicoid surfaces. *Opt. Commun.* **2023**, *533*, 129304. [CrossRef]
5. Sasian, J.; Reshidko, D. Method for the design of nonaxially symmetric optical systems using free-form surfaces. *Opt. Eng.* **2018**, *57*, 101704. [CrossRef]
6. Shi, H.; Jiang, H.; Zhang, X.; Wang, C.; Liu, T. Analysis of nodal aberration properties in off-axis freeform system design. *Appl. Opt.* **2016**, *55*, 6782–6790. [CrossRef] [PubMed]
7. Xu, C.; Lai, X.; Cheng, D.; Wang, Y.; Wu, K. Automatic optical path configuration variation in off-axis mirror system design. *Opt. Express* **2019**, *27*, 15251–15261. [CrossRef] [PubMed]
8. Zhong, Y.; Gross, H. Initial system design method for non-rotationally symmetric systems based on Gaussian brackets and Nodal aberration theory. *Opt. Express* **2017**, *25*, 10016–10030. [CrossRef] [PubMed]
9. Allain, G.; Thibault, S.; Johnson, R.B.; Mahajan, V.N.; Thibault, S. Distortion controlled optical design using orthogonal surface polynomials. In Proceedings of the Current Developments in Lens Design and Optical Engineering XXI, Online, 24 August–4 September 2020.
10. Thomas, G.C.; Lacy, G.C. Design concept for a Landsat-class imaging spectrometer with well-corrected spectral fidelity. In Proceedings of the Optical Science and Technology, SPIE's 48th Annual Meeting, San Diego, CA, USA, 3–8 August 2003; pp. 90–97.
11. Meng, Q.; Wang, H.; Liang, W.; Yan, Z.; Wang, B. Design of off-axis three-mirror systems with ultrawide field of view based on an expansion process of surface freeform and field of view. *Appl. Opt.* **2019**, *58*, 609–615. [CrossRef] [PubMed]
12. Cao, C.; Liao, S.; Liao, Z.; Bai, Y.; Fan, Z. Initial configuration design method for off-axis reflective optical systems using nodal aberration theory and genetic algorithm. *Opt. Eng.* **2019**, *58*, 105101. [CrossRef]
13. Bakic, S.; Vasiljevic, D. Optimization of the double-Gauss objective with the various evolution strategies and the damped least squares. *Phys. Scr.* **2014**, *T162*, 014034. [CrossRef]
14. Cheng, X.; Wang, Y.; Hao, Q.; Isshiki, M. Global and local optimization for optical systems. *Optik* **2006**, *117*, 111–117. [CrossRef]
15. Mouroulis, P.; Green, R.O.; Chrien, T.G. Design of pushbroom imaging spectrometers for optimum recovery of spectroscopic and spatial information. *Appl. Opt.* **2000**, *39*, 2210–2220. [CrossRef] [PubMed]
16. Available online: http://www.opticalres.com (accessed on 21 April 2024).
17. Lin, P.D. Seidel primary ray aberration coefficients for objects placed at finite and infinite distances. *Opt. Express* **2020**, *28*, 12740–12754. [CrossRef] [PubMed]
18. Wang, J.; Guo, B.; Sun, Q.; Lu, Z. Third-order aberration fields of pupil decentered optical systems. *Opt. Express* **2012**, *20*, 11652–11658. [CrossRef] [PubMed]

19. Bilal; Pant, M.; Zaheer, H.; Garcia-Hernandez, L.; Abraham, A. Differential Evolution: A review of more than two decades of research. *Eng. Appl. Artif. Intell.* **2020**, *90*, 103479. [CrossRef]
20. Gui, L.; Xia, X.; Yu, F.; Wu, H.; Wu, R.; Wei, B.; Zhang, Y.; Li, X.; He, G. A multi-role based differential evolution. *Swarm Evol. Comput.* **2019**, *50*, 100508. [CrossRef]

photonics

MDPI

Article

Wavelength Dependence of Modal Bandwidth of Multimode Fibers for High Data Rate Transmission and Its Implications

Xin Chen *, Hao Dong, Hao Chen, Jason E. Hurley, Zoren D. Bullock and Ming-Jun Li

Corning Incorporated, 1 Riverfront Plaza, Corning, NY 14831, USA; dongh3@corning.com (H.D.); chenh1@corning.com (H.C.); hurleyje@corning.com (J.E.H.); bullockzd@corning.com (Z.D.B.); lim@corning.com (M.-J.L.)

* Correspondence: chenx2@corning.com

Abstract: Vertical-cavity surface-emitting laser (VCSEL)-based transmission over multimode fiber (MMF) has achieved data rates of 100G per lane and is progressing towards 200G per lane. Recently, high-data-rate MMFs derived from OM3 and OM4 have been proposed. These fibers exhibit higher effective modal bandwidths at 910 nm, leading to a different wavelength dependence compared to conventional OM3 and OM4 MMFs. Understanding the wavelength dependence of these fibers is crucial to address their utilization in a broader range of applications. Through Monte Carlo simulations, we have obtained the low-end boundary of the effective modal bandwidths (EMBs) for these fibers, revealing capability improvements over the existing OM3 and OM4. The high-data-rate OM4 performs the same as or better than OM5 from 840 nm to 920 nm, while also showing a high bandwidth for the 850–870 nm wavelength window, favoring VCSELs with center wavelengths shifted toward 860 nm. We also obtained the link bandwidth, which includes both modal bandwidth and chromatic dispersion contributions, and the transmission reaches for various types of transceivers. We find that for both high-data-rate OM3 and high-data-rate OM4, the link bandwidth stays above the value at 850 nm until around 910 nm, delivering a similar transmission performance from 850 to 910 nm without declining towards longer wavelengths, unlike the standard OM3 and OM4. This characteristic favors a wider range of wavelength choices for VCSELs and enables optimal deployments for various applications.

Keywords: multimode mode fiber; high data rate transmission; wavelength dependence; link bandwidth; transmission reach

Citation: Chen, X.; Dong, H.; Chen, H.; Hurley, J.E.; Bullock, Z.D.; Li, M.-J. Wavelength Dependence of Modal Bandwidth of Multimode Fibers for High Data Rate Transmission and Its Implications. *Photonics* **2024**, *11*, 667. https://doi.org/10.3390/photonics11070667

Received: 20 June 2024
Revised: 8 July 2024
Accepted: 15 July 2024
Published: 17 July 2024

1. Introduction

Multimode fibers (MMFs) have been extensively employed for short-reach communications, such as within buildings, campuses, or data centers [1,2]. MMFs, used with vertical-cavity surface-emitting lasers (VCSELs), form a system suitable for short-reach applications spanning distances up to 100 m or more. VCSELs offer cost-effective manufacturing and a lower power consumption when compared to traditional single-mode lasers, making them an economical and energy-efficient choice for short-range communications. In contrast to single-mode fibers, MMFs boast a large core diameter and high numerical aperture, enabling the use of lower-cost VCSEL light sources and connectors between fibers.

The past decade has witnessed a rapid advancement in optical transmission technology, fueled by the explosive growth in data traffic and the soaring demand for cloud computing services in data centers [3]. The single-lane data rates have undergone a remarkable evolution, transitioning from 10G to 25G and further to 100G. The rise of artificial intelligence (AI) and graphics processing unit (GPU)-based computing has indeed catalyzed this acceleration in data rate increases, and the latest GPU architectures are now paving the way towards 200G speeds [4].

To achieve higher overall data rates for transceivers, several approaches have been adopted, including higher data rates per lane, higher modulation formats such as four-level pulse amplitude modulation (PAM4), parallel optics utilizing multiple fibers, and wavelength division multiplexing (WDM). In particular, WDM technology has emerged over the past decade in two types of transceivers. One such transceiver is the bidirectional (BiDi) transceiver, named for its capability of transmitting data in both directions [5]. It primarily utilizes two wavelengths, nominally around 850 nm and 910 nm. The data rate of such transceivers has reached 100G per lane, enabling transceiver-level data rates of 800G and 1.6 T/s [6]. Another competing technology is short-wavelength division multiplexing (SWDM) [7,8], which leverages four wavelengths, nominally around 850 nm, 880 nm, 910 nm, and 940 nm. To date, SWDM transceivers have been offered with 40G and 100G data rates, with each wavelength operating at 10 Gb/s and 25 Gb/s, respectively.

VCSEL technology has advanced significantly over the years, with 100G and 200G transmission demonstrated [9–13]. As the data rate moves to 100G per lane, as standardized by IEEE 802.3db [14], the transmission reach limitation becomes more severe. For data rates based on 25Gbaud or lower, OM3 and OM4 MMFs have been widely used for transmission reaches of 70 m and 100 m, respectively. Many data centers have been designed around the transmission reach capabilities. Starting at 100G per lane data rate, for 400G SR4 applications, the OM3 transmission reach has been reduced to 60 m from 70 m. In Terabit BiDi applications, such as 800G SR4.2, standard OM3 and OM4 fibers can only reach distances of 45 m and 70 m, respectively. Only OM5 fiber [15,16], mainly defined for SWDM applications, is specified for 100 m. In response to the bandwidth limitation for many existing MMFs, we have proposed a sub-category of OM3 and OM4 fibers that can achieve longer transmission reaches than standard OM3 and OM4 fibers at high data rates (HDRs) [17–20]. In this work, we refer to these fibers as 'HDR OM3' and 'HDR OM4', both of which have been launched as products [21]. One version of the proposed MMFs, 'HDR OM4', has been adopted by another fiber manufacturer more recently with the same 850/910 nm EMBs [22]. The 'HDR MMFs' are specified to have effective modal bandwidth (EMB) at both 850 nm and 910 nm to address the needs of these two wavelengths. For wider use in WDM applications, the VCSEL wavelengths are not limited to these two wavelengths. Therefore, it is necessary to understand the wavelength dependence of these fibers to determine the transmission reaches for various wavelengths that may be used, which motivated the current work.

In this paper, we conduct detailed modeling to determine the wavelength dependence of EMB for 'HDR MMFs'. A Monte Carlo simulation is used to generate the refractive index profiles of 10,000 MMFs with a range of variations. The EMB values of these fibers at various wavelengths are calculated. Using the worst-case values, we can establish the low-end boundary of these fibers. By utilizing the EMB values, we are able to further calculate the link bandwidth, which includes both the contributions from EMB and chromatic dispersion (CD) interaction with the laser linewidth. The system reach can be determined for a wavelength of interest based on a required link bandwidth value. In Section 2, we present the information on new 'HDR MMFs', i.e., their EMB values that define such types of fibers. In Section 3, we provide details of the Monte Carlo simulation of 10,000 MMFs. A comparison is made with known EMB values for OM3 and OM4. In Section 4, we demonstrate the link bandwidth using the wavelength dependence of EMB and CD of the MMFs. We also convert the link bandwidth into transmission reach for a few transmission applications to illustrate how we can benefit from the wavelength dependence of the EMB information. Finally, we draw conclusions in Section 6.

2. Multimode Fibers for High Data Rate Transmission

MMFs with a 50-micron core have been predominantly used in data communication systems. These fibers are categorized as OM2, OM3, and OM4, based on their effective modal bandwidth (EMB) at a wavelength of 850 nm. The abbreviation 'OM' stands for 'optical multimode'. The OM3 fiber has an EMB of 2000 MHz·km, while the OM4 fiber has a higher EMB of 4700 MHz·km at 850 nm. They were standardized in 2002 and 2009, respectively. OM5 is a sub-category of OM4, sharing the same EMB at 850 nm but with an additional EMB requirement of 2470 MHz·km at 953 nm, which is the longest wavelength supported by shortwave-wavelength division multiplexing (SWDM) technology [8]. OM5 was first specified by Telecommunications Industry Association(TIA) in 2016 as TIA-492AAAE [15], followed by International Electrotechnical Commission(IEC) in 2017 [16].

In recent years, the data transmission rate of VCSEL-based transceivers has reached 25 Gbaud. Among these transceivers, 100G SR4 and 100G SWDM utilize a 25G non-return-to-zero (NRZ) modulation format, while 100G BiDi transceivers employ a 50G PAM4 modulation format across two wavelengths. Meanwhile, higher speed 400G transceivers, such as 400G SR8 and 400G SR4.2, all utilize the PAM4 modulation format.

As mentioned in the Introduction, to address the needs of high data rate transmission, we have proposed the concept of 'HDR MMFs' for future-ready 100G-per-lane multimode applications. Essentially, 'HDR MMFs' are defined by re-categorizing a subset of OM3 or OM4 fibers. The EMB values of 'HDR MMF' are specified in Table 1 and are defined at two wavelengths of interest: 850 nm and 910 nm for 100G-per-lane transmission. Therefore, these fibers can support VCSEL transceivers involving either an 850 nm wavelength only or both 850 nm and 910 nm wavelengths, with targeted transmission reaches of 80 m and 100 m, respectively.

Table 1. EMB values of 'HDR MMFs' for 100G-per-lane applications.

MMF Solution	EMB @ 850 nm (MHz·km)	EMB @ 910 nm (MHz·km)	Reach for 800G BiDi (m)
'HDR OM3'	2890	2220	80
'HDR OM4'	4700	3100	100

3. Wavelength Dependence of EMB for 'HDR MMF' through Monte Carlo Simulation

3.1. Generation of Refractive Index Profiles for Monte Carlo Simulation

The relative refractive index profile, or delta profile of an MMF with a graded-index profile, is often referred to as the alpha profile. It takes the following form:

$$\Delta(\mathrm{r}) = \Delta_0[1 - (\mathrm{r}/\mathrm{a})^{\alpha}], \quad (1)$$

where Δ_0 is the relative refractive index in the center of the core defined as $\Delta_0 = \left(n_0^2 - n_1^2\right)/\left(2n_0^2\right)$, wherein n_0 is the refractive index in the center of the core, n_1 is the refractive index of the cladding, a is the core radius, and α is the profile shape parameter, also called the alpha parameter or the alpha shape parameter. As illustrated in Figure 1, the fiber has a 1% core delta, and a ~50 μm core diameter. Note that when the alpha value is two, the profile takes a parabolic shape. The MMF is most optimized when the alpha parameter has a value of around 2.1, at which point the MMF's modal bandwidth peaks at a wavelength near 850 nm. A trench with a negative delta value is placed outside the core to enhance the bending performance of the fiber [23]. The cladding is typically pure silica and is used as a reference with zero delta value.

Due to the nature of MMF fabrication, each individual MMF differs from another. The idea behind the current work is to generate an ensemble of MMFs with a range of variation types that can occur in real fibers. Figure 2 schematically shows the wavelength dependence of modal bandwidth for an individual MMF. The MMF follows a curve that can reach a peak or maximal modal bandwidth at a wavelength called the peak wavelength. Each MMF can have its own maximum modal bandwidth value and peak wavelength value. Therefore, they can have different modal bandwidth values at a given wavelength of interest, such as 850 nm. An MMF with an ideal refractive index profile can have a higher maximum bandwidth than other MMFs, but its peak wavelength can vary over a range, resulting in different wavelength dependence for each individual MMF. Below, we describe several types of variations that can affect the fiber modal bandwidth and its wavelength dependence.

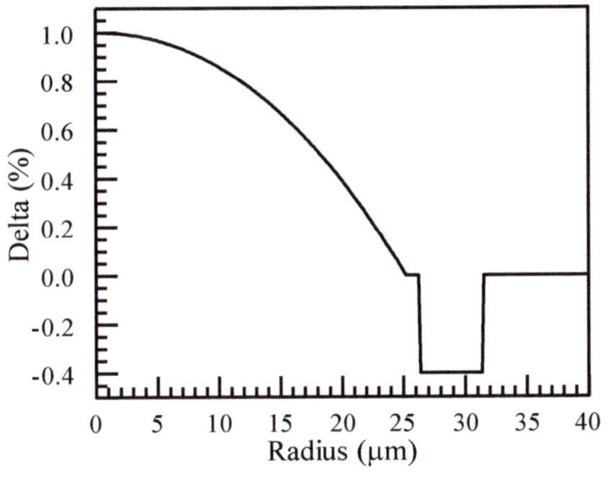

Figure 1. Delta profile of an MMF with 50 μm-diameter core.

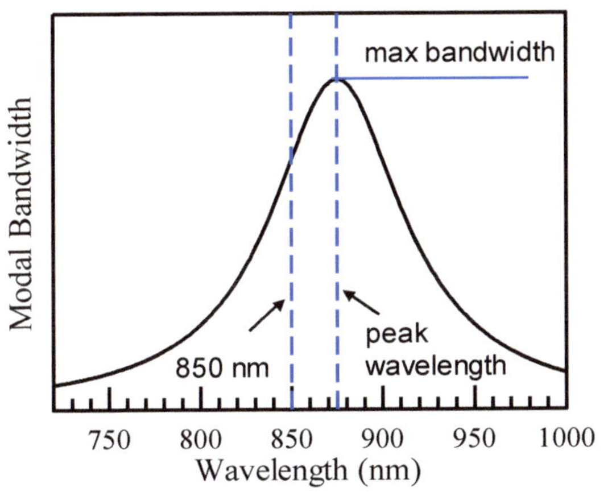

Figure 2. The schematic of wavelength dependency of modal bandwidth of an individual MMF.

One of the most important parameters to consider is the alpha parameter, as it determines where the fiber modal bandwidth peaks, as illustrated in Figure 2. When the alpha parameter is altered, the wavelength at which the modal bandwidth peaks also changes. Therefore, to emulate actual fiber, we would alter the alpha value within a range of over 200 nm. Figure 3 shows the relationship between alpha and peak wavelength. When the alpha value is decreased, the peak wavelength shifts to a higher value.

The second type of variation is the fluctuation of the alpha profile itself on top of the perfect alpha profile, which is referred to as non-alpha error. In the modeling, we expand such errors using Fourier series, as follows:

$$\Delta n(r) = \sum_{i=1}^{N} \left[a(i) \cdot \sin\left(\frac{i\pi r}{a}\right) + b(i) \cdot \cos\left(\frac{i\pi r}{a}\right) \right]. \tag{2}$$

In our actual generation of the Monte Carlo data, N is chosen to be six.

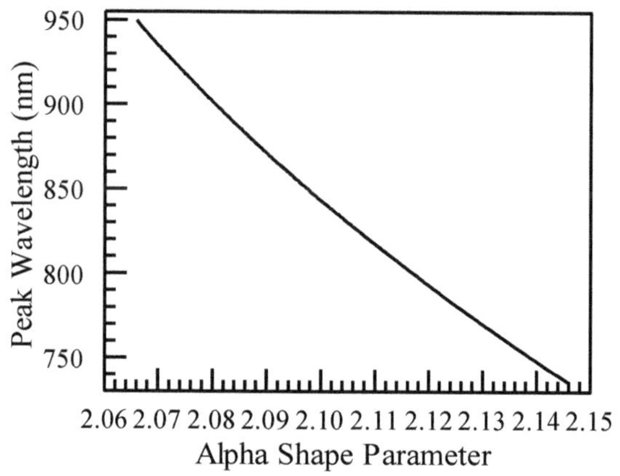

Figure 3. The peak wavelength as a function of alpha for MMF with 1% delta and 50 μm core diameter.

Another type of error that can occur in the center portion of the core involves a small spike or dip in the refractive index, leading to split pulses, which can degrade system performance [24]. In our modeling, we introduce such errors within the first several microns of the core radius. One last type of error to consider is the variation of the trench. In this case, we generate random offsets in the Monte Carlo data set. Alternatively, one can alter the depth of the trench to achieve similar performance variations.

3.2. Wavelength Dependence of EMB of MMFs

Using the refractive index profiles, the EMB of a particular fiber can be obtained from its differential mode delay (DMD) chart, which is the sequence of output pulses resulting from single-mode pulse launches at various radial offsets at the input. In reality, the DMD chart is measured using a DMD bench, while, in our case, it is calculated based on the refractive index profile [25]. In the standard, 10 VCSEL launch conditions were used [26], as the modal bandwidth depends on the launch condition. The DMD weight functions associated with the 10 launch conditions are shown in Figure 4. Using the measured DMD chart, the output pulses over ten VCSEL launch conditions are assembled as weighted by the ten DMD weight functions. The modal bandwidth associated with each weight or launch condition can be calculated from the 3 dB drop of the frequency response, S_{21}, and is referred to as calculated effective modal bandwidth (EMBc). The frequency response of a particular launch condition is calculated from the Fourier transform of the input pulse ($P_{in}(t)$) and output pulses $P_{out,i}(t)$ as follows:

$$S_{21,i} = \left| \frac{\mathcal{F}(P_{out,i}(t))}{\mathcal{F}(P_{in}(t))} \right|, \quad (3)$$

where $\mathcal{F}(P_{out,i}(t))$ and $\mathcal{F}(P_{in}(t))$denote the Fourier transform operations of the output pulse and input pulse respectively. The range of launch conditions ensures that essentially all conditions are covered, so the EMB reflects not just the outcome of one launch condition but the modal bandwidth capability of the specific fiber. With the 10 EMBcs obtained, the EMB can be further obtained. We also note that the integration of the DMD weight function over radius, which represents integrated power over radius, is referred to as encircled flux. The standard requires that, at the 4.5 μm radius, the integrated power is less than 30% of the total power and, at the 19 μm radius, the integrated power is over 86% of the total power.

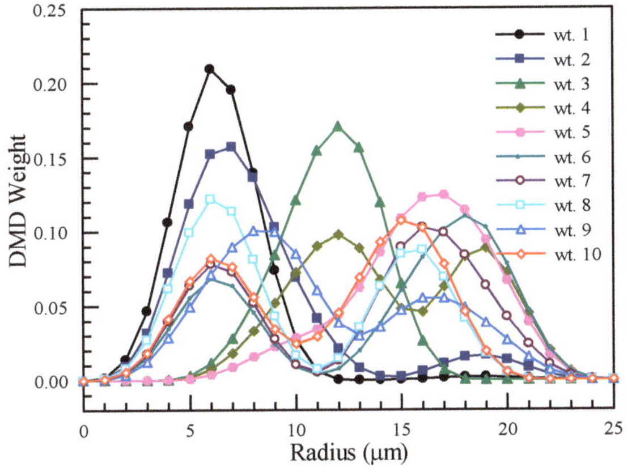

Figure 4. Ten weight functions associated with ten VCSEL launch conditions.

In the Monte Carlo study, we generated 10,000 cases of refractive index profiles to cover a wide range of profile variabilities, as described in Section 3.1, with a mixture of all types of profile errors. The magnitude of each type of error was also randomized within a range to yield MMFs with EMB spread over a range of interest. The purpose of the Monte Carlo set is to generate refractive index profiles with various possibilities that can affect the modal bandwidth under different launch conditions. For each wavelength, we calculated the EMB from the profiles as described above. By combining all the wavelengths, we can generate the wavelength dependence curve of one individual MMF. However, MMFs are primarily categorized as OM3 and OM4, so the interest lies in the low-end values of each type of MMF. IEC 60793-2-10 [16] published the low-end EMB values for OM3 and OM4 over the range of 840 nm to 953 nm. In the simulation presented in the current paper, we have repeated the work to ensure that the methodology yields consistent results. Our main objective, as described in Section 1, is to study the EMB wavelength dependence for 'HDR MMFs', which are further broken down into 'HDR OM4' and 'HDR OM3'. For most of the calculations, the low-end EMB values were obtained at the 99.9th percentile, meaning there is a 0.1% chance that the values can fall below the threshold. We believe this is a reasonable practice, as the truly worst case or lowest value can be overly pessimistic and, therefore, not a good representation of the product's practical capability. Later in the paper, we will discuss the impact of choosing such a percentile to put the results in perspective. In some cases, we will refer to the EMB of a particular type of MMF at its low-end value or 'worst' value without explicitly stating that these are the low-end values.

We first show the EMB of OM4 and OM3 in Figure 5a,b, referring to them as 'IEC OM4' and 'IEC OM3', respectively, as they are based on IEC guidance. The curves with discrete dots were obtained from Monte Carlo simulations for OM4 and OM3. They were labeled as 'Monte Carlo OM4' and 'Monte Carlo OM3' to differentiate them from the IEC-guided OM4 and OM3 EMBs. As the wavelength moves away from 850 nm, the EMB values for both OM4 and OM3 decrease, because the EMB value at 850 nm can be achieved by MMFs with peak wavelengths higher or lower than 850 nm. The agreement between the IEC guidance and the current Monte Carlo simulations is excellent. Note that the low-end EMB values in the simulation were obtained from the 99.9th percentile instead of using the 100th percentile or absolute lowest value, in order to be more realistically aligned with real-world situations. The agreement also gives us confidence that the methodology used in the current work would be valid when applied to 'HDR OM4' and 'HDR OM3'.

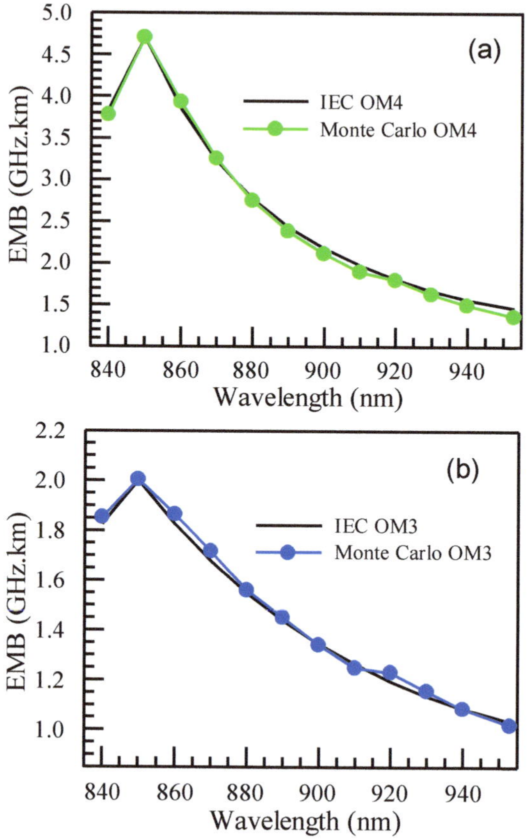

Figure 5. (**a**) The EMB vs. wavelength for OM4; (**b**) the EMB vs. wavelength for OM3.

In Figure 6, we show the wavelength dependence of EMB for 'HDR OM4' obtained from the Monte Carlo simulation with a comparison to two related types of MMFs, namely OM4 and OM5. The EMB values for OM4 and OM5 are based on IEC guidance. As mentioned in Section 1, OM5 was established for supporting SWDM-based VCSEL transceivers with the highest wavelength at 953 nm. SWDM is a type of WDM that includes more wavelengths and, therefore, more data traffic within one transceiver, which is in contrast to BiDi transceivers, which use 850 nm and 910 nm. However, in the past decade, BiDi transceivers have seen more widespread use and have now reached the Terabit age [6]. MMFs that can align with the underlying transceivers would allow for more economical deployments of such applications in larger volumes. It can be found that the EMB values of 'HDR OM4' are roughly equal to or above those of OM5 from 840 nm to 920 nm. Therefore, it has the same capability as OM5 for the wavelengths of interest for BiDi transceivers. Even for wavelengths from 920 to 953 nm, where the EMB of 'HDR OM4' falls below that of OM5, the EMB values are still significantly better than those of OM4, enabling better transmission reach when SWDM wavelengths are used. On the other hand, OM4's EMB decreases rapidly as the wavelength increases from 850 nm, which explains the limitations encountered for transmission at longer wavelengths, such as those used in BiDi transmission.

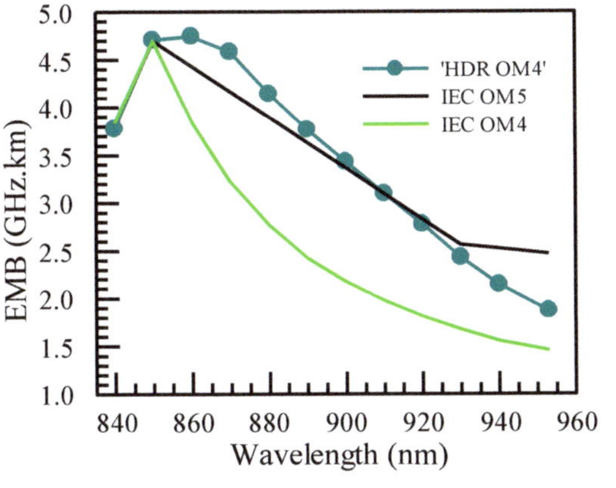

Figure 6. The EMB as a function of wavelength for 'HDR OM4' as compared to those for OM4 and OM5 defined by IEC.

In Figure 7, we show the wavelength dependence of EMB for 'HDR OM3' obtained from the Monte Carlo simulation, as compared to those of OM3 and OM4 based on IEC guidance. The EMBs of 'HDR OM3' over the whole wavelength range are higher than those of OM3, since the EMB at 850 nm is chosen to support an 80 m reach for 100G-per-lane transmission. Coupled with the corresponding EMB at 910 nm, the EMB over the entire wavelength range has been increased, not just for the 850 nm and 910 nm wavelengths where the fiber EMBs are defined. Also, when compared to OM4, the EMB of 'HDR OM3' becomes higher, starting at 883 nm. The overall higher EMB over the wavelength range can enable 'HDR OM3' to achieve a system reach of 80 m for 800G BiDi applications, even exceeding the specified reach for OM4 at 70 m, as will be shown in the next section.

In the Monte Carlo simulation presented above in Figures 6 and 7, we have used the 99.9th percentile for obtaining the low-end EMB. We would note that the low-end boundary can be affected by how EMB is distributed. To illustrate how the change in percentile can affect the EMB, we show the results obtained from two thresholds in Figure 8. Some differences are shown for wavelengths between 850 and 900 nm, but smaller differences are observed at higher wavelengths. In particular, in the 99.6th percentile, the EMB of 'HDR OM4' exceeds 4700 MHz·km over the range of 850 nm to 870 nm. The actual low-end EMB values can be fine-tuned to align with realistic situations with the support of actual measurement results.

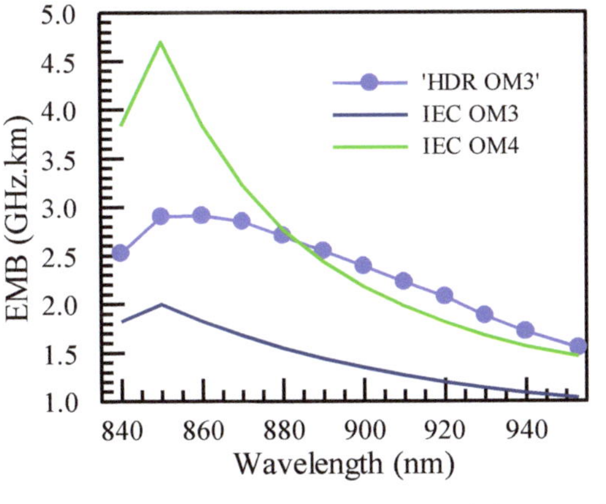

Figure 7. The EMB as a function of wavelength for 'HDR OM3' as compared to those for OM3 and OM4 defined by IEC.

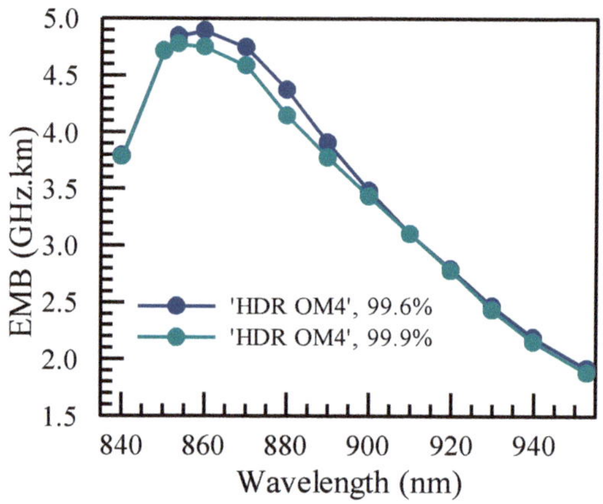

Figure 8. The EMB as a function of wavelength for 'HDR OM4' with two different percentiles.

4. Link Bandwidth and Transmission Capabilities of Various Applications as Related to the Wavelength Dependence of EMB

4.1. The Role of Chromatic Dispersion (CD) and Link Bandwidth

EMB is one important gauge of MMF quality. However, at the link level, the fiber-contributed bandwidth also includes chromatic dispersion and laser linewidth-contributed bandwidth [13,24], BW_{CD}. Therefore, the bandwidth at the link level, BW_{link}, is given by the following equation:

$$BW_{link} = (\frac{1}{EMB^2} + \frac{1}{BW_{CD}^2})^{-1/2}. \quad (4)$$

BW_{CD} is related to both the chromatic dispersion value and the root mean square (RMS) linewidth [16,27]. For most of the VCSEL transmission applications, the RMS linewidth is primarily specified at 0.6 nm, with some specified at different values, such as 0.58 nm and 0.65 nm. Since, currently, the commercially deployed VCSELs are multimoded, the existence of several modes results in a laser linewidth of around 0.6 nm. One future direction of sustaining the VCSEL transmission reach is to reduce the laser linewidth by adopting SM VCSEL [9,28,29], which can reduce the laser linewidth to around 0.1–0.2 nm. In Figure 9, we show the CD as a function of wavelength and the corresponding BW_{CD} in the same wavelength range as a standard-based 50 μm-core MMF. It can be observed that, as the wavelength increases, the CD drops, resulting in a higher BW_{CD}. This means that, as the wavelength increases, a lower EMB value is sufficient to achieve the same link bandwidth. Since link bandwidth is the measure that affects the transmission performance, one would resort to its value, not solely to the EMB value, to judge the link performance. The effect of EMB varies depending on the wavelength.

4.2. Link Bandwidths of Various MMFs at Lengths of Interest

The standardization of 100G-per-lane VCSEL transmission started with the IEEE 802.3db standard [14], which specifies the required link bandwidth for MMFs. The transmission is broken into two sub-categories, 'SR' and 'VR', wherein the 'SR' type has a longer reach than the 'VR' type. In this paper, unless otherwise specified, we refer to the 'SR' type by default for 100G-per-lane transmission. For 100G-per-lane transmission, the required link bandwidth, taking into account both EMB and CD, is 18 GHz, defined at a 3 dBe drop of the frequency response from zero frequency level. The VCSEL transmission defined by IEEE 802.3db is an 850 nm-only transmission, where the 850 nm window for the 'SR' type is specified to be 844–863 nm. Within this wavelength window, the link bandwidth is required to stay at 18 GHz or above. The same requirement is inherited by Terabit BiDi MSA for its 'SR' type of applications, which further includes the 910 nm wavelength, in addition to an 850 nm wavelength window. 'HDR OM3' and 'HDR OM4' were defined to meet 18 GHz in both 850 nm and 910 nm windows. The Monte Carlo study in Section 3 allows us to show the link bandwidth over the entire wavelength range, relative to the 18 GHz requirements for 100G-per-lane applications at a given transmission reach.

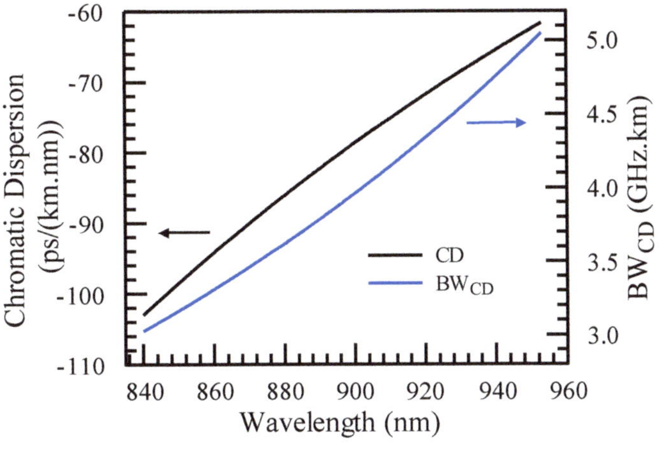

Figure 9. CD as a function of wavelength for MMF from 840 to 953 nm, shown as a black curve using left side axis scale pointed by the black arrow, and BW_{CD} as a function of wavelength for MMF from 840 to 953 nm, assuming 0.6 nm VCSEL RMS linewidth, shown as a blue curve using right side axis scale pointed by the blue arrow.

In Figure 10, we show the link bandwidth for 100 m 'HDR OM4', 100 m OM4, 100 m OM5, and 70 m OM4. The 18 GHz threshold is shown as a dashed line. For both 'HDR OM4' and OM5, the EMB stays above 18 GHz from 844 nm to 910 nm, ensuring that they both meet the link bandwidth requirements for the corresponding transmission reach for different types of transceivers. The drawback of OM4 is that its EMB drops rapidly once the wavelength exceeds 850 nm, and it has insufficient EMB at 910 nm for 100 m transmission. One has to substantially shorten the fiber length to restore the link bandwidth at 910 nm, as shown in the case of 70 m OM4. OM5 also meets the 18 GHz threshold between 844 nm and 910 nm. However, we note that the link bandwidth of 100 m OM5 drops below the critical 18 GHz threshold at wavelength above 910 nm, even though such fiber is specified for holding higher modal bandwidth up to 953 nm. The link bandwidth of 100 m OM5 is insufficient to support 100 m transmission once the wavelength is above 910 nm. Since OM5 was standardized around 2016–2017, predating the IEEE standard 802.3db for 100G per lane, it is not aligned with the 100G-per-lane transmission requirements. This highlights one limitation of OM5.

In Figure 11, we show the link bandwidth for 80 m 'HDR OM3', 70 m OM3, 45 m OM3, and 70 m OM4. For 80 m 'HDR OM3', the link bandwidth stays above the 18 GHz line from 850 nm to 910 nm, similar to 100 m 'HDR OM4' and 100 m OM5. To bring the link bandwidth back to 18 GHz for OM3 and OM4, their lengths would need to be reduced to 45 m and 70 m, respectively. We also observed that, for 70 m OM3, which has been used for 25 Gbaud transmission, the link bandwidth does not meet the 18 GHz threshold at any wavelength. This highlights the need to define 'HDR OM3' to restore transmission capabilities above 70 m.

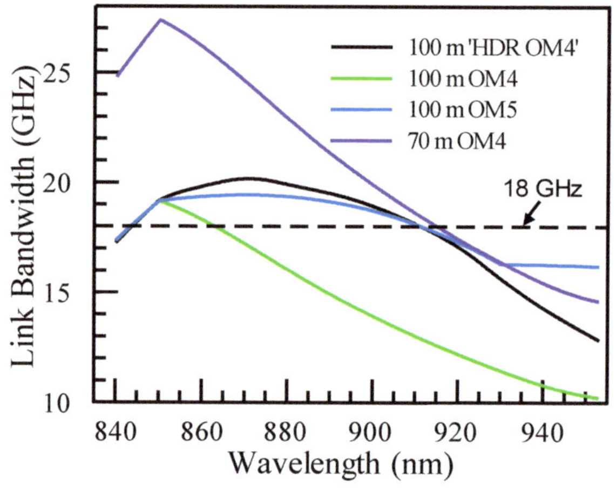

Figure 10. Link bandwidth of 100 m 'HDR OM4', 100 m OM4 and OM5, and 70 m OM4.

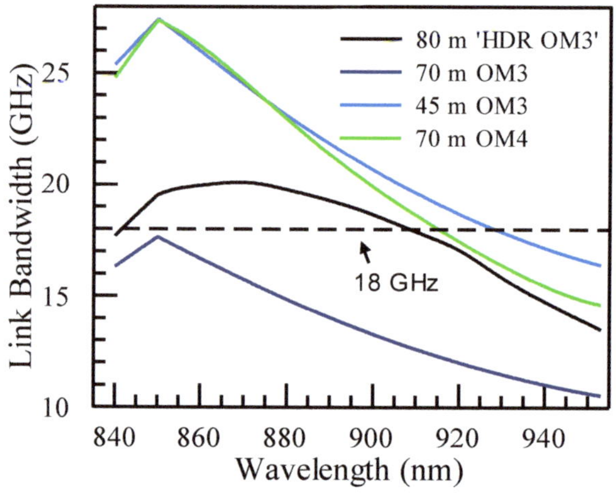

Figure 11. Link bandwidth of 80 m 'HDR OM3', 70 m OM3, 45 m OM3, and 70 m OM4.

4.3. Transmission Reaches for Various Types of Transceivers

The transmission reach is essentially the maximum length of an MMF that can meet the link bandwidth requirements. Table 2 lists the transmission reach of the transceivers with 100G-per-lane data rates at either the 850 nm or both 850/910 nm wavelength windows. Note that the focus here is on the 'SR' type of transmission. For the 850 nm window, the wavelength range is 844–863 nm, based on both IEEE 802.3db standard and Terabit BiDi MSA, while, for the 910 nm window, the wavelength range is 900–916 nm, based on Terabit BiDi MSA. One can see a notable limitation for OM3 in 400G SR4 and 800G SR8 transmission, as the reach has dropped to 60 m from the 70 m that people have grown accustomed to. The reaches of OM3 and OM4 for 800G SR4.2 or 800G BiDi are 45 m and 70 m, respectively, in contrast to the 70 m and 100 m people are accustomed to. The issue is alleviated with 'HDR OM3' and 'HDR OM4', which reach 80 m and 100 m distances, respectively. One subtle detail to note is that using a 0.6 nm RMS linewidth at a 910 nm window would result in the link bandwidth dropping below 18 GHz above 910 nm. This issue has been addressed in the MSA [6] by using a 0.58 nm laser linewidth for the wavelength window of 900–916 nm. The calculations in Figures 9 and 10 use 0.6 nm for simplicity when showing results. When a 0.58 nm laser linewidth is used, the link bandwidth over 900–916 nm would stay above 18 GHz for 100 m 'HDR OM4', 100 m OM5, and 80 m 'HDR OM3'. The fact that 'HDR OM3' can transmit over 80 m at 100G per lane is very attractive, as it not only carries over the benefit of OM3 from the 25 Gbaud age but also exceeds OM4 reach for 800G BiDi. It is especially beneficial for customers requiring 70–80 m distances for the majority of their links. Experimental testing using an 800G BiDi transceiver has also been conducted in [19,20], using 100 m 'HDR OM4' and 80 m 'HDR OM3', respectively, with good transmission performance demonstrated.

Table 2. The transmission reaches for 100G-per-lane-based transceivers.

Transceiver Protocol	MSA/Standard	Transmission Reach (m)				
		OM3	OM4	OM5	'HDR OM3'	'HDR OM4'
400G SR4	IEEE 802.3db	60	100	100	80	100
800G SR8	IEEE 802.3df	60	100	100	80	100
800G SR4.2	Terabit BiDi MSA	45	70	100	80	100

The 'HDR MMFs' can also serve well for 25 Gbaud-based VCSEL transmission. We list the calculated transmission reaches for various transceivers in Table 3. For 850 nm-based 25G SR, 100G SR4, and 400G SR8 transceivers, 'HDR OM3' can transmit up to 87 m, longer than OM3 at 70 m. For 100G BiDi and 100G SWDM4, 'HDR OM3' can transmit 120 m and 112 m, respectively, even exceeding OM4. For 'HDR OM4', it matches the reach of OM4 and OM5 for 25G SR, 100G SR4, 400G SR8, and 100G BiDi. For 100G SWDM4, 'HDR OM4' has a shorter reach than OM5 at 130 m, but its value is still significantly greater than that of OM4. Since commercially available SWDM-type transceivers are capped at 100G so far, with the underlying data rate at 25G NRZ per wavelength, while other types of transceivers have reached a 100G-per-lane data rate, the value of OM5, with its key strength aligned with a 953 nm EMB, is put in doubt [30,31].

Table 3. The transmission reaches for 25Gbaud-based transceivers.

Transceiver Protocol	MSA/Standard	Reach (m)				
		OM3	OM4	OM5	'HDR OM3'	'HDR OM4'
25G SR	IEEE 802.3by	70	100	100	87	100
100G SR4	IEEE 802.3bm	70	100	100	87	100
400G SR8	IEEE 802.3cm	70	100	100	87	100
400G SR4.2/100G BiDi	IEEE 802.3cm/proprietary	70	100	150	120	150
100G SWDM4	SWDM MSA	75	100	150	112	130

4.4. Implications outside 850 nm and 910 nm Wavelength Windows

VCSEL transmission data rates have increased from 10G to 25G and now reach 100G per wavelength. To fabricate VCSELs with higher bandwidth, indium has been added to the quantum well to improve the differential gain, resulting in the operating wavelength shifting higher [32]. This situation has occurred since the adoption of 25 Gbaud VCSEL transceivers. In Figure 12, we show the optical spectrum, as measured from one 850 nm channel of a 400G SR4.2 transceiver (LMQ8811B-PC+ from Hisense, Qingdao, China). Additionally, 400G SR4.2 transceivers adopt two wavelengths around 850 nm and 910 nm, with each channel operating with 50G PAM4 modulation. The center wavelength is 859.5 nm in this case. Although the nominal wavelength for the 850 nm window is defined as 844–863 nm in IEEE 802.3db for 100G per lane, it is known in the industry that, for much of the 850 nm window, 100G VCSELs have center wavelengths around 860 nm, with the range between 850 and 870 nm. The higher range is outside the current 850 nm window defined by the IEEE standard. While accepting this de facto situation, we show that 'HDR MMFs' are ideally positioned for such wavelength shifts of 850 nm VCSELs to maintain desired transmission reaches. In Figure 13, we zoom in on several curves from Figure 10 between 845 and 875 nm. As seen, for OM4, the link bandwidth drops below 18 GHz at 863 nm and beyond. Therefore, for the wavelength range between 863 nm and 870 nm, the transmission reach is below 100 m, with the lowest reach at 870 nm being 96 m. On the other hand, both 100 m 'HDR OM4' and 100 m OM5 retain a link bandwidth above 19 GHz, exceeding the 18 GHz threshold.

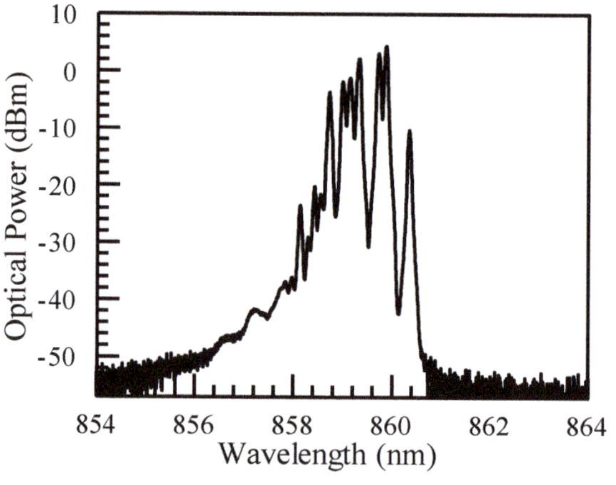

Figure 12. Optical spectrum of one channel from a 400G SR4.2 transceiver.

In addition to the benefits of 'HDR MMFs' for wavelength-shifted VCSELs toward 860 nm, we observe that the link bandwidth for such fibers would stay above the values at 850 nm, ensuring that VCSEL transmission at higher wavelengths would not suffer from lower bandwidth, as is the case for OM3 and OM4. This brings flexibility to VCSEL manufacturing and opens the door to using 'HDR MMFs' for broader benefits over the whole wavelength range from 850 nm to 910 nm, thereby not limited to the two wavelengths at which they are defined. Even at wavelengths above 910 nm, they still have better bandwidth than OM3 and OM4 while trailing OM5, which can be seen in Figures 9 and 10. Using 'HDR MMFs', transceivers can be implemented at any wavelength between 850 nm and 910 nm, with the benefits of achieving higher data rates for VCSEL transceivers, similar to what can be achieved at 850 nm.

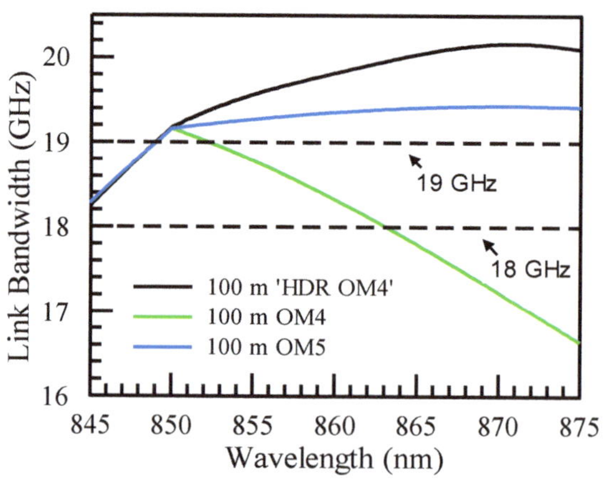

Figure 13. The link bandwidth as a function of wavelength for 100 m 'HDR OM4', 100 m OM4, and 100 m OM5.

5. Discussion

In this paper, we have studied the wavelength dependence of modal bandwidth for the 'HDR MMFs' we have recently proposed [17–21]. We have shown detailed results in terms of EMB and link bandwidth, which are directly tied to transmission performance. The 'HDR MMFs' were proposed to address the needs for VCSEL-based 100G-per-lane transmission over a wide wavelength range, with the highest performance from 850 nm to 910 nm. To better understand the benefits as they relate to wavelength, we discuss and highlight the capabilities from the following perspectives.

- Benefit for the 850 nm wavelength window: By definition, 'HDR OM3' and 'HDR OM4' are defined at 850 nm and 910 nm wavelengths. However, since we have taken into consideration the limitation of OM3 for 100G-per-lane transmission with a reach of 60 m, we have chosen the 850 nm EMB for 'HDR OM3' to support an 80 m reach. This ensures that 'HDR OM3' can also benefit 850 nm-only applications. It alleviates the issue of OM3's reach dropping below 70 m, a distance that many customers have grown accustomed to and around which many data centers have been designed.
- Addressing the wavelength shifting toward 860 nm in 850 nm window: As the VCSEL data rate increases to 100G per lane and will reach 200G per lane very soon, some implementations of VCSELs have shown a shift in the wavelength, more centered around 860 nm instead of 850 nm. The 'HDR MMFs' offer the benefit of having enhanced performance for the entire 850–870 nm wavelength window. Taking into account the CD contribution, the link bandwidth over this wavelength range stays above the level at 850 nm, meeting the high bandwidth demands despite the wavelength shift.
- Comparison to OM3 and OM4: One major limitation of OM3 and OM4 is that their EMBs decrease rapidly as the wavelength moves away from 850 nm, particularly when moving toward higher wavelengths up to 910 nm. Even though the 850 nm EMBs of these fibers remain adequate in many cases, the limitation at 910 nm restricts their applications for longer wavelength use, including 860 nm-centered applications and BiDi applications involving 850/910 nm dual wavelengths.
- Enabling VCSELs with more wavelengths: Currently, most VCSELs are implemented at 850 nm and 910 nm, with the exception of SWDM VCSELs, which work up to 953 nm but are so far limited to 100G data rates at the transceiver level. The current study shows that 'HDR MMFs' can have uniform link bandwidths from 850 nm to 910 nm. This opens up the feasibility of implementing VCSELs at any wavelength from 850 nm to 910 nm to achieve similar performance.
- Benefits for SWDM applications: even though SWDM applications have not been deployed in large volumes like other types of VCSEL transceivers and have not reached higher data rates, 'HDR MMFs' still demonstrate their advantages over OM3 and OM4 by delivering significantly better transmission reaches, although they trail somewhat behind OM5.
- Comparison to OM5: Unlike OM5, which was defined based on EMBs at 850 nm and 953 nm, 'HDR MMFs' were defined based on EMBs at 850 nm and 910 nm. 'HDR OM4' has the same EMBs as OM5 at both 850 nm and 910 nm. The wavelength dependence study in Section 3 shows that 'HDR OM4' has an EMB, from 850 nm to 910 nm, equal to or higher than that of OM5 guided by IEC. This implies that 'HDR OM4' can cover all the needs of OM5 from 850 nm to 910 nm and can be a more cost-effective and higher-volume solution for a wide range of applications using the relevant wavelengths.

6. Conclusions

VCSEL-based transmission over MMFs has been widely deployed. When the data rate for VCSEL-based transmission reaches 100G per lane, the modal bandwidth of OM3 and OM4 becomes limiting for certain wavelengths. In light of such limitations, we have proposed sub-categories of OM3 and OM4 fibers to address the needs of HDR transmission [17–21]. These fibers exhibit a higher EMBs at 910 nm, leading to a different wavelength dependence compared to conventional OM3 and OM4 MMFs. Understanding the wavelength dependence of these fibers is crucial to address their utilization in a broader range of applications. We adopted Monte Carlo simulations as the primary approach for the current study. We have generated refractive index profiles of 10,000 MMFs. By calculating EMB of each fiber over a range of wavelengths of interest, from 840 nm to 953 nm, we obtained the low-end boundary of the EMB at each wavelength. These values were compared with known cases of OM3 and OM4, revealing the capability differences.

We observe that both 'HDR OM4' and 'HDR OM3' offer higher EMBs than OM4 and OM3 for wavelengths longer than 850 nm. In particular, our findings indicate that the 'HDR OM4' performs the same or better than OM5 from 840 nm to 920 nm. Due to their higher bandwidth capability at longer wavelengths, these fibers are more suitable for high-data-rate transmission. We also obtained the link bandwidth that includes both the modal bandwidth contribution and the CD-contributed bandwidth, which is directly tied to the system level performance. We find that, for both 'HDR OM3' and 'HDR OM4', the link bandwidth stays above the value at 850 nm until around 910 nm. This means these fibers deliver similar transmission performances from 850 to 910 nm without declining towards longer wavelengths, unlike standard OM3 and OM4. In the application space, we further list the transmission reaches for various transceiver applications, both for 100G-per-lane-based and older 25 Gbaud-based transceivers. To better illustrate various aspects of 'HDR MMFs', in the discussion section, Section 5, we presented several comparisons. We believe the current study provides insights about the benefits of high-data-rate MMFs, enabling more VCSEL wavelength choices and optimal deployment in diverse scenarios.

Author Contributions: Conceptualization, X.C., H.D., H.C., Z.D.B. and M.-J.L.; methodology, X.C. and M.-J.L.; validation, X.C. and J.E.H.; formal analysis, X.C.; investigation, X.C. and J.E.H.; writing—original draft preparation, X.C.; writing—review and editing, X.C., H.D., H.C. and Z.D.B.; visualization, X.C. All authors have read and agreed to the published version of the manuscript.

Funding: This research received no external funding.

Institutional Review Board Statement: Not applicable.

Informed Consent Statement: Not applicable.

Data Availability Statement: Data are contained within the article, as shown in the figures and as-sociated descriptions.

Conflicts of Interest: Xin Chen, Hao Dong, Hao Chen, Jason E. Hurley, Zoren D. Bullock and Ming-Jun Li were employed by the company Corning Incorporated. The authors declare that the research was conducted in the absence of any commercial or financial relationships that could be construed as a potential conflict of interest.

References

1. Chen, X.; Bickham, S.R.; Abbott, J.S.; Coleman, J.D.; Li, M.-J.; Peng, G.D. Multimode fibers for data centers. In *Handbook of Optical Fibers*; Springer Nature Singapore Pte Ltd.: Berlin/Heidelberg, Germany, 2018; pp. 1–57.
2. Li, M.-J. MMF for High Data Rate and Short Length Applications. In Proceedings of the Optical Fiber Communication Conference, OFC 2014, San Diego, CA, USA, 24–28 March 2024; pp. 1–3.
3. Cisco Annual Internet Report (2018–2023) White Paper (Updated: March 9, 2020). Available online: https://www.cisco.com/c/en/us/solutions/executive-perspectives/annual-internet-report/index.html (accessed on 14 July 2024).
4. InfiniBand Network Application in DGX Cluster. Available online: https://www.naddod.com/blog/infiniband-network-application-in-dgx-cluster (accessed on 14 July 2024).
5. Cisco 40GBASE QSFP Modules Data Sheet. Available online: https://www.cisco.com/c/en/us/products/collateral/interfaces-modules/transceiver-modules/data_sheet_c78-660083.html (accessed on 14 July 2024).
6. Terabit BiDi MSA. Available online: https://terabit-bidi-msa.com/ (accessed on 14 July 2024).
7. Tatum, J.A.; Gazula, D.; Graham, L.A.; Guenter, J.K.; Johnson, R.H.; King, J.; Kocot, C.; Landry, G.D.; Lyubomirsky, I.; MacInnes, A.N.; et al. VCSEL-based interconnects for current and future data centers. *J. Light. Technol.* **2015**, *33*, 727–732. [CrossRef]

8. Shortwave Wavelength Division Multiplexing (SWDM) Alliance. Available online: https://www.swdm.org/ (accessed on 14 July 2024).
9. Chorchos, L.; Ledentsov, N.; Makarov, O.; Shchukin, V.A.; Kalosha, V.; Turkiewicz, J.P. Multi Aperture High Power 100G Single Mode 850nm VCSEL for Extended Reach 800G Ethernet. In Proceedings of the Optical Fiber Communications Conference and Exhibition (OFC), San Diego, CA, USA, 5–9 March 2023; pp. 1–3.
10. Zuo, T.; Zhang, T.; Zhang, S.; Liu, L. 850-nm VCSEL-based single-lane 200-Gbps PAM-4 transmission for datacenter intra-connections. *IEEE Photonics Technol. Lett.* **2021**, *33*, 1042–1045. [CrossRef]
11. Murty, M.V.R.; Wang, J.; Jiang, S.; Dolfi, D.; Wang, T.K.; Vaughan, D.; Feng, Z.; Leong, N.; Sridhara, A.; Joyo, S.T.; et al. Toward 200G per Lane VCSEL-based Multimode Links. In *Proceedings of the Optical Fiber Communication Conference (OFC) 2024, San Diego, CA, USA, 24–28 March 2024*; Technical Digest Series; Paper M2D.3; Optica Publishing Group: Washington, DC, USA, 2024.
12. Tirelli, S.; Corti, E.; Duda, E.; Pissis, A.; Hönl, S.; Hoser, M.; Paul, M.; Zibik, E. Lithographic Aperture VCSELs Enabling Beyond 100G Datacom Applications. In *Proceedings of the Optical Fiber Communication Conference (OFC) 2024, San Diego, CA, USA, 24–28 March 2024*; Technical Digest Series; Paper M2D.1; Optica Publishing Group: Washington, DC, USA, 2024.
13. Wettlin, T.; Lin, Y.; Stojanovic, N.; Calabrò, S.; Wang, R.; Zhang, L.; Kuschnerov, M. 200 Gb/s VCSEL transmission using 60 m OM4 MMF and KP4 FEC for AI computing clusters. *arXiv* **2024**, arXiv:2403.17275.
14. IEEE Standard for Ethernet—Amendment 3: Physical Layer Specifications and Management Parameters for 100 Gb/s, 200 Gb/s, and 400 Gb/s Operation over Optical Fiber Using 100 Gb/s Signaling. Available online: https://standards.ieee.org/ieee/802.3db/10283/ (accessed on 14 July 2024).
15. *TIA-492AAAE*; Detail Specification for 50-µm Core Diameter/125-µm Cladding Diameter Class 1a Graded-Index Multimode Optical Fibers with Laser-Optimized Bandwidth Characteristics Specified for Wavelength Division Multiplexing. TIA: Arlington, VA, USA, 2016.
16. *IEC 60793-2-10:2019*; Optical Fibres—Part 2–10: Product Specifications—Sectional Specification for Category A1 Multimode Fibres. Edition 7.0. IEC: Geneva, Switzerland, 2019.
17. Dong, H.; Chen, X. The Road to 800G with Multimode Fiber. Corning Incorporated. Available online: https://www.corning.com/catalog/coc/documents/articles/CO8000_Article_Road%20to%20800G%20multimode%20fiber.pdf (accessed on 14 July 2024).
18. Chen, H.; Wang, C.; Chen, X.; Dong, H.; Bullock, Z. Optimized Multimode Fiber for 100G/land and BiDi technologies. In Proceedings of the 2023 China Communication Academic Annual Conference, Dalian, China, 10–12 August 2023.
19. Chen, X.; Dong, H.; Chen, H.; Mu, J.; Zheng, L.; Yang, S.; Hurley, J.E.; Wood, W.A.; Bullock, Z.; Li, M.-J. Concept of BiDi Optimized OM4 Multimode Fiber for High Date Rate Short Reach VCSEL Transmission. In Proceedings of the 2023 Opto-Electronics and Communications Conference (OECC), Shanghai, China, 2–6 July 2023; pp. 1–5.
20. Chen, X.; Dong, H.; Huang, O.; Mu, J.; Chen, H.; Hurley, J.E.; Bullock, Z.; Li, M.-J. Concept of OM3 Multimode Fiber Optimized for 100G/lane and BiDi Based VCSEL Transmission. In Proceedings of the 2024 Opto-Electronics and Communications Conference (OECC), Melbourne, Australia, 30 June–4 July 2024. Paper #45.
21. Corning® ClearCurve® OM3 XT and OM4 XT Optical Fibers—Product Information. Available online: https://www.corning.com/media/worldwide/coc/documents/Fiber/product-information-sheets/PI-1568.pdf (accessed on 14 July 2024).
22. LaserWave® Dual-Band OM4+ Multimode Optical Fiber. Available online: https://fiber-optic-catalog.ofsoptics.com/LaserWave-Dual-Band-OM4-Multimode-Optical-Fiber-10033 (accessed on 14 July 2024).
23. Li, M.-J.; Tandon, P.; Bookbinder, D.C.; Bickham, S.R.; Wilbert, K.A.; Abbott, J.S.; Nolan, D.A. Designs of bend-insensitive multimode fibers. In Proceedings of the 2011 Optical Fiber Communication Conference and Exposition and The National Fiber Optic Engineers Conference, Los Angeles, CA, USA, 6–10 March 2011; pp. 1–3.
24. Bourdine, A.V.; Praporshchikov, D.E.; Yablochkin, K.A. Investigation of defects of refractive index profile of silica graded-index multimode fibers. *Proc. SPIE* **2011**, *7992*, 799206.
25. Lenahan, T.A. Calculation of modes in an optical fiber using the finite element method and EISPACK. *Bell Syst. Tech. J.* **1983**, *62*, 2663–2694. [CrossRef]
26. *IEC 60793-1-49*; Optical Fibres—Part 1-49: Measurement Methods and Test Procedures—Differential Mode Delay. Edition 3.0. IEC: Geneva, Switzerland, 2018.
27. Bottacchi, S. *Multi-Gigabit Transmission over Multimode Optical Fibre: Theory and Design Methods for 10 GbE Systems*; John Wiley & Sons, Ltd: Hoboken, NJ, USA, 2006.
28. Li, M.-J.; Li, K.; Chen, X.; Mishra, S.K.; Juarez, A.A.; Hurley, J.E.; Stone, J.S.; Wang, C.-H.; Cheng, H.-T.; Wu, C.-H.; et al. Single-Mode VCSEL Transmission for Short Reach Communications. *J. Lightwave Technol.* **2021**, *39*, 868–880. [CrossRef]
29. Ledentsov, N.N.; Makarov, O.Y.; Shchukin, V.A.; Kalosha, V.P.; Ledentsov, N.; Chrochos, L.; Sanayeh, M.B.; Turkiewicz, J.P. High Speed VCSEL Technology and Applications. *J. Lightwave Technol.* **2022**, *40*, 1749–1763. [CrossRef]
30. Understanding the Differences Between OM4 and OM5 Multimode Fiber. Available online: https://www.cisco.com/c/en/us/products/collateral/interfaces-modules/transceiver-modules/diff-om4-om5-multimode-fiber-wp.html (accessed on 14 July 2024).

31. Available online: https://blog.siemon.com/infrastructure/om5-no-reach-advantage-cisco-multimode-transceivers#:~:text=Most%20of%20Cisco's%20multimode%20transceivers,realize%20significant%20benefits%20from%20OM5 (accessed on 14 July 2024).
32. Chang, Y.-C.; Coldren, L.A. Optimization of VCSEL structure for high-speed operation. In Proceedings of the 2008 IEEE 21st International Semiconductor Laser Conference, Sorrento, Italy, 14–18 September 2008; pp. 159–160.

photonics

MDPI

Article

Lights off the Image: Highlight Suppression for Single Texture-Rich Images in Optical Inspection Based on Wavelet Transform and Fusion Strategy

Xiang Sun [1,2], Lingbao Kong [1,*], Xiaoqing Wang [2], Xing Peng [3] and Guangxi Dong [1]

[1] Shanghai Engineering Research Center of Ultra-Precision Optical Manufacturing, School of Information Science and Technology, Fudan University, Shanghai 200438, China; 3313@ecjtu.edu.cn (X.S.); 19210720013@fudan.edu.cn (G.D.)
[2] School of Electrical and Automation Engineering, East China Jiaotong University, Nanchang 330013, China; 2020048125300060@ecjtu.edu.cn
[3] College of Intelligent Science and Technology, National University of Defense Technology, Changsha 410073, China; pengxing22@nudt.edu.cn
* Correspondence: lkong@fudan.edu.cn

Abstract: A wavelet-transform-based highlight suppression method is presented, aiming at suppressing the highlights of single image with complex texture. The strategy involves the rough extraction of specular information, followed by extracting the high-frequency information in specular information based on multi-level wavelet transform to enhance the texture information in the original images by fusion strategy, and fusing with the same-level specular information to achieve the highlight suppression image. The experimental results demonstrate that the proposed method effectively removed large-area highlights while preserving texture details, and demonstrated the authenticity of the highlight estimation and the 'lights off' effect in the highlight-suppressed images. Overall, the method offers a feasibility for addressing the challenges of highlight suppression for visual detection image with rich texture and large-area highlights.

Keywords: highlight suppression; single image; wavelet transform; fusion strategy

Citation: Sun, X.; Kong, L.; Wang, X.; Peng, X.; Dong, G. Lights off the Image: Highlight Suppression for Single Texture-Rich Images in Optical Inspection Based on Wavelet Transform and Fusion Strategy. *Photonics* **2024**, *11*, 623. https://doi.org/10.3390/photonics11070623

Received: 21 May 2024
Revised: 26 June 2024
Accepted: 27 June 2024
Published: 28 June 2024

1. Introduction

In optical inspection systems, extreme external illumination can cause specular reflection on highly reflective surfaces like metal parts and smooth plastics, leading to overexposure and highlight regions that obscure the original target information [1], and it can further lead to causing a decrease in image quality and reducing the reliability of detection. Highlight suppression is critical for improving the quality of images captured in optical inspection systems. By effectively suppressing or eliminating highlights, it is possible to reveal the true texture and features of the inspected targets. This leads to more accurate defect detection and better-quality control. So, the suppression of highlights is particularly important in industries such as manufacture, medical equipment, and robots, where precision and reliability are paramount. Numerous methods have been proposed over the years to tackle this challenge.

The Dichromatic Reflection Model (DRM) describes reflected light as linear combinations of diffuse and specular reflection, where the diffuse reflection reflects the targets' texture information, and the specular reflection reflects the external lighting information, which masks the target texture and color features [2]. Based on the DRM, Klinker et al. [3] found that the diffuse and specular reflection components of the RGB channel in the image have a T-shaped distribution; they treated the dark channel of the image as approximately specular highlight information, and the color vectors of the diffuse and light sources were fitted by conducting principal component analysis on the diffuse and specular reflection regions, so the maximum a posteriori probability (MAP) method was used to obtain highlight removal images by calculating the probability distribution of the initial specular reflection components.

Moreover, several researchers have built upon these foundational concepts. Chen et al. [4] used a skin color model instead of a camera color calibration method to improve the removal effect of highlights. The maximum fraction of the diffuse color component in diffuse local patches is in smooth distribution [5]; based on this observation, Yang et al. regarded the specular component as image noise, so that the bilateral filter is employed to smooth the maximum fraction of the color components of the image to suppress the specular component; however, the method relies on the estimation of the maximum diffuse reflection. Zhao et al. [6] proposed a highlight separation method based on the local structure and chromaticity joint compensation (LSCJC) characteristic of an object, removed texture distortion in the estimated diffuse reflection component using the diffuse component gradient-magnitude similarity map, and compensated for the chromaticity in specular reflection region based on local chromaticity correlation. Subsequently, the model continued to expand, achieving effective separation of diffuse reflection components and specular reflection components [7].

Additionally, the sparse representation techniques can be applied to the separation of diffuse and specular components, enhancing the precision of highlight suppression. Akashi et al. [8], based on a modified version of sparse non-negative matrix factorization, and spatial prior, accurately separated the body colors and specular components, but this method may easily cause some pixels with zero value in strong highlight areas, so texture distortion is inevitable. Fu et al. [9] employed a linear combination of a few basis colors with sparse encoding coefficients to suppress the highlights with a small area, the main limitation of this method is may fail to restore the subtle textures from large specular highlight regions.

Most existing highlight suppression methods are mainly based on color images, analyzing the chromaticity or polarization of lighting to recover lost information. However, there are many grayscale images that exist in practice detection. The absence of color information can easily distort image intensity during highlight suppression, posing a significant challenge.

Ragheb et al. [10] proposed a MAP method for estimating the proportions of Lambertian and specular reflection components, accurately obtained the proportion of specular reflection, and realized a good surface normal reconstruction in the proximity of specular highlights, but the iterative conditional model in this method has local convergence and cannot obtain the global optimal solution, which can easily cause some residual highlights. Ma et al. [11] calculated the MAP of each reflection composition under simulated annealing based on the surface normal, and implemented image highlight detection and highlight removal, combined with the assumption of curvature continuity, but it does not take the impact of visual geometric features and edge smoothing on specular reflection regions into account, which can easily lead to excessive smoothness of the color and steep boundary brightness in specular reflection region. To address this issue, Yin et al. [12] established a model of diffuse and specular reflection components, then calculated the MAP to detect and recover the specular reflection region based on a Bayesian formula and BSCB (Bertalmio, Sapiro, Caselles, Ballester) model [13]; however, this method easily causes significant distortion when encountering specular reflection region.

The research mentioned above demonstrated that the specular reflection information can be removed. Direct separation of the specular component in the spatial domain is challenging, but it can be effectively achieved in the frequency domain by exploiting the gradient properties of highlight pixels. Chen et al. [14] proposed a method for removing specular highlights in natural scene images, whereby based on frequency domain analysis, they decomposed the input image into reflectance and illumination components, and estimated the specular reflection coefficient based on smooth features and chromaticity space, subtracting the specular component to achieve highlight removal. However, the method may not accurately recover information damaged by highlights. Zou et al. [15]

used the spatial-frequency image enhancement and region growing algorithm to realize the tile surface defects. The median filtering and local histogram equalization were used to implement it, followed by frequency domain enhancement using 2-D Gabor filter, and then use the region growing algorithm to realize the automatic segmentation of defect regions. But the method may cause lower accuracy for pockmark and chromatic aberration defects based on the observation that the maximum fraction of the diffuse color component changes smoothly in local patches.

These studies collectively demonstrate the diversity and effectiveness for image highlight suppression, for spatial domain methods, while intuitive and straightforward, often struggle with grayscale images where color information is absent, leading to potential distortions in image intensity. On the other hand, frequency domain methods can provide more robust suppression of highlights without significantly affecting the underlying texture. However, among the currently published methods, they may still face challenges in accurately extracting highlights and restoring textures, especially in regions with intricate textures. The processed highlight regions, usually with weak texture features, lack the highlight suppression for strong texture features, so there are still some limitations in image highlight removal for targets with a complex texture, while the sources commonly cause strong highlight regions in visual detection systems [16], which blocks the texture information. The highlights extracted by existing highlight suppression methods may easily contains much texture information, which may cause texture distortion when directly eliminating this highlight component.

Therefore, to solve this problem, this paper presents an image highlight suppression method based on wavelet transform to realize the image highlight suppression and texture restoration, and MATLAB (Version R2019a) is employed to implement the algorithm. First, the specular information is preliminarily separated based on a specular-free image, due to the complex texture, the extracted specular information still contains some residual texture information. So, the multi-stage wavelet transform is used to separate the low-frequency and high-frequency information of the extracted specular information and original image respectively. Finally, the low- and high-frequency information is processed by setting filters and a gain strategy, and the inverse wavelet transform is used to reconstruct image, so the highlight removal image and specular information can be obtained, where the concealed texture information in specular information is extracted to restore the texture.

2. Image Highlight Suppression Strategy

2.1. Rough Extraction of Specular Information

The reflected lights can be decomposed into diffuse and specular components [3], with the diffuse components reflecting the texture information of target, and the specular components reflecting the specular information [17,18].

The obtained image $Img_T(u,v)$ of size $m \times n$ can be reshaped into an array $M(i)$ ($i \in [1, m \times n]$) of length $m \times n$. The discrimination threshold T_M for the diffuse reflection component is then calculated from M, as follows,

$$T_M = \text{Mean}[M(i)] + \eta \cdot \text{Std}[M(i)] \tag{1}$$

where $\text{Mean}[M(i)]$ represents the average value, $\text{Std}[M(i)]$ represents the standard deviation, and η represents the intensity threshold of the specular information, which can be adjusted according to the actual situation of the image to obtain the specular information of different areas. When $M(i)$ is less than T_M, the corresponding pixels can be seen as the diffuse information. Take the stripe-projection image as an example, the intensity threshold η, which determines the area of extracted highlights, is set to 1.05, and the discrimination threshold, T_M, is equal to 196.77. Thus, the geometric factor $\beta_s(i)$ of the specular information can be calculated as follows,

$$\beta_s(i) = \begin{cases} M(i) - T_M & , M(i) > T_M \\ 0 & , otherwise \end{cases} \tag{2}$$

$\beta_s(i)$ is a one-dimensional array of length $m \times n$, and can be seen as the proximity between pixels and specular information. Large proximity implies that the pixels' value is close to the specular pixels. Conversely, a low proximity value suggests that the pixels are highly likely not specular information. Then, the obtained $\beta_s(i)$ can be reorganized into $Img_{Rou}(u,v)$ to represent the roughly extracted highlight pixels, which is normalized and then binarized to obtain the specular pixels, then according to the area of specular pixels, extract the largest connected area $\Gamma_{Lar}(u,v)$, which is a logical matrix. Extracting the largest area can enhance the robustness and accuracy by avoiding noise in small specular areas, reducing computational load, ensuring representative specular features, preventing overprocessing for natural visuals, and providing a stable intensity recovery. Here, the dilatation algorithm is employed to process the $\Gamma_{Lar}(u,v)$ to obtain the $\Lambda_{Lar}(u,v)$, which comprises pixels in the largest specular area and its surrounding diffuse reflection pixels. Then, perform the difference operation, and take the absolute value to obtain $Img_{Dif}(u,v)$, which is the diffuse information near maximum specular region, as follows,

$$Img_{Dif}(u,v) = |\Lambda_{Lar}(u,v) - \Gamma_{Lar}(u,v)| \tag{3}$$

Then, the $Img_{Dif}(u,v)$ is transformed into a logical matrix $\Gamma_{Dif}(u,v)$, where the non-zero elements in $Img_{Dif}(u,v)$ were converted to a logical value of 1 (true), and zeros were converted to a logical value of 0 (false). Then, use $\Gamma_{Dif}(u,v)$ and $\Gamma_{Lar}(u,v)$ as indexes to extract the corresponding pixels $Dif_M(l)$ and $Lar_M(l)$ from $M(i)$ respectively, which represents the intensity of the specular region and its nearby diffuse region. Using the $\Gamma_{Dif}(u,v)$ and $\Gamma_{Lar}(u,v)$ as indexes to extract $Dif_\beta(l)$ and $Lar_\beta(l)$ from $\beta_s(i)$, where the $Dif_\beta(l)$ is the specular information of the specular region, and $Lar_\beta(l)$ is the specular information of the nearby diffuse region. Variable l denotes the indices of elements in arrays $Dif_M(l)$, $Lar_M(l)$, $Dif_\beta(l)$, and $Lar_\beta(l)$. Thus, based on the principle of smooth transition, we have,

$$\overline{Dif_M(l)} - k \times \overline{Dif_\beta(l)} = \overline{Lar_M(l)} - k \times \overline{Lar_\beta(l)} \tag{4}$$

where k is a scale factor used to adjust the intensity of specular information, which makes the diffuse information smooth and natural in image. The $\overline{Dif_M(l)}$, $\overline{Lar_M(l)}$, $\overline{Dif_\beta(l)}$, and $\overline{Lar_\beta(l)}$ are the mean intensities of $Dif_M(l)$, $Lar_M(l)$, $Dif_\beta(l)$, and $Lar_\beta(l)$, respectively. Then, the value of pixels $S(i)$ of the specular information can be calculated by k, which is expressed as,

$$S(i) = k \times \beta_s(i) \tag{5}$$

For the stripe-projection image, according to Equations (2)–(4), k is 1.44. After the separation, the specular information is reorganized into a matrix $Img_S(u,v)$. Figure 1 shows a solved example of stripe-projection image, different η determines the extracted highlight areas, and further determines the extent of the subsequent highlight reduction. It can be seen from Figure 1 that there is much concealed texture information in the rough-extraction specular information Img_S, and direct difference operations on the image will eliminate both specular and texture information, which is unsuitable. Since the gradient variation in pixels in the specular region is much smaller than that of texture information, there is a significant difference in frequency domain [19]. In this paper, a frequency domain method is presented to improve the image highlight suppression.

2.2. First Fusion Strategy

Wavelet transform [20,21] is more suitable for the target image with complex features compared with Fourier transform, because the orthogonal wavelet reconstruction has better stability, and the symmetric wavelet bases can eliminate phase distortion, which can better reconstruct the decomposed images. Furthermore, wavelet bases with strong compact support have a higher attenuation speed so that can better detect fine features in images, and good smoothness can make it more beneficial for improving the frequency resolution during image decomposition and reducing distortion during image reconstruction.

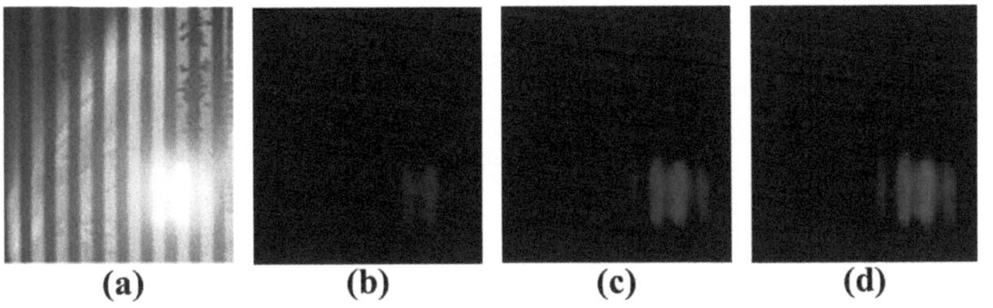

Figure 1. Rough-extraction specular information Img_S under different intensity thresholds of η: (**a**) original image Img_T; (**b**) $\eta = 1.05$; (**c**) $\eta = 0.85$; (**d**) $\eta = 0.65$.

In this paper, the 2-D discrete wavelet transform (2-D DWT) is used to perform multi-level decomposition of the images along the row, column, and diagonal directions. In a 2-D DWT, an image is divided into four sub-images because the process involves applying wavelet filters in both horizontal and vertical directions. (1) Horizontal filtering: The image is initially filtered along the rows (horizontal direction) using low-pass and high-pass wavelet filters, and then downsampled to create two sets of coefficients: low-frequency and high-frequency. (2) Vertical filtering: These coefficients are then filtered and downsampled along the columns (Vertical direction) using the low-pass and high-pass wavelet filters. The composite filtering produces four sub-bands: (1) Low-Low pass filtering: Results from low-pass filtering in both directions, obtaining the approximation of the image (or the lowpass filtered image). (2) Low-High pass filtering: Results from low-pass filtering horizontally and high-pass filtering vertically, obtaining horizontal edge details. (3) High-Low pass filtering: Results from high-pass filtering horizontally and low-pass filtering vertically, obtaining vertical edge details. (4) High-High pass filtering: Results from high-pass filtering in both directions, obtaining diagonal edge details.

For multi-level 2-D DWT, the object of each decomposition is the low-frequency information after the previous level 2-D DWT, as Figure 2 shows,

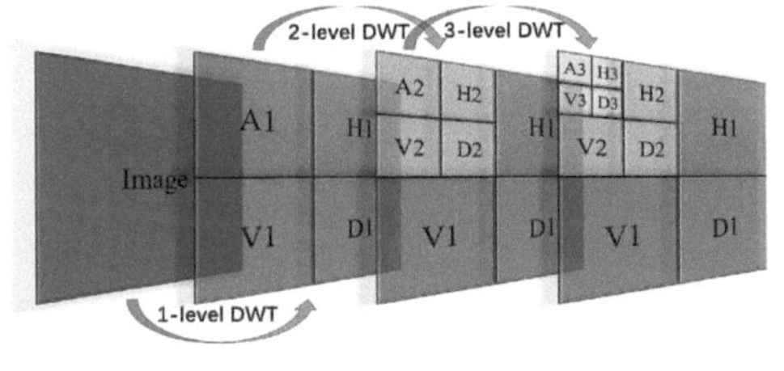

Figure 2. The 3-level decomposition of image by 2-D DWT.

Thus, the combination of these horizontal and vertical filtering steps generates four sub-images, each highlighting different frequency components and spatial characteristics of the original image. In this paper, a Symlet wavelet is employed to suppress the highlights based on single-image, two-fusion strategies are employed to restore the texture in highlight areas and suppress the specular information, respectively. The first fusion combines high-frequency information from imgs with imgt to enhance texture details, it is crucial for improving the visibility of fine details obscured by highlights. The second fusion, on the other hand, focuses on integrating the texture restoration image from first fusion back with the low-frequency information of Img_S, this ensures that the specular information can be removed while the texture of the image is preserved. Describing both fusions separately is essential because they serve different purposes within the highlight suppression framework and contribute uniquely to the final outcome.

Firstly, in order to extract the concealed texture information in Img_S to enhance the texture feature of original image Img_T, a multi-level DWT employed to decompose the low-frequency and high-frequency information of Img_S and Img_T respectively, the next-level decomposition object is the low-frequency information of previous-level decomposition. The technical route is as Figure 3 shows.

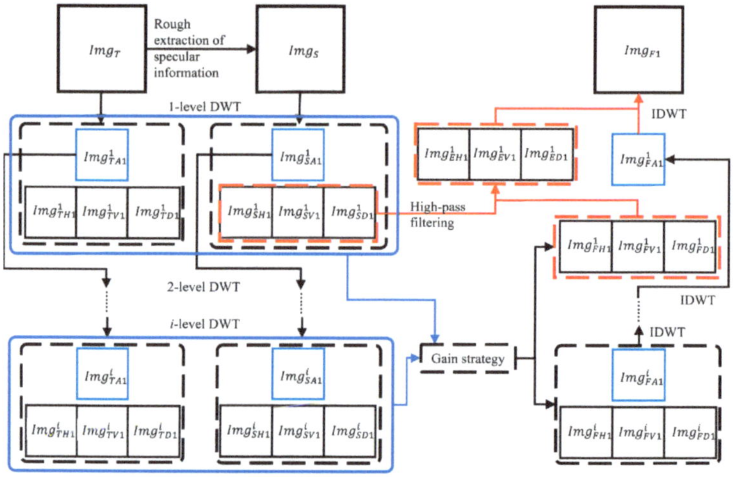

Figure 3. Technical route of enhancement for original image Img_T.

In Figure 3, Img^i_{SA1} denotes the low-frequency information of Img_S at different levels, while Img^i_{SH1}, Img^i_{SV1}, and Img^i_{SD1} denote the horizontal, vertical, and diagonal components of high-frequency information at different levels. Similarly, Img^i_{TA1} denotes the low-frequency information of Img_T at different levels, where the "A" denotes the "Approximation", it means the low-frequency information, Img^i_{TH1}, Img^i_{TV1}, Img^i_{TD1} denote the horizontal, vertical, and diagonal components of high-frequency information at different levels, where the "*H*" denotes the "Horizontal direction", "*V*" denotes the "Vertical direction", "*D*" denotes the "Diagonal direction", and the subscript "1" denotes the first 2-D DWT and fusion strategy.

After the multi-level DWT operation to the Img_S and Img_T respectively, the low-frequency information (Img^i_{SA1}, Img^i_{TA1}) and the high-frequency information at different levels (Img^i_{SH1}, Img^i_{SV1}, Img^i_{SD1}, Img^i_{TH1}, Img^i_{TV1}, Img^i_{TD1}) are obtained. The decomposition results and reconstruction images are shown in Figure 4.

Figure 4 illustrates that a single wavelet decomposition of *ImgS* does not completely separate low- and high-frequency information. The high-frequency information $\{Img^1_{SH1}, Img^1_{SV1}, Img^1_{SD1}\}$ is not visually obvious due to weak texture features in Img_S. However, after deeper wavelet decomposition, the high-frequency information becomes clearer, with stripes gradually becoming apparent, while residual texture features in low-frequency information Img^i_{SA1} are gradually eliminated. The texture features are fully eliminated in Img^4_{SA1}, accurately reflecting the specular information. For Img_T, its low and high frequency information are effectively separated by DWT, where the low-frequency information Img^1_{TA1} still contains the texture information, but interestingly, from the beginning of 2-level decomposition, the high-frequency features in Img^2_{TA1} are removed effectively, and the subsequent decompositions further separate the fine features, which proves the difference between highlight information and texture information in the frequency domain. After extracting the concealed texture information from Img_S by multi-level DWT, the original images Img_T are first processed to obtain texture restoration images Img_{F1}.

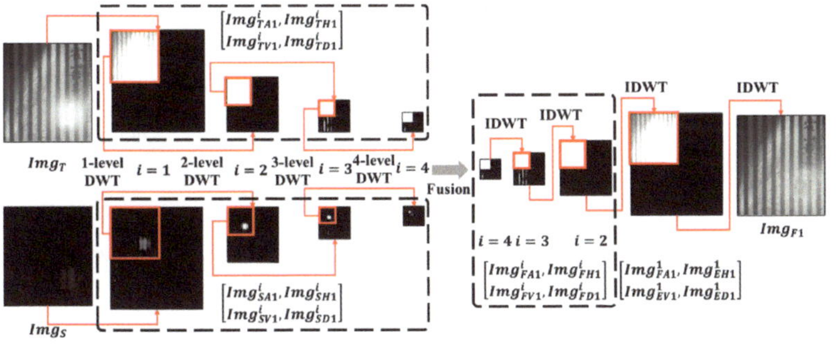

Figure 4. The decomposition results of Img_S and Img_T and reconstruction images at different levels.

The high-frequency information $\{Img^i_{SH1}, Img^i_{SV1}, Img^i_{SD1}\}$ in Img_S is extracted to enhance the high-frequency information $\{Img^i_{TH1}, Img^i_{TV1}, Img^i_{TD1}\}$ in Img_T. Then, the information sets $\{Img^i_{SA1}, Img^i_{SH1}, Img^i_{SV1}, Img^i_{SD1}\}$ and $\{Img^i_{TA1}, Img^i_{TH1}, Img^i_{TV1}, Img^i_{TD1}\}$ are subjected to hierarchical gain and fusion. The fusion strategies are expressed as follows,

$$\begin{bmatrix} Img^i_{FA1} \\ Img^i_{FH1} \\ Img^i_{FV1} \\ Img^i_{FD1} \end{bmatrix} = \begin{bmatrix} Gain^i_{TA1} \cdot Img^i_{TA1} \\ Gain^i_{TH1} \cdot Img^i_{TH1} \\ Gain^i_{TV1} \cdot Img^i_{TV1} \\ Gain^i_{TD1} \cdot Img^i_{TD1} \end{bmatrix} + \begin{bmatrix} Gain^i_{SA1} \cdot Img^i_{SA1} \\ Gain^i_{SH1} \cdot Img^i_{SH1} \\ Gain^i_{SV1} \cdot Img^i_{SV1} \\ Gain^i_{SD1} \cdot Img^i_{SD1} \end{bmatrix} \ i \geq 1 \tag{6}$$

where Img^i_{FA1}, Img^i_{FH1}, Img^i_{FV1}, and Img^i_{FD1}, respectively, represent the low-frequency and high-frequency information at different levels of fusion image Img_{F1}. The $Gain^i_{TA1}$, $Gain^i_{SA1}$, $Gain^i_{TH1}$, $Gain^i_{SH1}$, $Gain^i_{TV1}$, $Gain^i_{SV1}$, $Gain^i_{SD1}$, and $Gain^i_{TD1}$, respectively, represent the gain coefficients at different levels of the first wavelet decomposition information. Then, the low-frequency information of fusion image can be obtained by inverse discrete wavelet transform, as follows,

$$Img^{i-1}_{FA1} = \text{IDWT}\left(Img^i_{FA1}, Img^i_{FH1}, Img^i_{FV1}, Img^i_{FD1}\right) \ i \geq 2 \tag{7}$$

where IDWT represents the inverse discrete wavelet transform (IDWT) operation, through the IDWT, and the information set $\{Img^i_{FA1}, Img^i_{FH1}, Img^i_{FV1}, Img^i_{FD1} | i \geq 2\}$ is reconstructed into the low-frequency information Img^{i-1}_{FA1} up one level.

Since the high-frequency information $\{Img^i_{SH1}, Img^i_{SV1}, Img^i_{SD1} | i \geq 1\}$ reflects the residual texture information in Img_S, it needs to be retained, for Img_T, the high-frequency information $\{Img^i_{TH}, Img^i_{TV}, Img^i_{TD} | i \geq 1\}$ reflects the real texture information, here, an autocorrelation-function-based algorithm is used to calculate the gain coefficients of high-frequency information, as follows,

$$Gain^i = \langle \sum_{m-\omega}^{m+\omega} \sum_{n-\omega}^{n+\omega} \mu \cdot \lambda \cdot Img^i(x,y) Img^i(x-\varepsilon, y-\alpha) \rangle \Big/ \langle \sum_{m-\omega}^{m+\omega} \sum_{n-\omega}^{n+\omega} \left[Img^i(x,y)\right]^2 \rangle \tag{8}$$

where x and y are the variables, $m \times n$ are the size of each high-frequency information, μ is the reward factor, λ is an indicator to indicate whether the calculation condition is met, ε and α are the displacements in the x and y direction, respectively, which determine which two pixels in the image are used for comparison, and ω is the variable of offset, where $2\omega + 1$ is the size of a rectangular window. It means the computation between each pixel within a $(2\omega + 1) * (2\omega + 1)$ window and the pixel that is displaced by ε and α in the x and y direction, respectively. Img^i represents the high-frequency information of Img_T and Img_S. Taking the stripe-projection image as an example, in the fusion for constructing Img_{F1}, the gain coefficient $\{Gain^i_{SA1} | i \in [1,4\}$ is $\{0.05, 0.02, 0.02, 0.01\}$, $\{Gain^i_{SH1} | i \in [1,4\}$ is $\{1.01, 0.62, 0.54, 0.50\}$, $\{Gain^i_{SV1} | i \in [1,4\}$ is $\{0.98, 0.49, 0.44, 0.39\}$, $\{Gain^i_{SD1} | i \in [1,4\}$

is {0.97, 0.46, 0.41, 0.33}, and the $Gain^i_{TA1}$ $Gain^i_{TH1}$, $Gain^i_{TV1}$, and $Gain^i_{TD1}$ were equal to 1 due to the preserve texture principle. After multi-level reconstruction by IDWT, the first-level information set $\{Img^1_{FA1}, Img^1_{FH1}, Img^1_{FV1}, Img^1_{FD1}\}$ is obtained, then a high-pass filter with 3×3 convolution kernel is employed to extract the edge information of residual texture in $\{Img^1_{SH1}, Img^1_{SV1}, Img^1_{SD1}\}$, as follows,

$$\begin{bmatrix} Img^1_{EH1} \\ Img^1_{EV1} \\ Img^1_{ED1} \end{bmatrix} = \begin{bmatrix} 0 & a & 0 \\ a & b & a \\ 0 & a & 0 \end{bmatrix} * \begin{bmatrix} Img^1_{SH1} \\ Img^1_{SV1} \\ Img^1_{SD1} \end{bmatrix} + \begin{bmatrix} Img^1_{FH1} \\ Img^1_{FV1} \\ Img^1_{FD1} \end{bmatrix} \tag{9}$$

where the a is negative number, together with positive number b form the convolution kernel, in order to ensure that the extracted edge information does not destroy the overall texture of the original image due to excessive enhancement; here, a reference value is provided, where a is -1, and b is 0.7. Img^1_{EH1}, Img^1_{EV1}, and Img^1_{ED1} denote the horizontal, vertical, and diagonal component high-frequency information in first level after edge enhancement, the operator "$*$" denotes the convolution operation. The filter results are added to the first-level high-frequency information set for enhancement. Then, the obtained Img^1_{EH1}, Img^1_{EV1}, and Img^1_{ED1} are used to replace the corresponding high-frequency information Img^1_{FH1}, Img^1_{FV1}, and Img^1_{FD1}, IDWT is used to reconstruct them, as follows,

$$Img_{F1} = \text{IDWT}(Img^1_{FA1}, Img^1_{EH1}, Img^1_{EV1}, Img^1_{ED1}) \tag{10}$$

From Figure 4, texture information of all levels is enhanced compared with original image Img_T after wavelet decomposition. Img_{F1} a is a texture restoration image, and on this basis, the highlight suppression images Img_{F2} are obtained by removing the low-frequency information of Img_S.

2.3. Second Fusion Strategy

Img_{F1} needs to be decomposed by multi-level wavelet transform and fused with low-frequency information of Img_S, and the technical route is as shown in Figure 5.

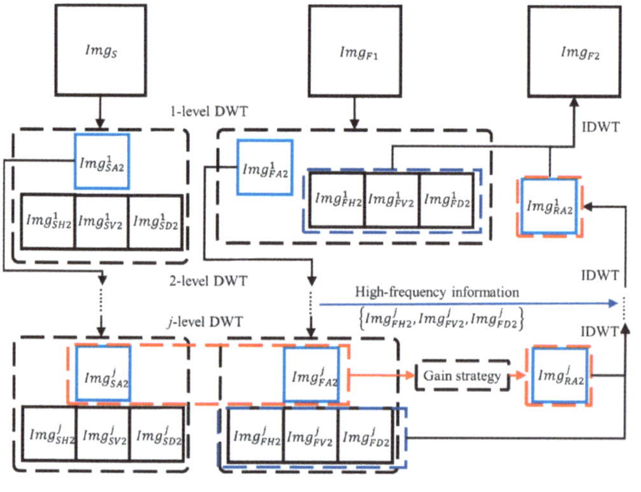

Figure 5. Technical route of highlight suppression for texture restoration image Img_{F1}.

In Figure 5, Img^j_{SA2} denotes the low-frequency information of Img_S at different levels, Img^j_{SH2}, Img^j_{SV2}, Img^j_{SD2}, denote the horizontal, vertical, and diagonal components of high-frequency information at different levels. Similarly, Img^j_{FA2} denotes the low-frequency information of Img_{F1} at different levels, Img^j_{FH2}, Img^j_{FV2}, and Img^j_{FD2}, denote the horizontal, vertical, and diagonal components of high-frequency information at different levels, and the subscript "2" denotes the second 2-D DWT and fusion strategy. As Figure 6 shows, in order to suppress the concealed specular information, and retain the texture details, the highest-level low-frequency information $Img_{SA2}^{\max(j)}$ of Img_S is selected to fuse with the same-level low-frequency information $Img_{FA2}^{\max(j)}$ of Img_{F1}. The decomposition results and reconstruction images at different levels, as well as the fusion strategy, are as follows,

$$\begin{cases} Img_{RA2}^{\max(j)} = Img_{FA2}^{\max(j)} + Gain_{SA2} \cdot Img_{SA2}^{\max(j)} \\ Img_{RA2}^{j-1} = IDWT(Img_{FA2}^{j}, Img_{FH2}^{j}, Img_{FV2}^{j}, Img_{FD2}^{j}) \end{cases} \quad j \geq 2 \tag{11}$$

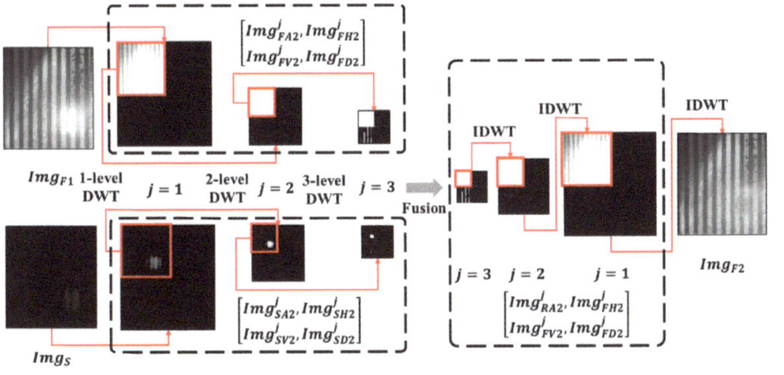

Figure 6. The decomposition results of Img_S and Img_{F1} and reconstruction images at different levels.

Since the low-frequency information Img_{FA2}^{j} will be replaced by higher-level decomposition information during wavelet reconstruction, based on the principle of highlight suppression, and considering that highest-level specular information is recalculated twice ($Img_{FA2}^{\max(j)}$ and $Img_{SA2}^{\max(j)}$), the gain coefficient $Gain_{SA2}^{\max(j)}$ should therefore be at least equal to -1.

Next, fuse it with information set $\{Img_{FA2}^{j}, Img_{FH2}^{j}, Img_{FV2}^{j}, Img_{FD2}^{j} | j \geq 2\}$, as shown in Figure 6. The calculated Img_{RA2}^{1} is used to replace the Img_{FA2}^{1} and then reconstructed by IDWT to obtain the highlight suppression image Img_{F2}, as follows,

$$Img_{F2} = \mathrm{IDWT}(Img_{RA2}^{1}, Img_{FH2}^{1}, Img_{FV2}^{1}, Img_{FD2}^{1}) \tag{12}$$

3. Results and Analysis

Figure 7 depicts some solved examples, which are respectively the stripe-projection image, the steel plate with diagonal stripes and the internal threads of the metal pipeline. Since they are grayscale images, they exclusively encompass intensity information. The targets mentioned above contain rich texture features and obvious image highlights, where the image highlights are mainly focused on the areas with rich textures, and it should be noted that large-area highlights exist in the images of steel plate with diagonal stripes and the internal threads of the metal pipeline.

This algorithm is inevitably sensitive to the intensity threshold η, whereby different input images often have a different optimal threshold η, which can be adjusted according to the area and intensity of highlights. Figure 7c clearly shows that the highlight suppression image Img_{F2} can significantly improves the original images Img_T. This is evident in the suppression of the main specular information without significant distortion in edge and intensity information, and it means that the restorations of texture information are natural and smooth, and display a good visual effect, which effectively validates the proposed method. The proposed method performs well in processing the highlights with strong texture features, as the texture and intensity in highlight regions are reasonably well restored, bringing clear stripes, reasonable intensity of diagonal striped, and relatively comprehensive thread information. Figure 7d depicts the highlight extraction images Img_H, which are obtained by the difference operation between Img_{F2} and Img_{F1}. Img_H do not contain obvious residual texture information, and while they accurately reflect the distribution and intensity of specular information, when preserving the original texture features of the target, the highlight-suppressed image Img_{F2} obtained after removing specular information Img_H exhibits a 'lights off' effect. Image histograms are calculated to objectively describe the highlight suppression performance, Figure 8 depicts the histogram comparisons of Img_T and Img_{F2}, and Table 1 shows the comparison of the highlight pixels' number of Img_T and Img_{F2}.

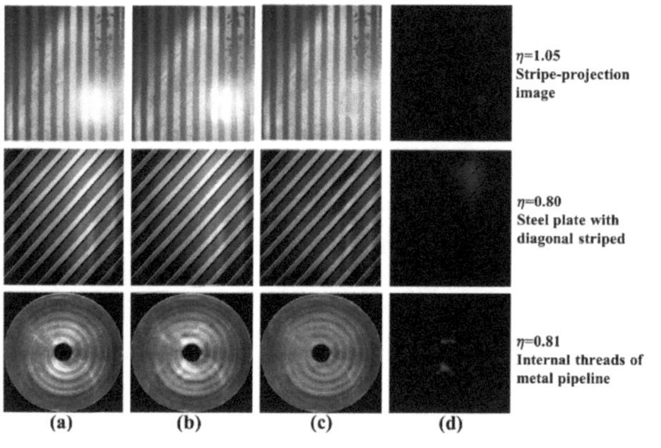

Figure 7. Solved examples. (**a**) Original image Img_T; (**b**) texture restoration image Img_{F1}; (**c**) highlight suppression image Img_{F2}; (**d**) extracted highlights Img_H.

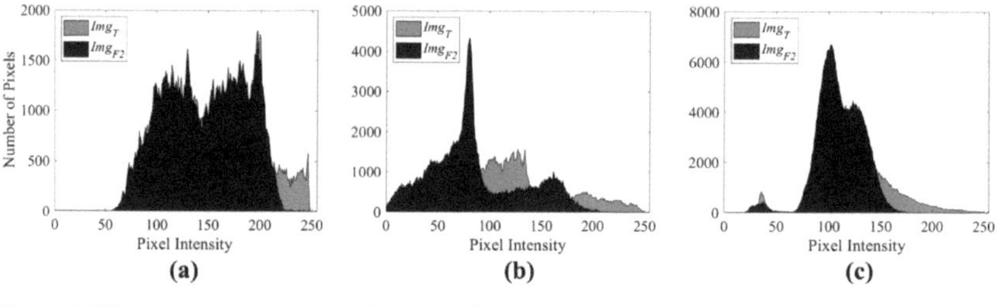

Figure 8. Histogram comparisons of Img_T and Img_{F2}. (**a**) Stripe-projection image; (**b**) steel plate with diagonal stripes; (**c**) internal threads of metal pipeline.

Due to the different area and intensity of image highlights, three different thresholds for highlight pixels are given in Table 1, which are 225, 190, and 191 respectively. The highlight pixels are mainly distributed in the range with high intensity, and it can be observed from Figure 8 and Table 1 that, compared with original images, the number of highlight pixels in three highlight suppression images are obviously decreased, and the removed highlight pixels have moved to lower intensity ranges. In order to further validate the proposed method, four classical methods (MSF method [17,18], Arnold's method [22], Fu's method [9], and Akashi's method [8]) are used for comparison. Figure 9 shows the highlights extracted by five different methods.

Table 1. Comparison of the number of highlight pixels.

Image	Stripe-Projection Image	Steel Plate with Diagonal Striped	Internal Threads of Metal Pipeline
Img_T	8640 [1]	16,953 [2]	9429 [3]
Img_{F2}	195	1536	8

[1] Highlight threshold ≥ 225. [2] Highlight threshold ≥ 190. [3] Highlight threshold ≥ 191.

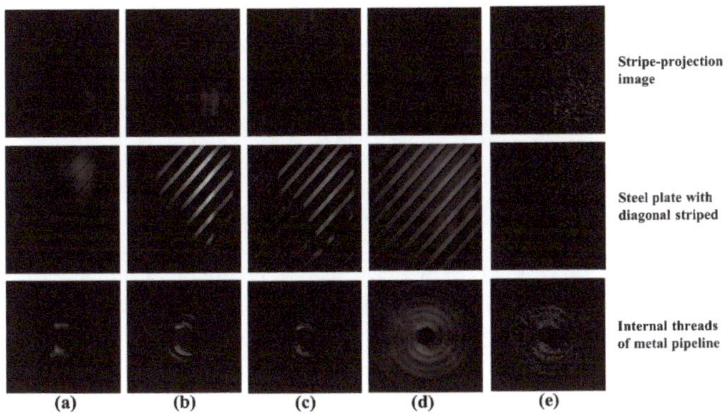

Figure 9. The extracted highlights. (**a**) Proposed method; (**b**) MSF method [17,18]; (**c**) Arnold's method [22]; (**d**) Fu's method [9]; (**e**) Akashi's method [8].

Figure 9 shows that, compared to classical methods, the highlights extracted by the proposed method do not exhibit significant edge information, and therefore does not include the texture information, thereby reducing image distortion after removing highlights, while the highlights extracted by classical methods both contain much texture features, which may cause the loss of edge information and distortion of intensity information. From an ocular perception standpoint, the highlights extracted by the proposed method provide a more authentic estimation of highlights, reflecting the specular information generated by the target's intrinsic high-reflectance properties and illumination. Figure 10 depicts the comparison of highlight suppression image.

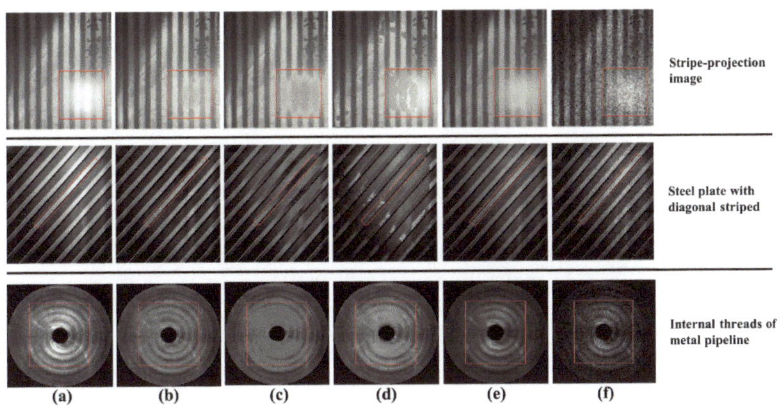

Figure 10. Highlights suppression images. (**a**) Original images; (**b**) proposed method; (**c**) MSF method [17,18]; (**d**) Arnold's method [22]; (**e**) Fu's method [9]; (**f**) Akashi's method [8].

In Figure 10, the main highlight region is marked in red boxes; in order to compare the texture information in the highlight region, the restoration results of different methods are displayed for images. Figure 11 depicts the intensity spatial display of Figure 10.

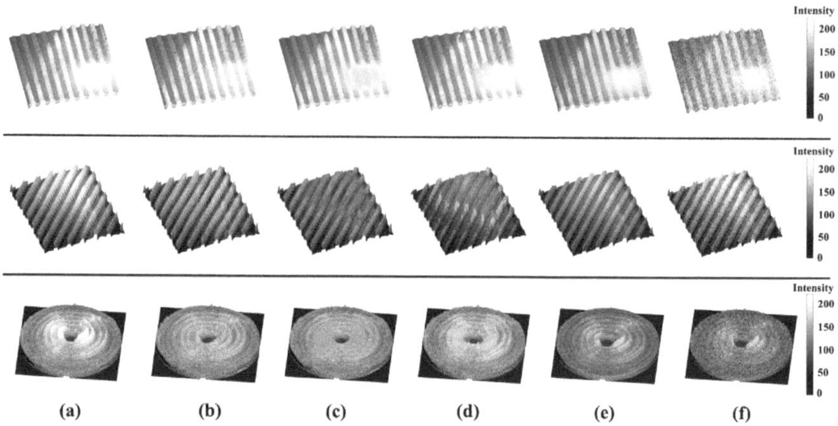

Figure 11. Intensity spatial display of Figure 10. (**a**) Original images; (**b**) proposed method; (**c**) MSF method [17,18]; (**d**) Arnold's method [22]; (**e**) Fu's method [9]; (**f**) Akashi's method [8].

From Figures 10 and 11, a false feature occurred in Figure 11c,d. For the MSF method [17,18], the highlights suppression image of stripe-projection and internal threads of the metal pipeline have obvious losses of texture information; the stripes and the threads in the highlight regions have almost disappeared, while there is some intensity distortion in the image of a steel plate with diagonal stripes, and the pixel intensity in the highlight regions have significant differences from other regions. For Arnold's method [22], the highlight areas are filled proportionally using Gaussian smoothing with the original image, which may cause the poor edge preservation, so it brings obvious losses of texture information in the image of internal threads of the metal pipeline, or even false texture information and intensity in the image of stripe-projection and the steel plate with diagonal stripes. Fu's method [9] shows a better performance in texture distortion compared with Figure 11c,d, but it still has significant residual highlights and loses some texture details, due to the smoothness balance issue between the diffuse layers and the highlight layers. Compared with the methods mentioned above, the proposed method has a significant improvement in the restoration of texture and intensity information in highlight regions. The effect of image highlight suppression by Akashi's method [8] seems limited due to the lack of color information and because the image exhibits a significant decrease in quality, and it is also prone to noise when processing grayscale images.

4. Discussion

Since the concealed texture information extracted from highlight areas is used in the texture restoration of the original images, the restoration results achieved by the proposed method are therefore comparatively more authentic. The results obtained demonstrate the efficacy of the proposed wavelet transform and fusion strategy in suppressing highlights for single texture-rich images. Arnold's method [22] and the BSCB method [13] focused on generating pixel information to restore the texture, and as the BSCB method [13] is a pure texture restoration algorithm, it needs to combine highlight extraction algorithms to achieve the suppression of highlights, it is more suitable for texture restoration with small area.

Other previous methods, such as the MSF Method [17,18] and those based on the DRM, have shown effectiveness in separating diffuse and specular reflection components in color images. However, these methods may struggle with grayscale images where color information is absent, leading to potential distortions in image intensity and changes in signal-to-noise ratio, like Akashi's method [8] and Fu's method [9]. The local structure and chromaticity joint compensation (LSCJC) [6] method has been successful in separating diffuse and specular components by leveraging local chromaticity correlations, but this method is limited when applied to single-channel grayscale images, where chromaticity information is not available. In contrast, the proposed method's reliance on frequency domain analysis rather than spatial domain chromaticity features allows it to overcome this limitation, providing a more versatile solution for various imaging scenarios. This represents a significant advancement in highlight suppression technology, particularly for industrial applications where grayscale imaging is prevalent. The multi-level decomposition of the image into low-frequency and high-frequency information through wavelet transform allows for precise manipulation of the images' frequency content. This capability enables the suppression of highlights while preserving essential texture details, which is a notable improvement over methods that may inadvertently smooth out these details. For instance, the method by Yang et al. [5], which employs edge-preserving low-pass filtering, effectively removes specular highlights but can sometimes over-smooth the image, losing critical texture information. The proposed method mitigates this issue by targeting only the high-frequency components associated with highlights. Furthermore, the fusion strategy employed in our method is critical for combining the processed high-frequency and low-frequency components to reconstruct a highlight-suppressed image. Thus, this method ensures that the texture details are preserved and enhanced, leading to a more visually coherent and accurate representation of the inspected surface. This is particularly beneficial when dealing with images of metal parts or smooth plastics, where surface texture is crucial for quality assessment.

However, since the texture restoration relies on the weak texture concealed in highlight regions, the proposed method may cause some limitations in texture restoration when the pixel intensity reaches 255. Furthermore, the number of decomposition levels is not a constant for every case. The texture of images can affect the level of wavelet decomposition. Certainly, for highlight suppression of general images in optical inspection, the level of wavelet decomposition will not be too high; in this paper, the wavelet decomposition levels of several solved examples are all within 5. For different information in images, such as specular information, a solution to this problem is to first set a high wavelet decomposition level, then calculate the gain coefficient of specular information for each level; if the value is close to 0, its impact on the results will also tend toward nonexistent. Future research could explore the possibilities for improving the automation level of this method, analyzing the impact of image resolution and wavelet filters on the decomposition level, and breaking the limitation and the integration of this method with other image processing techniques, such as machine learning algorithms, to further enhance its robustness and adaptability. Additionally, testing the method across a broader range of materials and surface conditions would help in fine-tuning the algorithm for specific industrial applications.

5. Conclusions

The targets with high reflectance properties may bring the image highlights, which is detrimental to subsequent detection tasks, and if there exist complex texture features in highlight regions, the highlights extracted by traditional highlight suppression methods may contain a significant amount of texture features, which results in a loss of substantial target information after removing the highlight components. The main difficulty in processing these images lies in effectively suppressing highlights while restoring complex texture features. This paper introduces a highlight suppression method that utilizes wavelet transform and a fusion strategy to effectively suppress highlights in single images. The experimental results show that the real specular information is eliminated effectively, while the texture information is reasonably restored. Compared with classical methods, the proposed method does not exhibit significant texture information in extracted highlights, it provides more authentic estimation of highlights, and it effectively processes the images with rich texture and large-area highlights, with the obtained highlight-suppressed images after specular information removal exhibiting a 'lights off' effect. This method can lead to more reliable optical inspections, and can be integrated directly into vision inspection/measurement (e.g., fringe projection profilometry), medical imaging, and video processing. Due to its simplicity and hardware-free nature, it attracts a lower cost in improving existing visual equipment, and it can enhance image quality, improve object detection and recognition, facilitate accurate diagnosis, and optimize the viewing experience by reducing glare and overexposure.

Author Contributions: Conceptualization, X.S. and L.K.; methodology, X.S.; software, X.S.; validation, X.W. and X.P.; formal analysis, X.W. and G.D.; investigation, X.S. and L.K.; resources, X.S. and L.K.; data curation, X.W.; writing—original draft preparation, X.S. and L.K.; writing—review and editing, X.S. and L.K.; visualization, X.S.; supervision, L.K.; project administration, X.S. and L.K.; funding acquisition, X.S. and L.K. All authors have read and agreed to the published version of the manuscript.

Funding: This research was funded by Natural Science Foundation of Jiangxi Province (20224BAB214053), Shanghai Science and Technology Innovation Program (23ZR1404200), National Key R&D Program (2022YFB3403404), and Science and Technology Research Project of Education Department of Jiangxi Province (GJJ210668).

Institutional Review Board Statement: Not applicable.

Informed Consent Statement: Not applicable.

Data Availability Statement: Data underlying the results presented in this paper are not publicly available at this time but may be obtained from the authors upon reasonable request.

Conflicts of Interest: The authors declare no conflicts of interest.

References

1. Chen, H.H.; Luo, L.J.; Guo, C.S.; Ying, N.; Ye, X.Y. Specular Highlight Removal Using a Divide-and-Conquer Multi-Resolution Deep Network. *Multimed. Tools Appl.* **2023**, *82*, 36885–36907. [CrossRef]
2. Feng, W.; Li, X.H.; Cheng, X.H.; Wang, H.H.; Xiong, Z.; Zhai, Z.S. Specular Highlight Removal of Light Field Based on Dichromatic Reflection and Total Variation Optimizations. *Opt. Lasers Eng.* **2021**, *151*, 106939. [CrossRef]
3. Klinker, G.J.; Shafer, S.A.; Kanade, T. The Measurement of Highlights in Color Images. *Int. J. Comput. Vis.* **1988**, *2*, 7–32. [CrossRef]
4. Chen, L.; Lin, S.; Zhou, K.; Ikeuchi, K. Specular Highlight Removal in Facial Images. In Proceedings of the IEEE Conference on Computer Vision and Pattern Recognition, Honolulu, HI, USA, 21–26 July 2017; pp. 2780–2789.
5. Yang, Q.X.; Tang, J.H.; Ahuja, N. Efficient and Robust Specular Highlight Removal. *IEEE Trans. Pattern Anal. Mach. Intell.* **2015**, *37*, 1304–1311. [CrossRef] [PubMed]
6. Zhao, Y.Q.; Peng, Q.N.; Xue, J.Z.; Kong, S. Specular Reflection Removal Using Local Structural Similarity and Chromaticity Consistency. In Proceedings of the IEEE International Conference on Image Processing, Quebec City, QC, Canada, 27–30 September 2015.
7. Suo, J.L.; An, D.S.; Ji, X.Y.; Wang, H.Q.; Dai, Q.H. Fast and High-Quality Highlight Removal from a Single Image. *IEEE Trans. Image Process.* **2016**, *25*, 5441–5454. [CrossRef] [PubMed]
8. Akashi, Y.; Okatani, T. Separation of reflection components by sparse non-negative matrix factorization. In Proceedings of the Asian Conference on Computer Vision, Singapore, 1–5 November 2014; pp. 611–625.
9. Fu, G.; Zhang, Q.; Song, C.F.; Lin, Q.F.; Xiao, C.X. Specular Highlight Removal for Real-world Images. *Comput. Graph. Forum J. Eur. Assoc. Comput. Graph.* **2019**, *38*, 253–263. [CrossRef]
10. Ragheb, H.; Hancock, E.R. A Probabilistic Framework for Specular Shape-From-Shading. *Pattern Recognit.* **2003**, *36*, 407–427. [CrossRef]
11. Ma, J.Q.; Kong, F.H.; Zhao, P.; Gong, B. A Specular Shape-From-Shading Method with Constrained Inpainting. *Appl. Mech. Mater.* **2014**, *530–531*, 377–381. [CrossRef]
12. Yin, F.; Chen, T.; Wu, R.; Fu, Z.; Yu, X. Specular Detection and Removal for a Grayscale Image Based on the Markov Random Field. In Proceedings of the International Conference of Young Computer Scientists, Engineers and Educators, Harbin, China, 19–22 August 2016; pp. 641–649.
13. Bertalmio, M.; Sapiro, G.; Caselles, V. Image inpainting. In Proceedings of the Special Interest Group on Computer Graphics and Interactive Techniques, New Orleans, LA, USA, 23–28 July 2000; pp. 417–424.
14. Chen, H.; Hou, C.G.; Duan, M.H.; Tan, X.; Jin, Y.; Lv, P.L.; Qin, S.Q. Single Image Specular Highlight Removal on Natural Scenes. In Proceedings of the Chinese Conference on Pattern Recognition and Computer Vision, Zhuhai, China, 29 October–1 November 2021; pp. 78–91.
15. Zou, G.F.; Li, T.T.; Li, G.Y.; Peng, X.; Fu, G.X. A Visual Detection Method of Tile Surface Defects based on Spatial-Frequency Domain Image Enhancement and Region Growing. In Proceedings of the Chinese Automation Congress, Hangzhou, China, 22–24 November 2019; pp. 1631–1636.

16. Sun, X.; Liu, Y.; Yu, X.Y.; Wu, H.B.; Zhang, N. Three-Dimensional Measurement for Specular Reflection Surface Based on Reflection Component Separation and Priority Region Filling Theory. *Sensors* **2017**, *17*, 215. [CrossRef] [PubMed]
17. Shen, H.L.; Zheng, Z.H. Real-Time Highlight Removal Using Intensity Ratio. *Appl. Opt.* **2013**, *52*, 4483–4493. [CrossRef] [PubMed]
18. Ren, W.H.; Tian, J.D.; Tang, Y.D. Specular Reflection Separation with Color-Lines Constraint. *IEEE Trans. Image Process.* **2017**, *26*, 2327–2337. [CrossRef] [PubMed]
19. Cai, J.R.; Gu, S.H.; Zhang, L. Learning a Deep Single Image Contrast Enhancer from Multi-Exposure Images. *IEEE Trans. Image Process.* **2018**, *27*, 2049–2062. [CrossRef] [PubMed]
20. Donoho, D.L. De-Noising by Soft-Thresholding. *IEEE Trans. Inf. Theory* **1995**, *41*, 613–627. [CrossRef]
21. Makbol, N.M.; Khoo, B.E.; Rassem, T.H. Block-Based Discrete Wavelet Transform-Singular Value Decomposition Image Watermarking Scheme using Human Visual System Characteristics. *IET Image Process.* **2016**, *10*, 34–52. [CrossRef]
22. Arnold, M.; Ghosh, A.; Ameling, S.; Lacey, G. Automatic Segmentation and Inpainting of Specular Highlights for Endoscopic Imaging. *EURASIP J. Image Video Process.* **2010**, *2010*, 814319. [CrossRef]

photonics

MDPI

Article

Nanosensor Based on the Circular Ring with External Rectangular Ring Structure

Shuwen Chang [1,2,3], Shubin Yan [2,3,*], Yiru Su [2], Jin Wang [1,2,3], Yuhao Cao [1,2,3], Yi Zhang [2], Taiquan Wu [2] and Yifeng Ren [1]

1 School of Electrical and Control Engineering, North University of China, Taiyuan 030051, China; sz202215006@st.nuc.edu.cn (S.C.); sz202215044@st.nuc.edu.cn (J.W.); s202215001@st.nuc.edu.cn (Y.C.); renyifeng@nuc.edu.cn (Y.R.)

2 School of Electrical Engineering, Zhejiang University of Water Resources and Electric Power, Hangzhou 310018, China; 2022b03001@stu.zjweu.edu.cn (Y.S.); zhangyi@zjweu.edu.cn (Y.Z.); wutq@zjweu.edu.cn (T.W.)

3 Joint Laboratory of Intelligent Equipment and System for Water Conservancy and Hydropower Safety Monitoring of Zhejiang Province and Belarus, Hangzhou 310018, China

* Correspondence: yanshb@zjweu.edu.cn; Tel.: +86-186-3611-2255

Abstract: This paper presents a novel nanoscale refractive index sensor, which is produced by using a metal–insulator–metal (MIM) waveguide structure coupled with the circular ring with an external rectangular ring (CRERR) structure with the Fano resonance phenomenon. In this study, COMSOL software was used to model and simulate the structure, paired with an analysis of the output spectra to detail the effect of constructional factors on the output Fano curve as measured from a finite element method. After a series of studies, it was shown that an external rectangular ring is the linchpin of the unsymmetrical Fano resonance, while the circular ring's radius strongly influences the transducer's capability to achieve a maximum for 3180 nm/RIU sensitivity and a FOM of 54.8. The sensor is capable of achieving sensitivities of 0.495 nm/mgdL^{-1} and 0.6375 nm/mgdL^{-1} when detecting the concentration of the electrolyte sodium and potassium ions in human blood and is expected to play an important role in human health monitoring.

Keywords: Fano resonance; nanorefractive index sensor; sensitivity; figure of merit (FOM)

Citation: Chang, S.; Yan, S.; Su, Y.; Wang, J.; Cao, Y.; Zhang, Y.; Wu, T.; Ren, Y. Nanosensor Based on the Circular Ring with External Rectangular Ring Structure. *Photonics* **2024**, *11*, 568. https://doi.org/10.3390/photonics11060568

Received: 12 May 2024
Revised: 12 June 2024
Accepted: 13 June 2024
Published: 17 June 2024

1. Introduction

Surface plasmon polaritons (SPPs) on a metal surface are collective oscillations of free electrons with photons that produce an electromagnetic spectrum, which transpire as they migrate along the metal–dielectric junction [1]. The travel of the SPPs is in a direction paralleling the metal surface, and it is in the direction perpendicular to the metal surface that their electric field strength decays exponentially; thus, SPPs are bound to the metal surface and exhibit surface-localized properties [2,3]. When the incident light and the free electrons on the metal surface have the same frequency, the electromagnetic field energy can be well localized on the metal surface, which enhances the cooperation between transmitted light and other substances. Therefore, by utilizing this special optical property, a new path can be opened for the design of novel optical sensors, whose theory and related techniques have become a research field at the forefront of nanophotonics, with a widespread application in ultra-diffractive lithography, beamsplitters [4,5], spatial technology, optical switches [6], sensors [7], and highly integrated photonic circuits, among others.

SPPs in nano-optoelectronic devices have attracted the interest of many researchers by virtue of their great advantages over conventional electronic devices, leading to the design of a wide variety of waveguide structures, such as filters [8–10], rectangular waveguides based on MIM [11,12], nanoshells [13], nanowires [14,15], and farnesoidal resonances [16–19], among others. Among the various structures mentioned above, MIM waveguides have been recognized by researchers as having advantages such as a simple structure, easy processing and acceptable propagation length, and have been designed into various types of micronano-optoelectronic devices. One of the unique phenomena based on the MIM waveguide using the coupled cavity resonance, Fano resonance has aroused the interest of researchers from various countries because the Fano resonance phenomenon produces steep and sharp resonance transmission spectral lines, which are affected by much less compared to other micro-nano photonic devices. However, in the MIM waveguide structure, its structural composition dictates the existence of a slit, which ensures that SPPs can propagate efficiently in this slit, even in the subwavelength state. In this paper, the surface plasmon MIM waveguide is used as the sensing structure, and the Fano resonance effect is used as the sensing basis.

Fano resonance has been labeled as a phenomenon that produces a dispersive resonance with an unsymmetrical line formed by the integration of two scattering amplitudes, one for scattering in the continuous state (background-dependent) while the other is for excitation in the dispersive state (resonance-dependent). In close proximity to the resonance energy, the change in the amplitude of the backscattering with energy is usually gentle; however, the amplitude of the resonance scattering, both in magnitude and in phase, changes quite rapidly, leading to the onset of asymmetry [20,21]. The output waveform of the Fano resonance is extremely susceptible to fluctuation in the surrounding environment, and small changes in the environment can cause large wavelength drifts [22–24]. Therefore, the Fano resonance-based MIM waveguide metal structure can be applied to the sensing field to greatly improve the sensitivity and provide a better choice for high-precision measurements. Such sensors have now moved from the early stages of proof-of-concept demonstration and theoretical modeling to further device and system development and practical applications. Research efforts are increasingly focused on the detection of samples from more realistic and complex media. One such area is the development of portable rapid diagnostic devices for immediate care. Another important area is vapor detection [25].

Rakhshani et al., in 2018, designed a plasma nanosensor consisting of an array of nanorods inside a square resonant cavity coupled with two slot cavities. Their sensor properties were obtained by filling with a grape solution and varying its concentration, with a sensitivity of 892 nm/RIU [26]. Chen Ying et al., in 2019, devised a compact optical waveguide structure with a MIM waveguide coupled to a metal baffle SCRR, and their structure also obtained desirable sensing characteristics, in which the sensitivity was as high as 1120 nm/RIU [27]. A. Noual et al. proposed resonators consisting of a bus waveguide made of graphene coupled to resonators inserted along the waveguide, where each resonator consisted of a set of two coupled graphene nanoribbons (CGNR) [28]. Yamina Rezzouk et al. proposed a sensor connected to an infinite waveguide in a T-shaped cavity consisting of a short segment of length d_0 and two side branches of length d_1 and d_2. The whole system consists of a MIM plasmonic waveguide that operates in the telecommunication range with a sensitivity of 1400 nm/RIU [29]. The performance of the sensors presented herein is superior to the performance of the sensors presented in the literature [30–32]. And compared to the sensors proposed in the literature [33], although the sensitivity of the sensors proposed herein is slightly lower, the sensors proposed herein, however, can be adjusted with multiple geometrical parameters to suit different applications.

The structure of this new nanoscale refractive index sensor, designed in this paper, consists of a MIM waveguide with two rectangular baffles and a ring with a rectangular annular cavity, which is simple and easy to parameterize. Changing the geometrical parameters of the structure affects both the transmittance and the position of the Fano resonance peaks, and the sensor structure has a large number of adjustable geometrical parameters. In this paper, when the sensor was used to detect the concentrations of electrolyte sodium and potassium ions in human blood, it was found that the structure was sensitive to changes in the refractive index and was able to achieve sensitivities of 0.495 nm/mgdL^{-1} and 0.6375 nm/mgdL^{-1}. The structure has excellent sensing properties and can be used for concentration detection in solutions as well as for sensors at the nanoscale.

In this paper, based on existing studies and basic theories, a nanoscale sensor with a simple structure is proposed and investigated, which is derived from the coupling of a MIM and CRERR structure. Using COMSOL 5.4 software and in conjunction with the finite element method (FEM) [34], we have carried out an in-depth simulation study of the sensing characteristics of the designed sensor. This paper addresses the influence of various geometric parameters on the transmission characteristics, which include the exterior radius of the ring, the length of the baffle rectangle, the rotation angle of the external rectangular ring, the height of the outer rectangular ring, as well as the coupling distance among MIM waveguides in relation to the CRERR structure.

2. Structure and Methods

Given that the 2D model is more simplified than the 3D model, and the 2D model can reduce computational complexity while retaining key physical properties, it makes the study more efficient. Secondly, the height of the proposed structure is much larger than the skinning depth of the SPPs, so that the 2D model can be used instead of the 3D model for simple calculations. As shown in Figure 1, the entire sensor structure is symmetric along the centerline and comprises a MIM waveguide having a rectangular baffle coupled to a CRERR structure. The radius of the outside circle and the radius of the inside circle of the CRERR structure are denoted by R and r, respectively, and the values of R and r satisfy R = r + 50 nm. In addition to the above-mentioned parameters, the height of the rectangular stopper is precisely set to h, and g is the key value used to describe the doubling distance of the coupling between the MIM waveguide and the CRERR. Furthermore, α indicates the specific angle at which the rectangular annular cavity is rotated around the center of its torus, while L represents, in turn, the height of the rectangular annular cavity. Additionally, w serves as a common parameter that defines not only the width of the MIM waveguide but also the width of the annular cavity, the rectangular cavity and the rectangular baffle. When the value of w is in a small range, the SPPs in the up-and-down metal–dielectric interfaces will exhibit two dispersion modes due to mutual coupling. In the even-symmetric mode, the SPPs can maintain a long propagation distance and have a relatively low energy loss. In order to achieve efficient energy transmission and low loss in this study, we deliberately designed all the structures based on the even-symmetric mode. To achieve this, the width w is precisely set to 50 nm in the MIM waveguide structures to ensure that only one transmission mode, the even-symmetric mode, exists.

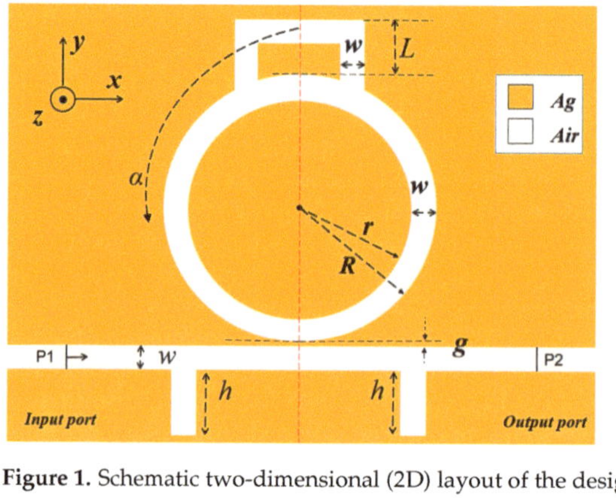

Figure 1. Schematic two-dimensional (2D) layout of the designed sensor fabric.

In order to have a better excitation of SPPs and because silver has a stronger electric force as well as less powerful consumption, metallic silver was used for the metal layer in this study. In Figure 1, the yellow area is for silver material, while the white area is for air, whose dielectric constant is 1. The dispersion model used in this paper to represent the dielectric constant of metallic silver uses the Debye–Drude dispersion equation as follows [35]:

$$\varepsilon(\omega) = \varepsilon_\infty + \frac{\varepsilon_s - \varepsilon_\infty}{1 + j\omega\tau} + \frac{\sigma}{j\omega\varepsilon_0} \tag{1}$$

where $\varepsilon_s = -9530.5$ is the quiescent dielectric constant, $\varepsilon_\infty = 3.8344$ is the relative permittivity at the wireless angular frequency, ε_0 is the permittivity in vacuum, $\tau = 7.35 \times 10^{-15}$ s is the relaxation time, and $\sigma = 1.1486 \times 10^7$ S/m is the electrical conductivity of silver.

In some special cases, such as on the surface of nanostructures or in waveguide structures with specific designs, the TE-polarized modes may be converted into TM-polarized modes through certain mechanisms (e.g., mode conversion or scattering), and thus, SPPs may be generated, which needs to be achieved by precisely designing the geometrical parameters and material properties of nanostructures. Therefore, in general, TE polarization modes do not directly generate SPPs, and TE polarization modes in nanostructures can generate SPPs under specific conditions [36]. Electromagnetic waves can excite SPPs only in the TM mode on the layer surface of the metal. Hence, this paper focuses on analyzing the TM mode, and its equation in the MIM waveguide can be described as [37]

$$\tanh(kw) = -\frac{2kp\alpha_c}{k^2 + p^2\alpha_c} \tag{2}$$

where w represents the waveguide width, and k is the wavevector in the waveguide, $\alpha_c = \left[k_0^2(\varepsilon_{in} - \varepsilon_m) + k\right]^{1/2}$, $p = \varepsilon_{in}/\varepsilon_m$. The wavevector k_0 in a vacuum is denoted as $k_0 = 2\pi/\lambda_0$; k is determined by Equation (2), and ε_m and ε_{in} denote the dielectric constants in the metal and dielectric.

Given the high cost of nanoscale device fabrication, simulation has emerged as an efficient means of allowing us the flexibility to adapt the structure of the sensor and ultimately maximize its performance. In this paper, COMSOL 5.4 software was used to successfully fabricate a 2D model structure of CRERR and subsequently perform an exhaustive simulation analysis. The overall simple simulation is described as follows: (1) Identify the appropriate physical field modules and interfaces. (2) The geometric drawing tool that comes with COMSOL 5.4 can be used to draw the design structure. (3) Before performing the simulation, the port locations of the model, the applicable material model and the corresponding fluctuation formulae first need to be carefully selected and configured. Subsequently, in order to ensure the accuracy and efficiency of the simulation, the entire simulation area will be meshed using a triangular meshing method. (4) Parametric scanning can be easily set up in COMSOL's simulation environment using its own solver. Specifically, specify a wavelength range from 1600 nm to 3000 nm, and set the step size to 1 nm. Once these settings are made, simply click the 'Calculate' button, and COMSOL will automatically perform a parametric scan to compute and simulate the response of the physical model point by point in this wavelength range.

In refractive index sensors, sensitivity (S) is generally defined as the amount of variation in the value of the resonance pole position experienced when there is a unit movement in the refractive index. This concept can be precisely expressed by the following mathematical formula: [38,39]:

$$S = \frac{\Delta\lambda}{\Delta n} \tag{3}$$

where Δn is the modification of the refractive index, and $\Delta\lambda$ is the change in resonance wavelength. The unit of S can be expressed as nm/RIU.

The FOM is a reflection of the holistic performance of the refractive index sensor and is another key parameter for measuring many sensors, which can be expressed by the formula

$$FOM = \frac{S}{FWHM} \tag{4}$$

where $FWHM$ denotes the half-height width.

3. Results and Analysis

In the preliminary study of this paper, we compared the performance of a single annular cavity structure with that of a CRERR structure sensor. Through careful comparative analyses, we found that the CRERR structure sensor demonstrated slightly higher sensitivity over a preset range of refractive index fluctuations, and its FOM value was superior. In view of these advantages, we decided to further investigate the properties and potential of the CRERR structure's sensor in depth. The initial parameters of the structure are as follows: R = 250 nm, g = 10 nm, h = 160 nm, $\alpha = 0°$ and L = 60 nm.

A series of exhaustive simulation analyses have been carried out in order to thoroughly investigate the formation mechanism of the Fano resonance and its spreading properties in the sensor. These analyses cover the overall structure of the system, the stand-alone CRERR structure, and the simplified double baffle composition. As shown in Figure 2, we can see and understand the complete system fabric that the transmittal frequency profile exhibits as an unsymmetrical sharp form, deriving its linear shape from the interference of two scattering amplitudes—one in the continuous state (associated with the background) and the other in the discrete state of excitation (associated with the resonance). The energy of the resonant state must be in the energy range of the continuous state (i.e., the background) for this effect to occur. In the vicinity of the resonance energy, the amplitude of the background scattering usually varies gently with energy; however, the amplitude of the resonance scattering, both in magnitude and in phase, varies quite rapidly, leading to the onset of asymmetry. In the structure containing only two baffles, the transmission spectral line presents an approximately straight line feature, and all points on this line possess high transmittance. However, when we introduced the CRERR structure into the system, the collection capability of the entire system for electric fields and the capture capability of SPPs were significantly improved. This improvement makes the system more advantageous for sensor performance evaluation.

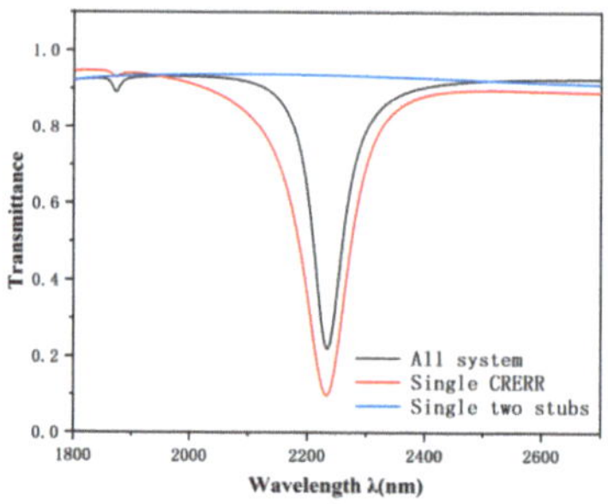

Figure 2. Transmission spectra of the entire structure (black line), a single CRERR structure (red line) and only two baffle structures (blue line).

In an attempt to explain the generation process of the Fano resonance in more detail, we further investigated the distributed magnetic field at the resonance inclination ($\lambda_{SingleCRERR} = 2232$ nm, $\lambda_{All\ system} = 2243$ nm) for a single CRERR structure and the whole system, as depicted in Figure 3. It is noteworthy from the figure, that here, the SPPs could pass through the waveguide and pair to the CRERR structure in both the individual CRERR structure and the whole system structure. The distributions of the naturalized magnetism are dominated by the CRERR fabric, which is sparsely observed in the bus waveguide, indicating that significant resonance occurs. Moreover, we can also see from the figure that the CRERR's upper and lower parts are inverted, and the overall structure is similar to the normalized magnetic field distributed by a single CRERR structure, but the addition of the rectangular baffle increases the propagation of the SPPs throughout the system, which further contributes to the formation of the Fano resonance.

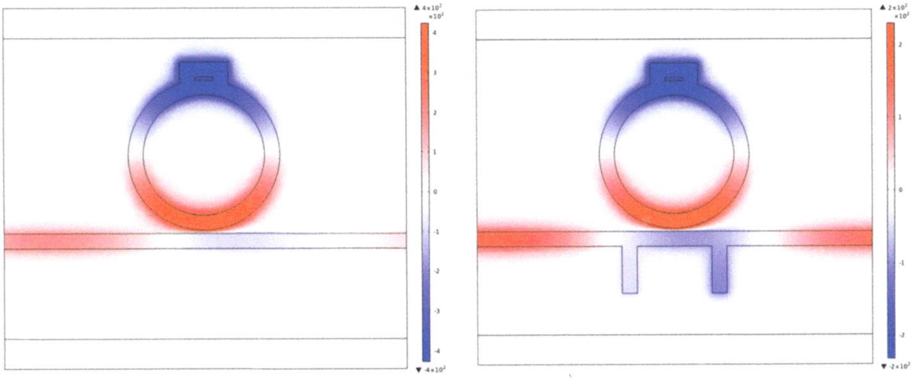

(a) (b)

Figure 3. (**a**) Normalized distribution of magnetic field at λ = 2232 nm for the CRERR structure; (**b**) normalized distribution of magnetic field at λ = 2243 nm for the whole system.

We systematically adjusted the values of the refractive indices, which were set to 1.00, 1.01, 1.02, 1.03, 1.04 and 1.05, and carried out an exhaustive comparative study of the same values. From Figure 4a, it can be seen that the shape of the transmission spectral curve is constant when the refractive index n is varied, and the curve undergoes an almost equidistant redshift when the refractive index n is increased. As illustrated in Figure 4b, the displacement of the inclination wavelength changes linearity according to the difference in the refractive index n. It is evident that, based on these features, this can be used as a refractive index transducer in our presented structure. After calculating the skewness of the sensitivity fit line, an optimal parameter for the construction was calculated, and the sensitivity of the transducer obtained reached 3180 nm/RIU and a FOM of 54.8, which is the optimal parameter for this structure.

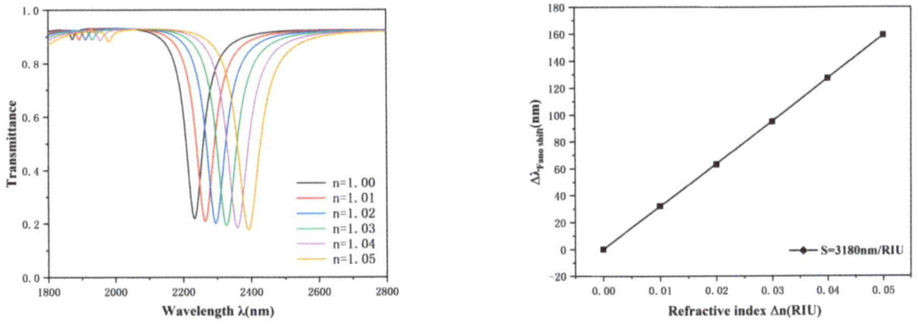

(a) (b)

Figure 4. (**a**) Transmission spectra at variable refractive indices; (**b**) sensitivity fitted lines at variable refractive indices.

We set the values of α to −120°, −60°, 120°, 90°, 60°, 30° and 0°, and as demonstrated in Figure 5, the transmission curves, when rotated at the same angle counterclockwise and clockwise, have curves with essentially the same shape, so it is sufficient for us to study the case where α is greater than 0°. As α changes from 0° to 120°, a shift of the curve to the left is observed; i.e., the S of the sensor decreases, and the FOM decreases. After that, we observed the magnetic field distribution in the lumen of CRERR from several perspectives, finding that the differential magnetic scene distribution is different at different angles, and the magnetic scene concentration is higher where the cavity is located, while the magnetic field distribution of these structures is symmetric near the centerline of the rectangular toroidal cavity. This is an indication of the fact that these structures are in the same resonance mode and that the location of the rectangular annular cavity leads to differences in these contrasting structures, with the rectangular annular cavity region being the structure that predominantly binds the energy. When α is 0° the transmission spectrum is an unsymmetrical curve, having an ultra-low transmit rate and a great range of wavelengths, which indicates that this is a Fano resonance of high sensitivity.

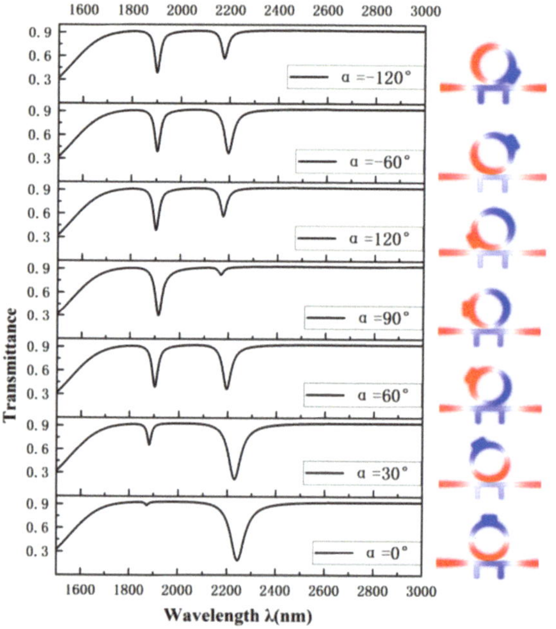

Figure 5. Transmission spectra of rectangular annular cavity rotated at different angles.

The above study shows that the transmission spectral curves of the sensor are basically the same for the same angle of counterclockwise and clockwise rotation, and the performance of the sensor improves as the angle of rotation of the rectangular annular cavity decreases. Thus, the structure when α is 0° is chosen as the main structure to be studied in the following discussion.

An in-depth study of the rectangular annular cavity was carried out by investigating its height L at 60 nm, 70 nm, 80 nm, 90 nm and 100 nm. As shown in Figure 6a, the transmission spectrum is red-shifted with increasing L. This is due to the fact that the increase in L causes an extension of the effective length of the structure, which induces a redshift of the resonance trough, but the change in the transmittance is not very significant. Figure 6b indicates that as the height of the rectangular annular cavity increases, there is no significant effect on the sensitivity change. From Figure 6c, it is observed that the FWHM becomes dramatically larger as the height L from 60 nm to 100 nm is added, and the FWHM is smallest at a rectangular annular cavity height of 60 nm, and the FWHM is 72 nm at a L of 100 nm. Since the effect on both the transmission spectrum and sensitivity is insignificant when L is varied, it can be concluded that the FOM is maximum when L is 60 nm, which is when the sensor's performance is best. Therefore, after comprehensive consideration, 60 nm is the most suitable height for the rectangular annular cavity.

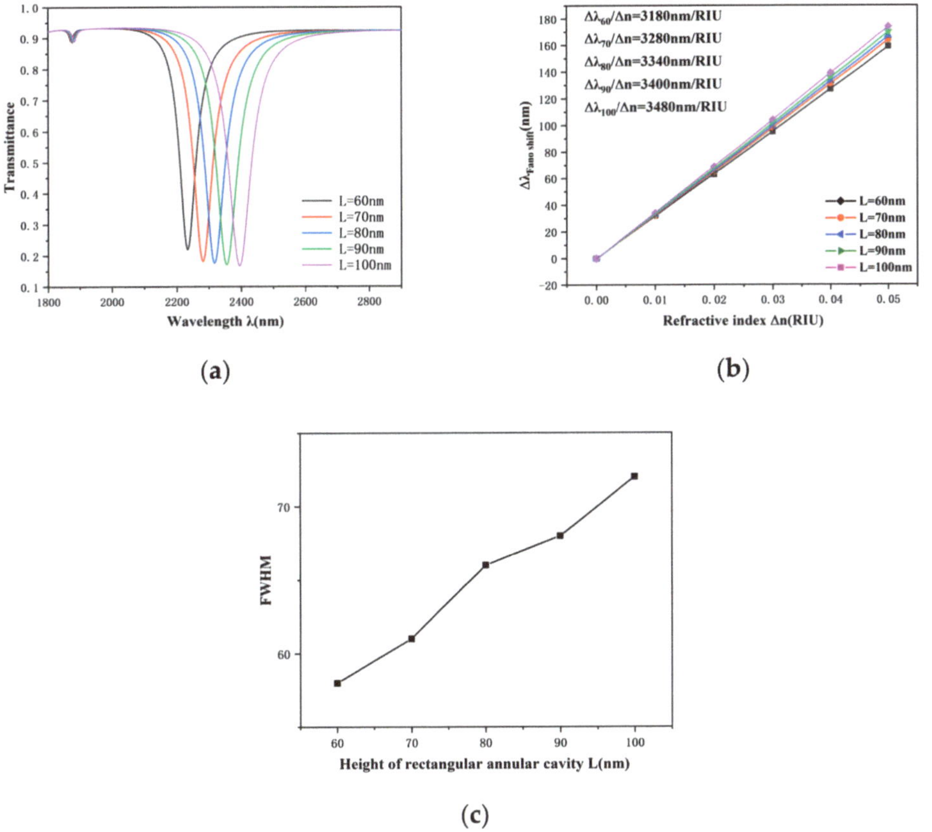

Figure 6. (**a**) Transmission spectra at different heights of the rectangular annular cavity; (**b**) sensitivity fitted lines at different heights of the rectangular annular cavity; (**c**) variation in FWHM values at different heights of the rectangular annular cavity.

We then estimated the impact of R on the entire system. The parameters of R are adjusted to 210 nm, 220 nm, 230 nm, 240 nm and 250 nm, illustrated in Figure 7a, which demonstrates that R has the lowest transmittance and maximum wavelength when it is 250 nm. As R increases, the transmission spectral profile shows a significant redshift, which is attributed equally to the expansion of the reactive length of the structure due to the increase in R, which leads to a redshift of the resonance trough. The sensitivity fit line for this structure is provided in Figure 7b, from which it can be observed that linearity between the varying refractive index and the varying wavelength exhibits an excellent linear relationship. In addition to this, the sensitivity of the CRERR structure is significantly improved as the outer radius R of the structure continues to increase. We also note that the growth of R increases susceptibility from 1880 nm/RIU to 3180 nm/RIU, indicating that the outer radius R is the main geometrical parameter for the structure to enhance sensitivity. In summary, we chose 250 nm as the outer radius of the structure in this paper, at which point the structure has a higher S and FOM sensitivity.

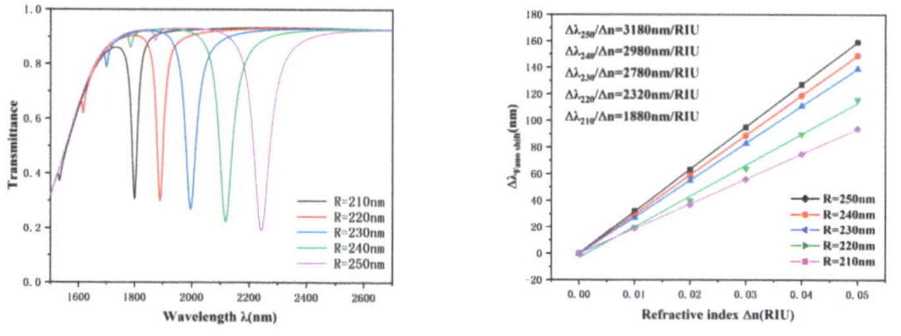

(a) (b)

Figure 7. (**a**) Transmission spectra of different Rs; (**b**) sensitivity fitted lines of different Rs.

Then, this paper evaluated the impact of the MIM waveguide's structural parameters on the transmission properties of this sensor. The two rectangular baffle heights h grew from 120 nm to 160 nm, with a step interval of 10 nm for five sets of data. As shown in Figure 8a, the tilted wavelength and transmittance of the Fano resonance do not significantly change when h is varied, and the degree of asymmetry of the transmission spectral curve does not change significantly. During Fano resonance production, a succession of broad bands is generated by the MIM waveguide. As shown in Figure 8b, the height of the rectangular baffle also has some effect on the FWHM, and it can be seen from the experiment that as the height h of the rectangular baffle increases, the corresponding FWHM value becomes smaller. The FWHM is minimized at 58 nm when h is 160 nm.

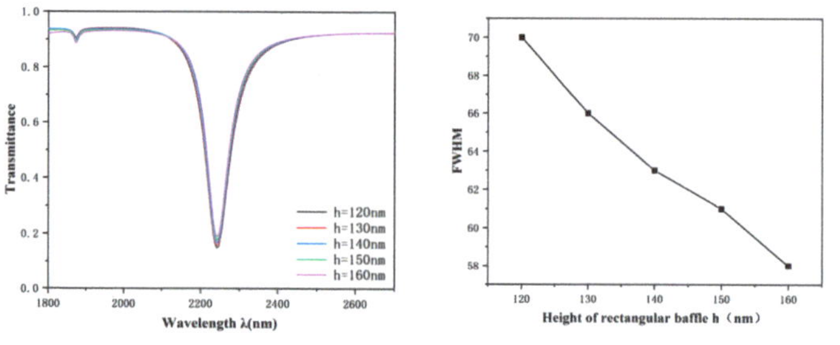

(a) (b)

Figure 8. (**a**) Transmission spectra of a rectangular baffle at variable heights; (**b**) variation in FWHM values at different heights of a rectangular baffle.

Next, we will deeply exploit the effect of the coupling distance between the waveguide and the CRERR on the transmission features. Specifically, we set the starting value of the coupling distance g to be 5 nm. Take this as a benchmark, and follow the increment of 5 nm up to 25 nm. By observing Figure 9a, we can distinctly see that the transmission spectral profile significantly shifts towards the short wavelength (i.e., blueshift) with the growth of the coupling distance, and the transmittance also shows an increasing trend. Further, Figure 9c shows that the FWHM value decreases with an increasing coupling distance, a phenomenon that indicates that, as the coupling gap between the waveguide and the CRERR structure widens, the coupling effect between them becomes weaker, while the electric field strength also weakens accordingly [40]. Therefore, by the choice of a suitable coupling distance g, a transmission spectrum with a lower transmittance can be obtained. When $g < 10$ nm, the corresponding FWHM value increases dramatically. As shown in Figure 9b, the sensitivity decreases as the coupling distance increases but decreases at coupling distances greater than 10 nm. The sensitivity at this point is maximized to 3620 nm/RIU, as g is equal to 5 nm, but the corresponding FWHM value is as high as 159 nm, and the FOM value is only 22.76, as shown in Equation (4). The larger FWHM clearly destroys the FOM of the structure, as the coupling distance is less than 10 nm. After comprehensive consideration, 10 nm was selected as the preferable coupling length for the submitted sensor. At this point, the sensitivity of the proposed sensor is 3180 nm/RIU with a FOM value of 54.8.

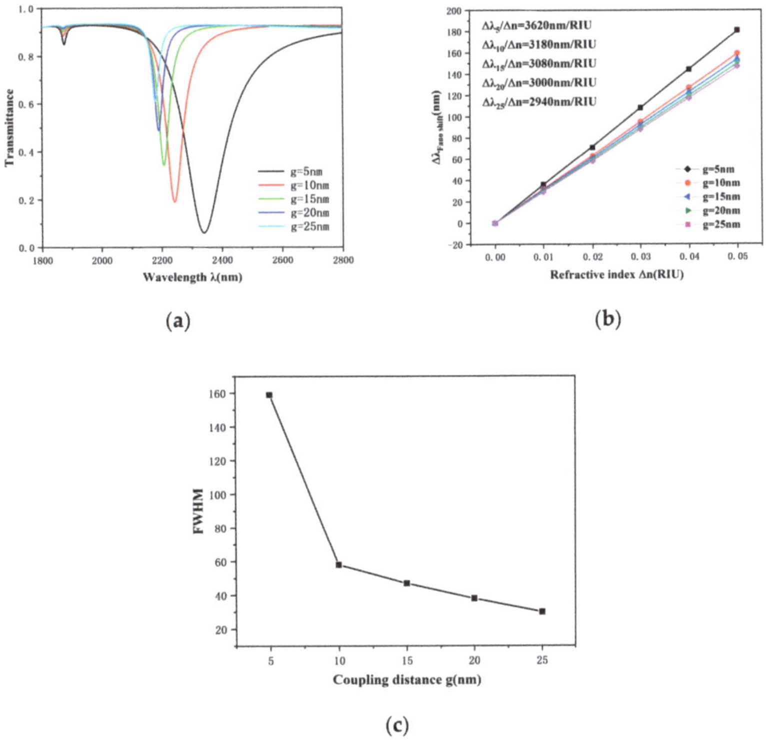

Figure 9. (**a**) Transmission spectra obtained with various coupling gaps; (**b**) sensitivity fitted lines with various coupling gaps; (**c**) variation in FWHM values with various coupling gaps.

4. Biological Sensing Applications

The sensor proposed in this work has a simple structure and high sensitivity; thus, we applied it as a sensor for detecting electrolyte concentrations in human blood. The balance between sodium (Na^+) and potassium (K^+) ions in electrolytes ensures the functioning of the body's cells, and the monitoring of Na^+ and K^+ concentrations plays an extremely important role in the control of cardiovascular and neurological diseases. The linear relationship between the refractive index and the Na^+ concentration and K^+ concentration at constant temperature is expressed as the following [41]:

$$n_{Na^+} = 1.3373 + 1.768 \times 10^{-3}\left(\frac{C_1 \times k_1}{393}\right) - 5.8 \times 10^{-6}\left(\frac{C_1 \times k_1}{393}\right)^2 \quad (5)$$

$$n_{K^+} = 1.3352 + 1.6167 \times 10^{-3}\left(\frac{C_2 \times k_2}{529.8}\right) - 4 \times 10^{-7}\left(\frac{C_2 \times k_2}{529.8}\right)^2 \quad (6)$$

where C_1 and C_2 denote the concentration of Na^+ and K^+, respectively, and k_1 and k_2 represent the concentration factors of Na^+ and Na^+, respectively, whose values are set to 30 and 50, respectively. Furthermore, n_{Na^+} and n_{K^+} are the refractive index values for different concentrations of Na^+ and K^+, respectively. The sensitivity of the sensor for detecting electrolyte concentrations can be expressed as

$$S_C = \frac{\Delta\lambda}{\Delta C} \quad (7)$$

where $\Delta\lambda$ represents the shift in the transmission spectrum, and ΔC represents the change in the concentration of the detected substance.

Because of the high optical activity of the surface plasmon excitations in this sensor system, it is theoretically possible to place small amounts of blood samples in the waveguide and resonator. In addition, a change in the concentration of the sample results in a change in the refractive index of the sensor, which indirectly reflects the change in concentration by calculating the shift in the transmission spectrum caused by the change in refractive index. The structural parameters of the proposed sensor were fixed at R = 250 nm, h = 160 nm, α = 0°, L = 60 nm and g = 10 nm; the concentrations of Na^+ were set to 200, 250, 300, 350 and 400 mgdL^{-1}; and the concentrations of K^+ were set to 0, 20, 40, 60 and 80 mgdL^{-1}.

Observing Figure 10a,b, we can see that the transmission spectra show a clear redshift as the concentration value of the substance to be detected increases. The sensitivity of this concentration detection sensor is shown in Figure 10c,d, which indicates that the structure exhibits a satisfactory linear fit, which ensures the accuracy of the measurements. The sensor is capable of achieving sensitivities of 0.495 nm/mgdL^{-1} and 0.6375 nm/mgdL^{-1} for the detection of electrolyte sodium and potassium ion concentrations in human blood. With the advantages of simple structure, fast response, high reliability and easy integration at the nanoscale, this sensor model provides a high-performance cavity option for surface plasma-based biosensor devices, which is expected to open up new opportunities in the field of medical detection.

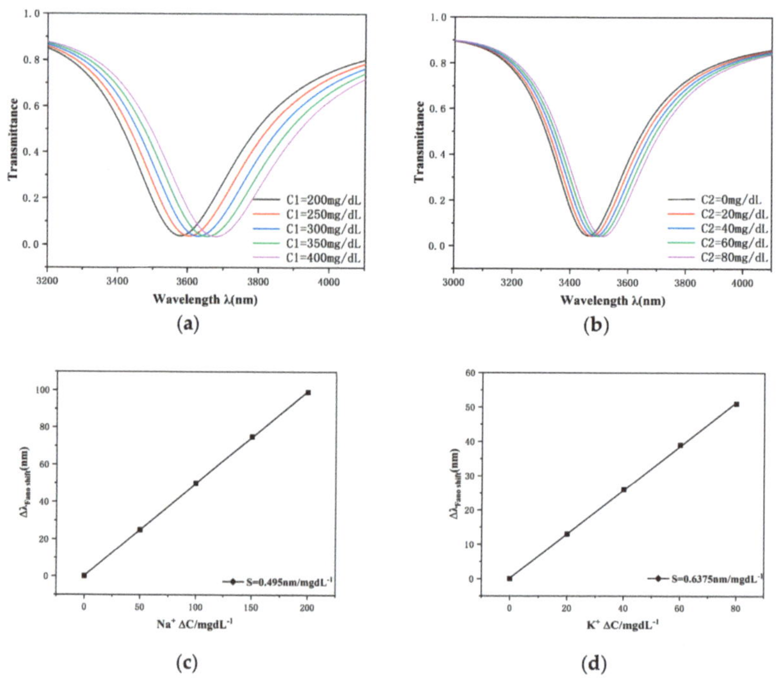

Figure 10. (**a**) Transmission spectra of different concentrations of Na^+; (**b**) transmission spectra of different concentrations of K^+; (**c**) sensitivity fit lines for different concentrations of Na^+; (**d**) sensitivity fit lines for different concentrations of K^+.

5. Conclusions

In this research, a new structure for a nanoscale diffraction rate transducer was developed, comprising a MIM waveguide having two rectangular baffles and a circular coupling with a rectangular annular cavity. We have analyzed the propagation characteristics of the overall structure by a series of experimental comparisons using the finite element method (FEM) to determine the optimal CRERR structure and obtain an excellent Fano resonance curve. A new CRERR structure is introduced in the narrowband mode, depending on the MIM waveguide, with a double rectangular baffle as the coupling foundation. The modeling results show a significant redshift of the wavelength of the Fano resonant dip with the expansion of the refractive index n and the radius of the outer circle R. However, the wavelength of the Fano resonance inclination produces a significant blueshift with the increase in the coupling distance. The variation in the height of the double rectangular baffle has no significant effect on the wavelength of the Fano resonance valley but has a large effect on the FWHM. When the height of the rectangular annular cavity increases, the sensitivity does not change significantly, but its FWHM increases significantly, while the sensing performance of the sensor becomes better when the rotational angle of the rectangular annular cavity declines. Under specific structural parameters, i.e., when R is set to 250 nm, h to 160 nm, α angle to 0°, L length to 60 nm and the coupling distance g to 10 nm, our proposal of the transducer structure exhibits its best sensing performance. This optimized transducer has a sensitivity of up to 3180 nm/RIU and a FOM of 54.8, demonstrating its excellent potential and performance in sensing applications. The sensor is capable of achieving sensitivities of 0.495 nm/mgdL^{-1} and 0.6375 nm/mgdL^{-1} when detecting concentrations of electrolyte sodium and potassium ions in human blood and is expected to play an important role in human health monitoring.

Author Contributions: Conceptualization, S.C., Y.C. and Y.S.; methodology, S.C.; software, Y.S. and Y.C..; validation, Y.S., J.W. and Y.C.; formal analysis, J.W.; investigation, J.W.; resources, Y.Z.; data curation, S.C.; writing—original draft preparation, S.C.; writing—review and editing, S.Y. and Y.R.; visualization, T.W. and Y.R.; supervision, T.W., Y.Z. and Y.R.; project administration, S.Y., Y.Z. and T.W.; funding acquisition, S.Y. All authors have read and agreed to the published version of the manuscript.

Funding: This research was funded in part by the National Natural Science Foundation of China under Grant No. 62374148 and in part by the Zhejiang Provincial Natural Science Foundation of China under Grant No. LD21F050001, the Key Research Project by the Department of Water Resources of Zhejiang Province under Grant No. RA2101, the Key Research and Development Project of Zhejiang Province under Grant No. 2021C03019, and the Funds for Special Projects of the Central Government in Guidance of Local Science and Technology Development under Grant No. YDZJSX20231A031.

Informed Consent Statement: Informed consent was obtained from all subjects involved in the study.

Data Availability Statement: The data used to support the findings of this study are available from the corresponding author upon reasonable request.

Conflicts of Interest: The authors declare no conflicts of interest.

References

1. Motil, A.; Danon, O.; Peled, Y.; Tur, M. Pump-Power-Independent Double Slope-Assisted Distributed and Fast Brillouin FiberOptic Sensor. *IEEE Photonics Technol. Lett.* **2014**, *26*, 797–800. [CrossRef]
2. Law, M.; Sirbuly, D.J.; Johnson, J.C.; Goldberger, J.; Saykally, R.J.; Yang, P. Nanoribbon waveguides for subwavelength photonics integration. *Science* **2004**, *305*, 1269–1273. [CrossRef] [PubMed]
3. Kresic, I.; Kruljac, M.; Ban, T.; Aumiler, D. Electromagnetically induced transparency with a single frequency comb mode probe. *J. Opt. Soc. Am. B* **2019**, *36*, 1758–1764. [CrossRef]
4. Chen, J.; Li, Z.; Lei, M.; Fu, X.; Xiao, J.; Gong, Q. Plasmonic Y-splitters of High Wavelength Resolution Based on Strongly Coupled-Resonator Effects. *Plasmonics* **2012**, *7*, 441–445. [CrossRef]
5. Chen, Z.; Wang, W.; Cui, L.; Yu, L.; Duan, G.; Zhao, Y.; Xiao, J. Spectral Splitting Based on Electromagnetically Induced Transparency in Plasmonic Waveguide Resonator System. *Plasmonics* **2015**, *10*, 721–727. [CrossRef]

6. Miroshnichenko, A.E.; Flach, S.; Kivshar, Y.S. Fano resonances in nanoscale structures. *Rev. Mod. Phys.* **2010**, *82*, 2257–2298. [CrossRef]
7. Lu, H.; Liu, X.; Mao, D.; Wang, G. Plasmonic nanosensor based on Fano resonance in waveguide coupled resonators. *Opt. Lett.* **2012**, *37*, 3780–3782. [CrossRef] [PubMed]
8. Hu, N.; Zhang, G.; An, H.; Shi, Y.; Gu, M. Design and Optimization of the Multifunctional Rectangular Cavity Band-Pass Filter Based on the Surface Plasmon Polariton. *Plasmonics* **2016**, *12*, 1457–1462. [CrossRef]
9. Wu, W.; Yang, J.; Zhang, J.; Huang, J.; Chen, D.; Wang, H. Ultra-high resolution filter and optical field modulator based on a surface plasmon polariton. *Opt. Lett.* **2016**, *41*, 2310–2313. [CrossRef]
10. Zhang, Q.; Huang, X.G.; Lin, X.S.; Tao, J.; Jin, X.P. A subwavelength coupler-type MIM optical filter. *Opt. Express* **2009**, *17*, 7549–7554. [CrossRef]
11. Wen, K.; Hu, Y.; Chen, L.; Zhou, J.; Lei, L.; Guo, Z. Fano Resonance with Ultra-High Figure of Merits Based on Plasmonic Metal-Insulator-Metal Waveguide. *Plasmonics* **2015**, *10*, 27–32. [CrossRef]
12. Zhang, Z.D.; Wang, R.B.; Zhang, Z.Y.; Tang, J.; Zhang, W.D.; Xue, C.Y.; Yan, S.B. Electromagnetically Induced Transparency and Refractive Index Sensing for a Plasmonic Waveguide with a Stub Coupled Ring Resonator. *Plasmonics* **2016**, *12*, 1007–1013. [CrossRef]
13. Khan, A.D.; Miano, G. Plasmonic Fano Resonances in Single-Layer Gold Conical Nanoshells. *Plasmonics* **2013**, *8*, 1429–1437. [CrossRef]
14. Bian, Y.; Zheng, Z.; Liu, Y.; Liu, J.; Zhu, J.; Zhou, T. Hybrid wedge plasmon polariton waveguide with good fabrication-error-tolerance for ultra-deepsubwavelength mode confinement. *Opt. Express* **2011**, *19*, 22417–22422. [CrossRef] [PubMed]
15. Gao, Z.; Zhang, X.; Shen, L. Wedge mode of spoof surface plasmon polaritons at terahertz frequencies. *J. Appl. Phys.* **2010**, *108*, 113104. [CrossRef]
16. Pang, S.; Huo, Y.; Xie, Y.; Hao, L. Fano resonance in MIM waveguide structure with oblique rectangular cavity and its application in sensor. *Opt. Commun.* **2016**, *381*, 409–413. [CrossRef]
17. Zafar, R.; Salim, M. Enhanced Figure of Merit in Fano Resonance-Based Plasmonic Refractive Index Sensor. *IEEE Sens. J.* **2015**, *15*, 6313–6317. [CrossRef]
18. Zafar, R.; Salim, M. Analysis of asymmetry of Fano resonance in plasmonic metal-insulator-metal waveguide. *Photonics Nanostruct.-Fundam. Appl.* **2017**, *23*, 1–6. [CrossRef]
19. Zhou, J.; Chen, H.; Zhang, Z.; Tang, J.; Cui, J.; Xue, C.; Yan, S. Transmission and refractive index sensing based on Fano resonance in MIM waveguide-coupled trapezoid cavity. *AIP Adv.* **2017**, *7*, 015020. [CrossRef]
20. Zhang, R.; Pu, S.; Li, X. Gold-Film-Thickness Dependent SPR Refractive Index and Temperature Sensing with Hetero-Core Optical Fiber Structure. *Sensors* **2019**, *19*, 4345. [CrossRef]
21. Han, Z.; Bozhevolnyi, S.I. Plasmon-induced transparency with detuned ultracompact Fabry-Perot resonators in integrated plasmonic devices. *Opt. Express* **2011**, *19*, 3251–3257. [CrossRef] [PubMed]
22. Yi, X.; Tian, J.; Yang, R. Tunable Fano resonance in plasmonic MDM waveguide with a square type split-ring resonator. *Optik* **2018**, *171*, 139–148. [CrossRef]
23. Ren, X.; Ren, K.; Cai, Y. Tunable compact nanosensor based on Fano resonance in a plasmonic waveguide system. *Appl. Opt.* **2017**, *56*, H1–H9. [CrossRef] [PubMed]
24. Gai, H.; Wang, J.; Qian, T. Modified Debye model parameters of metals applicable for broadband calculations. *Appl. Opt.* **2007**, *46*, 2229–2233. [CrossRef] [PubMed]
25. Yuze, S.; Xudong, F. Optical ring resonators for biochemical and chemical sensing. *Anal. Bioanal. Chem.* **2011**, *399*, 205–211.
26. Reza, M.R.; Alireza, T.; Ali, M.M. Design of a plasmonic sensor based on a square array of nanorods and two slot cavities with a high figure of merit for glucose concentration monitoring. *Appl. Opt.* **2018**, *57*, 7798–7804.
27. Chen, Y.; Xu, Y.; Cao, J. Fano resonance sensing characteristics of MIM waveguide coupled Square Convex Ring Resonator with metallic baffle. *Results Phys.* **2019**, *14*, 102420. [CrossRef]
28. Noual, A.; Amrani, M.; El Boudouti, E.H.; Pennec, Y.; Djafari-Rouhani, B. Terahertz multi-plasmon induced reflection and transmission and sensor devices in a graphene-based coupled nanoribbons resonators. *Opt. Commun.* **2019**, *440*, 1–13. [CrossRef]
29. Rezzouk, Y.; Khattou, S.; Amrani, M.; Noual, A.; El Boudouti, E.H.; Talbi, A.; Djafari-Rouhani, B. Bound States in the Continuum and Induced Resonances in a Simple Plasmonic Waveguide with Sensing Application. *Photonics* **2023**, *10*, 1284. [CrossRef]
30. Ciminelli, C.; Campanella, C.M.; Dell'Olio, F.; Campanella, C.E.; Armenise, M.N. Label-free optical resonant sensors for biochemical applications. *Prog. Quantum Electron.* **2013**, *37*, 51–107. [CrossRef]
31. Chen, J.; Gan, F.; Wang, Y.; Li, G. Plasmonic Sensing and Modulation Based on Fano Resonances. *Adv. Opt. Mater.* **2018**, *6*, 1701152. [CrossRef]
32. Zhang, Y.; Liu, W.; Li, Z.; Li, Z.; Cheng, H.; Chen, S.; Tian, J. High-quality-factor multiple Fano resonances for refractive index sensing. *Opt. Lett.* **2018**, *43*, 1842–1845. [CrossRef]
33. Wu, T.; Liu, Y.; Yu, Z.; Peng, Y.; Shu, C.; Ye, H. The sensing characteristics of plasmonic waveguide with a ring resonator. *Opt. Express* **2014**, *22*, 7669–7677. [CrossRef] [PubMed]
34. He, Y.-J.; Hung, W.-C.; Lai, Z.-P. Using Finite Element and Eigenmode Expansion Methods to Investigate the Periodic and Spectral Characteristic of Superstructure Fiber Bragg Gratings. *Sensors* **2016**, *16*, 192. [CrossRef] [PubMed]

35. Li, S.; Wang, Y.; Jiao, R. Fano resonances based on multimode and degenerate mode interference in plasmonic resonator system. *Opt. Express* **2017**, *25*, 3525–3533. [CrossRef]
36. Samusev, A.; Mukhin, I.; Malureanu, R.; Takayama, O.; Permyakov, D.V.; Sinev, I.S.; Baranov, D.; Yermakov, O.; Iorsh, I.V.; Bogdanov, A.A.; et al. Polarization-resolved characterization of plasmon waves supported by an anisotropic metasurface. *Opt. Express* **2017**, *25*, 32631–32639. [CrossRef]
37. Kekatpure, R.D.; Hryciw, A.C.; Barnard, E.S.; Brongersma, M.L. Solving dielectric and plasmonic waveguide dispersion relations on a pocket calculator. *Opt. Express* **2009**, *17*, 24112–24129. [CrossRef] [PubMed]
38. Qiao, L.; Zhang, G.; Wang, Z.; Fan, G.; Yan, Y. Study on the Fano resonance of coupling M-type cavity based on surface plasmon polaritons. *Opt. Commun.* **2019**, *433*, 144–149. [CrossRef]
39. Zhu, J.; Li, N. MIM waveguide structure consisting of a semicircu-lar resonant cavity coupled with a key-shaped resonant cavity. *Opt. Express* **2020**, *28*, 19978–19987. [CrossRef]
40. Noual, A.; Abouti, O.E.; El Boudouti, E.H.; Akjouj, A.; Pennec, Y.; Djafari-Rouhani, B. Plasmonic-induced transparency in a MIM waveguide with two side-coupled cavities. *Appl. Phys. A* **2017**, *123*, 49. [CrossRef]
41. Hassan, M.F.; Sagor, R.H.; Amin, M.R.; Islam, M.R.; Alam, M.S. Point of Care Detection of Blood Electrolytes and Glucose Utilizing Nano-Dot Enhanced Plasmonic Biosensor. *IEEE Sens. J.* **2021**, *21*, 17749–17757. [CrossRef]

photonics

MDPI

Article

Experimental Study on Evolution of Chemical Structure Defects and Secondary Contaminative Deposition during HF-Based Etching

Xiao Shen [1,2,3], Feng Shi [1,2,3,*], Shuo Qiao [1,2,3], Xing Peng [1,2,3] and Ying Xiong [1,2,3]

1 College of Intelligence Science and Technology, National University of Defense Technology, Changsha 410073, China; hillbert2009@163.com (X.S.); sqiao525@163.com (S.Q.); pengxing@fudan.edu.cn (X.P.); annie_xiong@163.com (Y.X.)

2 Hunan Key Laboratory of Ultra-Precision Machining Technology, National University of Defense Technology, Changsha 410073, China

3 Laboratory of Science and Technology on Integrated Logistics Support, National University of Defense Technology, Changsha 410073, China

* Correspondence: shifeng@nudt.edu.cn

Abstract: Post-processing based on HF etching has become a highly preferred technique in the fabrication of fused silica optical elements in various high-power laser systems. Previous studies have thoroughly examined and confirmed the elimination of fragments and contamination. However, limited attention has been paid to nano-sized chemical structural defects and secondary precursors that arise during the etching process. Therefore, in this paper, a set of fused silica samples are prepared and undergo the etching process under different parameters. Subsequently, an atomic force microscope, scanning electron microscope and fluorescence spectrometer are applied to analyze sample surfaces, and then an LIDT test based on the R-on-1 method is applied. The findings revealed that appropriate etching configurations will lead to certain LIDT improvement (from initial 7.22 J/cm^2 to 10.76 J/cm^2), and HF-based etching effectively suppresses chemical structural defects, while additional processes are recommended for the elimination of micron- to nano-sized secondary deposition contamination.

Keywords: HF-based etching; micro morphology; chemical structure defects; depositive reaction products

Citation: Shen, X.; Shi, F.; Qiao, S.; Peng, X.; Xiong, Y. Experimental Study on Evolution of Chemical Structure Defects and Secondary Contaminative Deposition during HF-Based Etching. *Photonics* **2024**, *11*, 479. https://doi.org/10.3390/photonics11050479

Received: 3 April 2024
Revised: 29 April 2024
Accepted: 6 May 2024
Published: 20 May 2024

1. Introduction

As the crucial terminal part of the ITER (International Thermonuclear Experimental Reactor), Final Optics Assembly (FOA) comprises fused silica optical elements capable of withstanding immense laser energy fluxes approaching the mega-joule level [1]. In operation, multiple laser beams traverse the FOA and converge upon a polyimide target capsule to necessitate the prior thermonuclear reaction, which induces subsequent nuclear fusion. In the National Ignition Facility (NIF) of the United States, 192 laser beams simultaneously transmit through the FOA, achieving an output energy of up to 1.9 mega-joules [2]; in France's Laser Megajoule facility (LMJ), the number of laser beams increases to 240, reaching a total energy output of 2.4 mega-joules [3]. Exposed to such high throughput, fused silica optical elements are vulnerable to various laser induced damages that can potentially result in rapid component failure.

A previous study revealed that the intrinsic LIDT (Laser-Induced Damage Threshold) of fused silica stands at 100 J/cm^2 [4]; however, in practice, the observed LIDT of fused silica optical components is notably lower than theoretical threshold. Numerous researchers have identified the primary factors contributing to this significant discrepancy between practical and theoretical LIDT values, such as damage precursors [5,6] derived from manufacturing integrated with grinding, polishing and post-treatment processing, including fragments, contamination and nano-scale defects. These defects exhibit a strong propensity to induce intense laser radiation absorption, resulting in significant energy accumulation, and once the energy deposition surpasses a certain limit, severe laser-induced damage ensues. Of all the precursors, fragments derived from material cutting, grinding and polishing are often exhibited as surface or subsurface cracks, pits, scratches and pocking marks, etc. Typically, the sizes of fragments range in size from sub-micron to several nanometers [7,8]; the occurrence of contamination is the result of residual polishing powder remaining in the superficial hydrolysis layer and subsurface defect layer, including particles such as CeO_2 and ZrO_2 during polishing and Cu and Fe particles during machining [9–11]. Nano-scale defects including chemical structure defects and trace amounts of salt deposits, unlike fragments and contamination, tend to cause laser-induced damage under low throughput and will evolve into precursors when exposed to high throughput. The former results from molecular bond breaking during machining [12,13], and the latter comes from post-processing aiming at removing surface and sub-surface precursors. Ultimately, all of these precursors can lead to the performance degradation of the FOA [14].

To mitigate various damage precursors, the wet etching process is also introduced [15–17]. This approach aims to suppress precursors while maintaining optical surface precision. The removal mechanism of surface and subsurface precursors during wet etching is the global chemical reaction between the etchant and the substrate. Scholars have conducted thorough research on HF-based wet etching processes; Wong investigated morphological evolution during the HF etching of cracks originating from the grinding process and found that the passivation of cracks can lead to a significant improvement in LIDT [16]; the Lawrence Livermore National Laboratory integrated HF etching with ultrasonic washing technology to establish AMP (Advanced Mitigation Processing), a method which has proven to be a reliable and effective means of removing fragments and contamination in post-treatment processes [15,18]. Through the latest AMP technology, Bude discovered that the initially scratched optical surface remained intact even under laser irradiation of 10 J/cm^2 (@351 nm, 5 ns) after passivation [19]. Zheng discovered that an appropriate etching process parameter can enhance LIDT and further identified a near-linear correlation between surface hardness and LIDT [20]. As research progressed, scholars also revealed that HF wet etching has a tendency to induce secondary precursors due to depositive reaction products, compromising surface accuracy if the etching parameter is not carefully controlled. Suratwala et al. observed that secondary pollutants adhere to the surface during HF wet etching, affecting the intrinsic characteristics of fused silica, including LIDT [15]. Bude revealed that nano-sized salt deposits resulting from post-treatment processes play an important role in laser-induced damage; when exposed to laser irradiation at 488 nm and 25 J/cm^2 for 5 ns, NaCl crystal deposits from deionized water caused significant damage to optical elements [13].

Since current post-processing based on HF etching on optical elements in high-power laser systems is a kind of "pollution first and treatment later" or "treatment during pollution" type of process in essence, the intrinsic surface of optical elements is theoretically inaccessible, and the final performance of the optical element is determined by post-processing to a large extent. However, the majority of existing studies on post-processing are primarily focused on crack passivation and pollution removal through HF etching. The impact of HF etching on chemical structure defects and the evolution and impact of secondary pollution introduced by HF etching have not been fully elucidated. Additionally, with HF being a highly toxic and corrosive compound, only by understanding the laws of the derivation and evolution of various precursors in HF etching can we achieve optical surfaces with superior performance in a more environmentally friendly and efficient manner. The paper is structured as follows: in Section 2, the sample preparation, testing, and characterization methods are introduced. Section 3 presents the research findings. Section 4 discusses the pertinent research results, and Section 5 summarizes the entire work presented in the paper.

2. Materials and Methods

As a weak acid, the ionization products of HF solution are H^+, F^-, $HF_2{}^-$ and H_2F_2 [21]. In HF wet etching, the corresponding chemical equation is given by the following [15]:

$$SiO_{2(solid)} + 3HF^-_{2(aq)} + H^+_{(aq)} = SiF^{2-}_{6(aq)} + 2H_2O_{(aq)} \tag{1}$$

The mechanism of the dissolution of SiO_2 in HF wet etching is clarified by Judge [22]: the reaction rate is determined by the adsorption process of HF molecules, $HF_2{}^-$ and H^+ ions on fused silica. Similar to SiO_2 crystals, silicon atoms and oxygen atoms together form a three-dimensional reticular structure, composed of SiO_4 tetrahedrons through covalent bonds. During etching, $HF_2{}^-$ and HF are adsorbed onto siloxane groups (≡Si-O-Si≡), while H^+ ions are attached to bridging oxygen atoms in siloxane bonds. The gathering effect of HF and $HF_2{}^-$ leads to an increase in electron density on bridging oxygen atoms, which strengthening the alkalinity of these oxygen atoms; subsequently, more H^+ ions are attracted, resulting in an elevated rate of silicon–oxygen bond breaking. The entire process encompasses two distinct steps, as outlined below:

Step one, the protonation of bridging oxygen atoms:

$$Si - O - X + H^+ \rightarrow Si - O(H)^+ - X \text{ (X represnts Si or H)} \tag{2}$$

Step two, the nucleophilic attack of electrophilic silicon atoms by HF_2^-:

$$Si - O(H)^+ - X + HF_2^- \rightarrow Si - F + HO - X + HF \tag{3}$$

The etching rate model can be defined as follows [21]:

$$R = \left(k_1\left[HF_2^-\right] + k_0[H_2F_2]\right)\frac{K_3\left[H^+\right]}{1 + K_3\left[H^+\right] + 1/K_4\left[H^+\right]} + k_2\left[HF_2^-\right]\frac{K_4\left[H^+\right]}{1 + K_4\left[H^+\right] + K_3K_4\left[H^+\right]^2} \tag{4}$$

The dissolution rate of SiO_2 in hydrogen fluoride solution is governed by the protonation and deprotonation of the surface reaction site; for a complete SiO_4 element, four rapid nucleophilic substitution reactions are essential to deprive one silicon atom away from the substrate. Therefore, it can be concluded that the presence of numerous chemical structure defects on fused silica substrate will certainly lead to an accelerated etching rate.

2.1. Sample Preparation

To study the evolution of damage precursors in HF etching, 6 samples are prepared, marked from #0 to #5, and the size of each sample is Ø 50 mm × 5 mm. The raw material of each sample is Suprasil 300 from the Heraeus company. Before etching, each sample goes through a smoothing process on a self-developed system [23] via pitch lap; the detailed parameters are shown in Table 1. Measurement via a Zygo interferometer indicates that the initial surface roughness is 0.28 nm, as illustrated in Figure 1.

Table 1. Smoothing parameters.

Parameters	Value
Polishing abrasive	CeO_2
Abrasive diameter	1.5 μm
Rotating rate	150 r/min
Feeding rate	300 mm/min
Smoothing pressure	0.02 Mpa
Smoothing duration	180 min

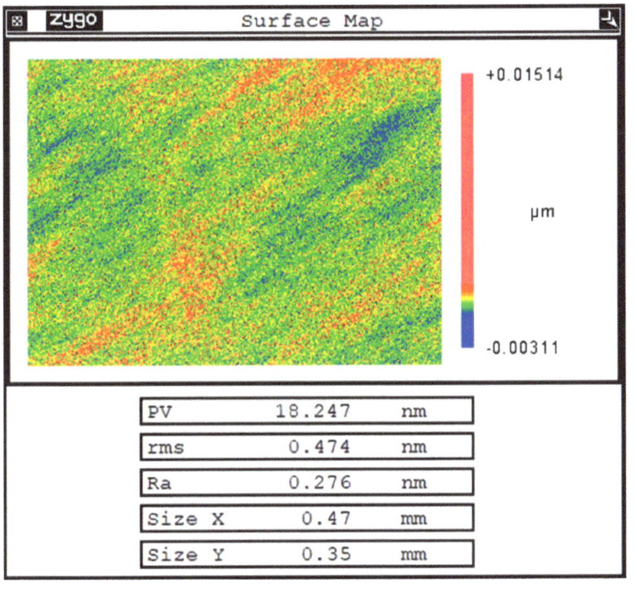

Figure 1. Surface measurement of initial samples.

2.2. Megasonic-Aided Etching Configuration

Current HF etching applications are usually performed in two ways: dynamic etching (or megasonic-aided etching) and static etching. In the extensive removal process via static etching technology, the rapid accumulation of reaction products leads to significant local precipitation easily, causing a profound effect on removal homogeneity and resulting in the deterioration of surface shape and roughness; consequently, the LIDT decreases drastically [24,25]. To avoid this problem, megasonic-aided etching, which combines the etching process with ultrasonic vibration, are more appealing to researchers. Assisted by an ultrasonic acoustic field, a thin boundary layer is generated adjacent to the reaction surface, as shown in Figure 2. According to the theory of boundary layer, the liquid vibration and surface material transfer efficiency caused by liquid vibration exhibit a negative correlation with the thickness of the acoustic layer. The thickness of the acoustic layer is determined by Equation (5), as stated in reference [21].

$$\delta = \sqrt{\frac{\mu}{\pi f \rho}} \tag{5}$$

where μ and ρ denote the viscosity and density of the liquid, respectively, and f represents the frequency of the megasonic.

In the process of megasonic-aided etching, the mass fraction of the HF solution is 5 wt. %, and the frequency is set to 1.3 MHz to ensure that the thickness of the acoustic layer caused by fluid viscosity remains below 0.5 µm [22]. Influenced by an ultrasonic acoustic field, the diffusion of micro reaction product deposition in solution will be more efficient, leading to a significant enhancement in the reaction rate between HF and SiO_2. For the 6 samples considered in this paper, Figure 3 illustrates the dynamic etching procedures and etching configurations. Prior to HF etching, deionized water rinsing is able to eliminate any disruptive contamination introduced by previous machining steps. The specific ultrasonic parameter setup for the rinsing is detailed in Table 2. All operations outlined in Figure 3 are carried out in a class 100 clean room, with a steady ambient temperature of 25 °C, to avoid external contamination during rinsing, etching and drying.

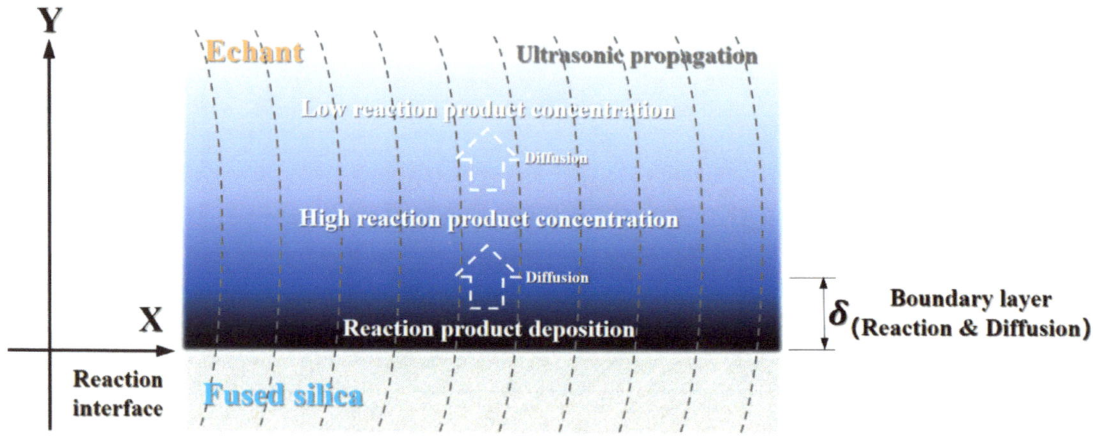

Figure 2. HF etching aided by megasonic acoustic field.

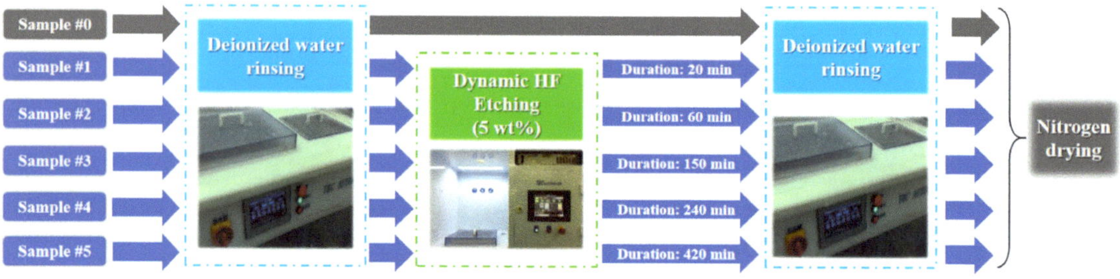

Figure 3. Etching schedule of each sample.

Table 2. Rinsing parameters.

Step	1	2	3	4	5	6	7
Frequency [kHz]	50	100	150	200	250	300	350
Duration [min]	5	5	5	5	5	5	5
Temperature [°C]				22			

2.3. Laser-Induced Damage Threshold Test Configuration

The laser induced damage test for each sample is carried out in the laboratory of advanced optical fabrication, National University of Defense Technology, and the test platform and the schematic are shown in Figure 4a,b, respectively. The test wavelength is 355 nm, the pulse width (FWHM) is 7 ns, the target spot shape is round, the laser spot diameter is 1.2 mm, and the degree of modulation is 3.2. During the test, the temperature is 22 ± 0.2 °C, and the humidity is 35 ± 5%. The measurement of threshold is based on the R-on-1 method; namely, by testing the same point with increasing laser flux, once laser damage occurs, the corresponding throughput is the local LIDT of the point.

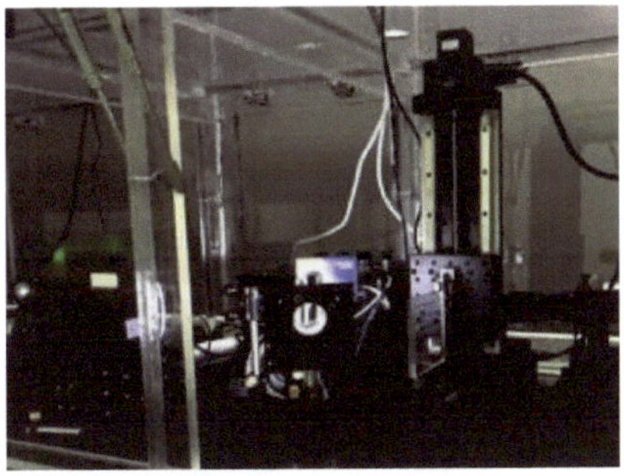

(**a**) Damage test platform

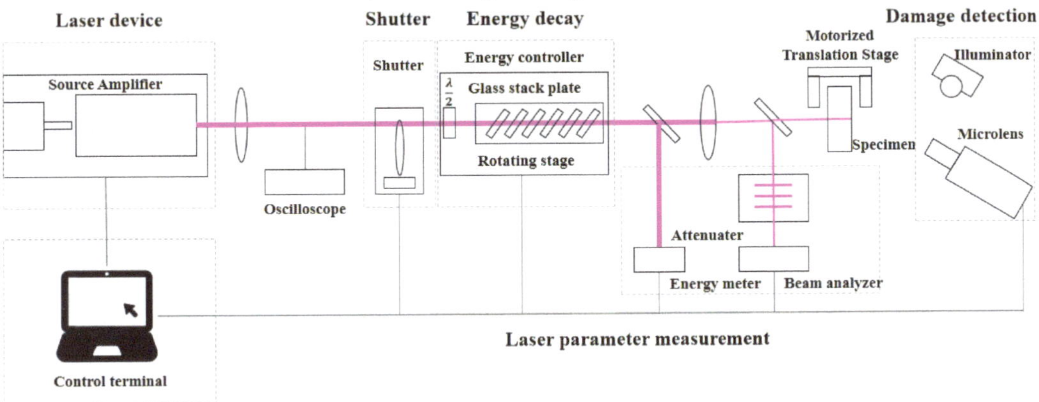

(**b**) Schematic of the platform

Figure 4. LIDT test platform.

3. Results

3.1. Evolution of Surface Micro Morphology

The atomic force microscope (AFM) of Bruker (detailed configurations are scanning frequency of 1.0 Hz and measuring area of 5 μm × 5 μm) is applied to assess the surface quality. Results on sample #0 revealed evident polishing marks left by the prior smoothing process on the initial surface, as depicted in Figure 5a. Additionally, a Form Talysurf PGI 1240 surface profiler is employed to detect the real-time etching depth.

In contrast, sample #0 remains unexposed to dynamic etching, maintaining its original roughness. Figure 5b–f depicts the microstructural evolution under varying etching durations, in 20 min etching period increments. Following the removal of the superficial hydrolytic layer, the numbers of subsurface defects are revealed. These defects, including scattered micro black pits and scratches, become evident, as shown in Figure 5b. As etching proceeds, the micro pits become increasingly shallower, and mottled gel-like depositions begin to emerge, appearing as white spots in the AFM images shown in Figure 5b–f. The figure also reveals that the quantity of white deposition spots peaks at 150 min of etching (sample #3), and subsequently, the count of white deposition spots decreases. Concurrently, the initial roughness deteriorates, increasing from 0.28 nm to 1.5 nm.

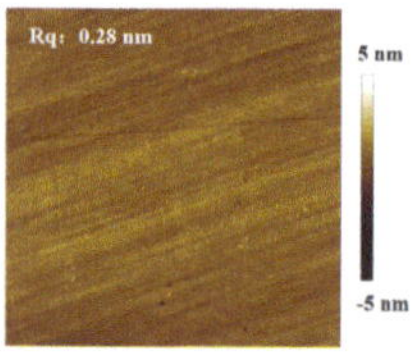

(**a**) Initial roughness (#0)

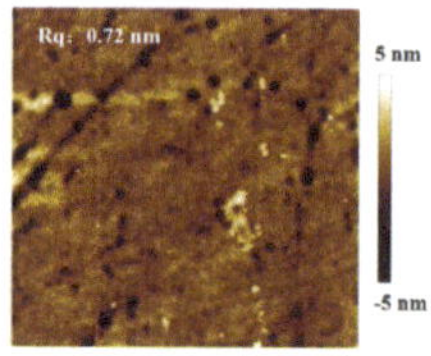

(**b**) Surface roughness after etching for 20 min (#1), etching depth 0.62 μm

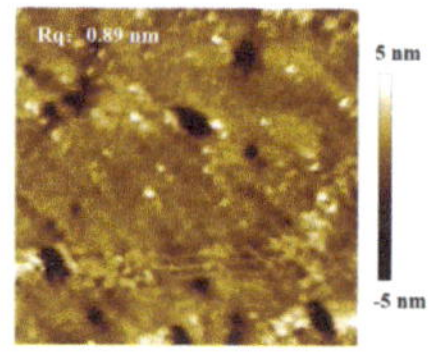

(**c**) Surface roughness after etching for 60 min (#2), etching depth 1.43 μm

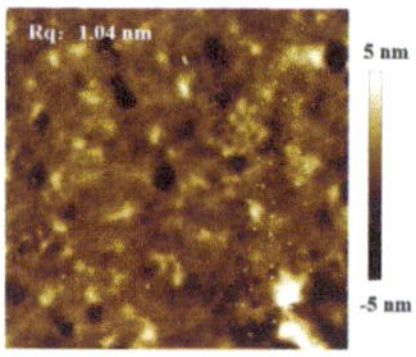

(**d**) Surface roughness after etching for 150 min (#3), etching depth 3.59 μm

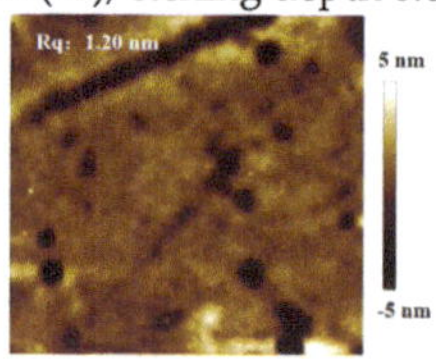

(**e**) Surface roughness after etching for 240 min (#4), etching depth 5.47 μm

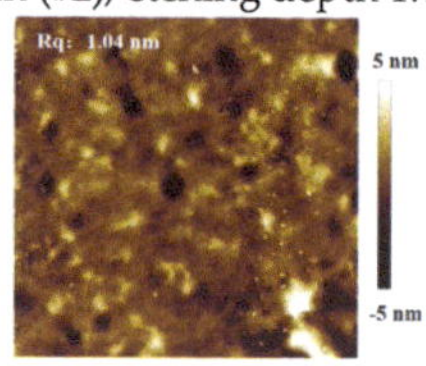

(**f**) Surface roughness after etching for 420 min (#5), etching depth 9.43 μm

Figure 5. Micro morphology of the samples under different etching depths.

3.2. Composition Identification of Micro Depositions

Using a PHENOM-ProX scanning electron microscope, a detailed examination is conducted on mottled sub-micron-scale residual depositions on sample #5. The elemental composition of two separate test points and surrounding areas are analyzed with an element mapping module. Figure 6b–e presents the energy spectrum scanning data of test point one, and it is evident that the distribution of impurity elements Na, Ca, Cl and C align precisely with the morphological distribution of residual depositions at test point one; the detailed composition ratio for each element is shown in Table 3.

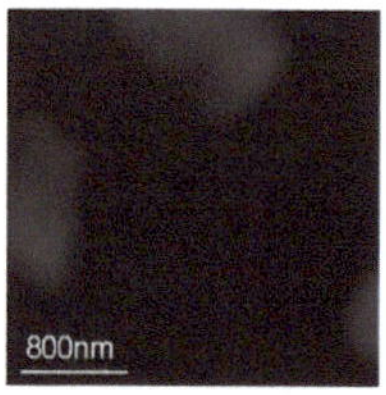

(**a**) Mottled deposits on sample #5 (point one)

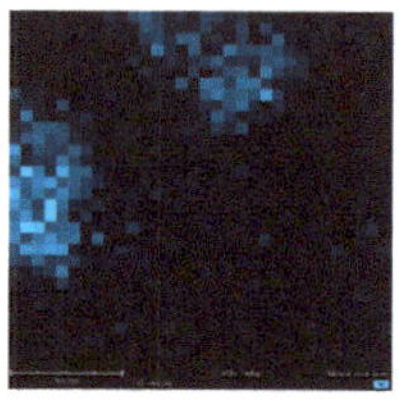

(**b**) Na element distribution

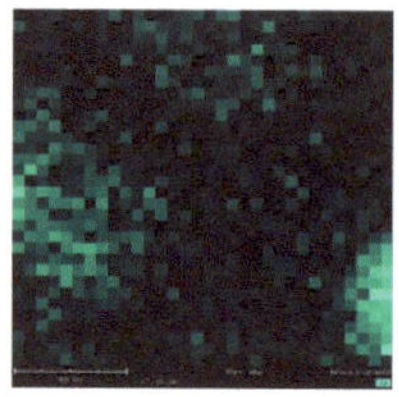

(**c**) Ca element distribution

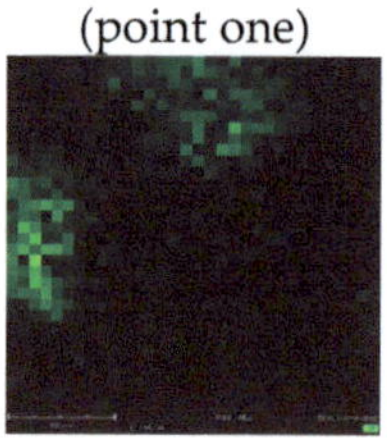

(**d**) Cl element distribution

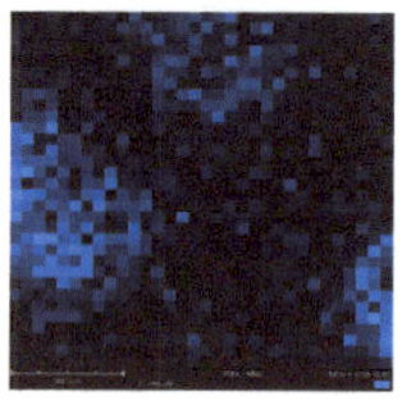

(**e**) C element distribution

Figure 6. Energy spectrum scanning results of test point one.

Table 3. Elemental composition of the deposits at test point one.

Element	Si	O	Na	C	Cl	Ca	Others
Composition ratio	50.48%	18.42%	1.13%	27.87%	0.51%	0.11%	1.48%

Under same image scale parameters, test point two is also observed, and the corresponding energy spectrum scanning results in Figure 7 demonstrate a shared distribution among residues and Na, K, Cl and C elements.

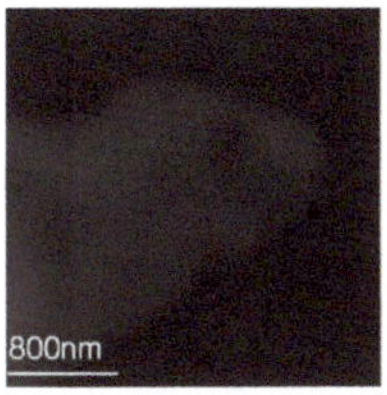

(**a**) Mottled deposits on sample #5 (Point two)

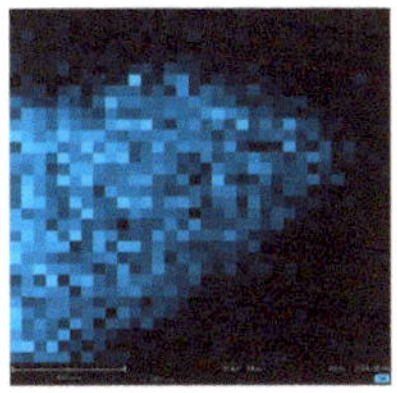

(**b**) Na element distribution

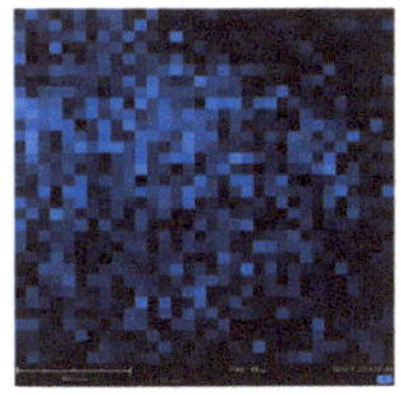

(**c**) K element distribution

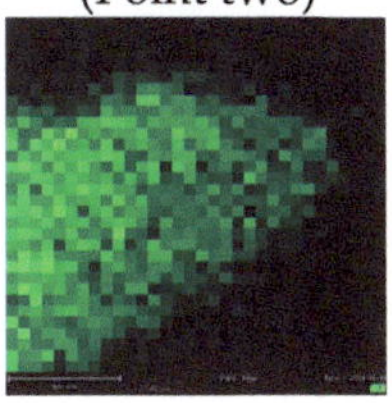

(**d**) Cl element distribution

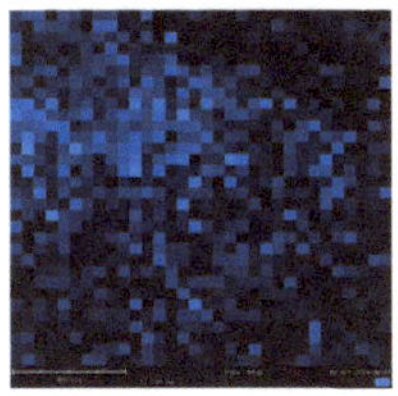

(**e**) C element distribution

Figure 7. Energy-spectrum scanning results of test point two.

Considering results above, besides Si and O elements originating from sample substrate, the residual deposition also contains C, Na, Cl, K and other elements, as shown in Tables 3 and 4. For both points, the C element content is significantly higher than other external elements. Furthermore, it is worth mentioning that the gathering-up effects of impure elements on residual depositions are particularly evident at both test points when compared to surroundings.

Table 4. Elemental composition of the deposits at test point 2.

Element	Si	O	Na	C	Cl	K	Others
Composition ratio	15.62%	39.33%	3.76%	35.48%	2.10%	0.24%	3.47%

3.3. Evolution of Chemical Structural Defects

In nano-scale fused silica substrate, there appears to be a 3D reticular amorphous structure composed of countless silicon–oxygen bonds. Previous research revealed that the chemical structure defects resulting from pre-machining processing are an oxygen-deficient center (ODC), a nonbridging oxygen hole center (NBOHC) and an E′-center (E′), as depicted in Figure 8a–c respectively. These imperfections, stemming from disrupted silicon–oxygen bonds, exert a significant constraint on the LIDT of optical elements [26,27].

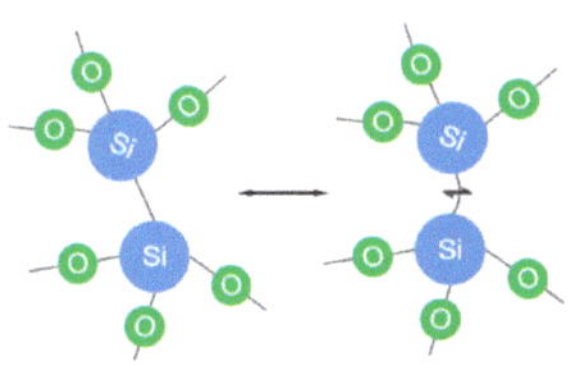

(**a**) Oxygen-deficient center

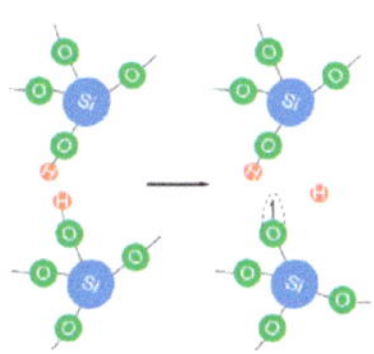

(**b**) Nonbridging oxygen hole center

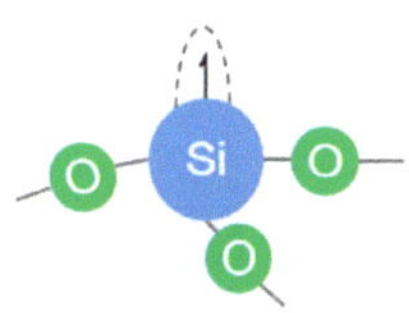

(**c**) E′-center

Figure 8. Chemical structure defects in fused silica substrate.

Chemical structural defects can be characterized via fluorescence spectroscopic detection. In this regard, an AJY TAU-3 fluorescence spectrometer is applied to test samples #0, #3 and #5. The testing parameters are specified as follows: the excitation wavelength is 248 nm, the spectral resolution is 0.01 nm, and a long-wavelength pass filter with a cutoff wavelength of 320 nm is also used to mitigate the interference of secondary excitation.

The fluorescence spectra of samples #0, #3 and #5 are presented in Figure 9. The initial surface exhibits two prominent absorption peaks. The peak centered around 440 nm originates from ODC defects, whereas the peak centered around 640 nm corresponds to NBOHC defects [28]. These structural defects arise from prior machining processes. During hydrolysis, the nano network of fused silica is broken first, enabling impurity elements such as Ce and H to penetrate the surface of the fused silica and form new bonding structures with Si and O elements; hence, ODC and NBOHC defects are introduced into sample #0. As depicted in Figure 9b, after 150 min, the fluorescence spectrum of ODC on sample #3 exhibits a significant reduction in characteristic peaks, and the intensity decreases from 1514.3 to 1031.0, which indicates a considerable decrease in the concentration of ODC defects; for NBOHC, the intensity level around 640 nm remains. However, the intensity level of noise near two characteristic absorption peaks changes significantly after etching; the reason is as follows: with the deprivation of oxygen atoms and silicon atoms from three-dimensional reticular structure during a chemical reaction, the residual part of an formerly complete SiO_4 tetrahedron will continuously induce new chemical structure defects; thus, the noise intensity level is strengthened.

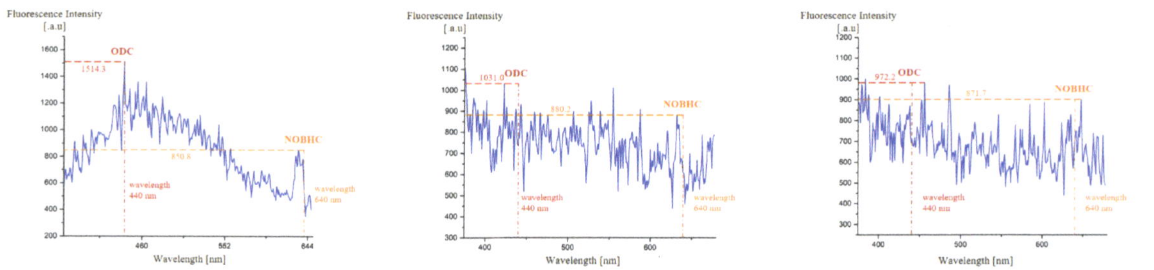

(**a**) Fluorescence spectrum of sample #0 (initial sample)

(**b**) Fluorescence spectrum of sample #3 (etching for 150 min)

(**c**) Fluorescence spectrum of sample #5 (etching for 420 min)

Figure 9. Fluorescence spectrum of the samples under different etching durations.

Upon further etching to 420 min, the fluorescence spectrum for #5's surface remains consistent with that of sample #3, which means that even excessive HF etching does not introduce any new characteristic peaks sourced from ODC. For NBOHC, the situation is quite different, as the intensity level indicates no noticeable change. Based on the results, HF etching technology exhibits a significant suppressive effect on ODC defects for fused silica-based optical elements.

3.4. LIDT Testing Results

The results of laser damage threshold testing on different HF etching depths are presented in Figure 10. After the polishing process, the initial LIDT is 7.22 J/cm^2. As the etching depth increases to 3.53 μm, the LIDT increases to 10.76 J/cm^2, which exhibits a notable upward trend. However, further increases in etching depth result in a significant decrease in LIDT. Specifically, at etching depths of 5.46 microns and 9.54 microns, the LIDT decreases by 13.75% and 19.61%, respectively, compared to the highest value ever achieved.

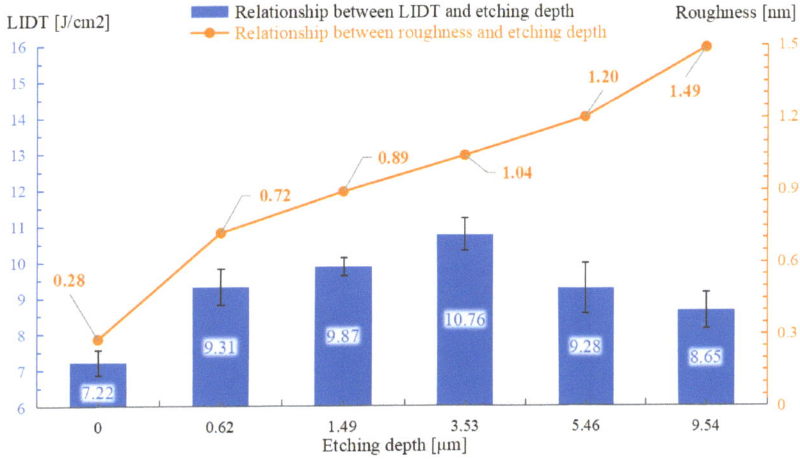

Figure 10. Relationship between LIDT, surface roughness and etching depth.

4. Discussion

Through measurements and LIDT tests in Section 4, the specific influence of hydrogen fluoride-based etching on the characteristics of a fused silica sample are revealed. Nonetheless, the underlying mechanism of phenomena observed during the tests merit further elucidation, and the enhancement of laser damage resistance performance also merits deeper exploration.

Over a duration of 420 min, the average removal rate is approximately 23 nm/min, compared to that of 30 to 170 nm/min in the grinding process and 2 to 8 nm/min in the polishing process [29]. Hydrogen fluoride-based etching is of high efficiency due to its feature of a global reaction effect with any exposed surface of the substrate.

The relationship between roughness and etching depth is also shown in Figure 11. Surface roughness decays rapidly first and then slows down, keeping a relatively stable decay rate, followed by a gradual slowing down, then maintains a relatively constant decay rate. Meanwhile, the etching rate exhibits a similar trend. This phenomenon can be attributed to the specific composition of the superficial hydrolysis layer (presented in Figure 12) on initial samples; the layer consists of an inhomogeneous and loosely packed mixture of micro fused silica fragments, silicic acid gel (hydrolysis relevant product, ≡Si—OH) and impurities originating from the deposition of hydrolysis during prior polishing and smoothing processes. The normal compression loads on a fused silica surface during these prior machining processes will enhance the chemical activity of the hydrolysis layer [30]. Additionally, some impurities do not react with HF. The combined effect of all these factors will certainly result in a rapid reaction rate between the hydrolysis layer and HF solution. Once the hydrolysis layer is removed, the reaction rate slows down.

Meanwhile, the etching product also deposits on an uneven reaction surface, presented as micron-scale white spots in Figure 5. Then, HF has to penetrate these gel-like depositions to react with the substrate, which leads to the further deterioration of surface roughness. It is also noteworthy that the etching rate within the hydrolysis layer exhibits a non-constant characteristic. The gradually decreasing roughness decay rate and etching rate observed in the figure suggest that the concentration of the hydrolysis layer diminishes with the increase in layer depth. As the etching process progresses, subsurface defects resulting from the prior processing stage are exposed, the surface becomes rougher and scratches transform from sharp contours to smoother arc-shaped contours. The increasingly rough micro morphology of the local surface gives rise to varying local etching rates. Additionally, the accumulation of a depositive etching product manifests as micron-scale white spots in Figure 5. On an uneven reaction surface, the homogeneity of the local reaction rate will be even worse, since HF has to penetrate these gel-like depositions to react with the substrate. This phenomenon consequently leads to a further deterioration in surface roughness.

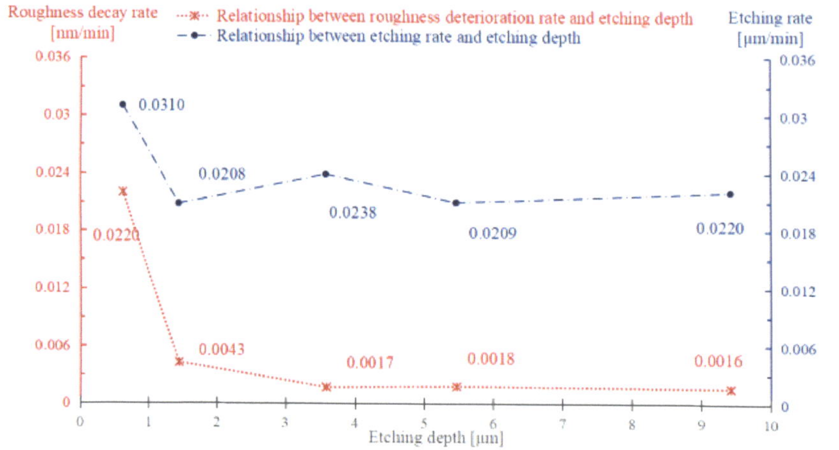

Figure 11. Relationship between roughness and etching depth.

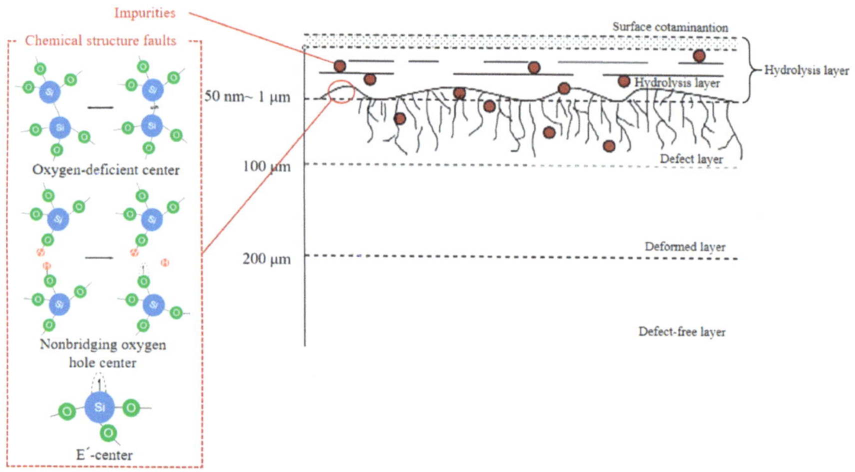

Figure 12. Specific composition of superficial hydrolysis layer.

Despite the application of megasonic-assisted etching, white mottled residual depositions can still be observed on sample #5 through SEM. These depositions are primarily composed of sodium, calcium and potassium salts, and of all the elemental constituents, silicon and oxygen elements originating from fused silica substrate dominate the elemental composition. Notably, a significant accumulation of impure elements exists within the residual depositions adhering to the surfaces, which is attributed to the adsorption of ions through a gel-like depositive reaction product during etching. The impure elements within these depositions may originate from two potential sources:

(1) From fused silica samples: in HF etching, the chemical network structure of silicane will be disassembled, causing the dissolution of the alkalis initially embedded in fused silica. For example, the contents of calcium, potassium and sodium elements in suprasil 300 produced by Heraeus that are used in this paper are 0.05 ppm, 0.01 ppm and 0.05 ppm, respectively [31];

(2) From deionized water: the alkali elements may also stem from deionized water in HF solution, as deionized water often contains trace amount of dissolved ionic impurities.

By analyzing the LIDT test results, it is evident that the appropriate selection of etching parameters can significantly enhance the LIDT. During first 150 min of etching (where etching depth reaches 3.59 μm), HF effectively reacts with the fused silica, removing fragment defects, chemical structure defects and impurities resulting from prior polishing and smoothing processes, leading to a certain improvement on LIDT. However, once the etching depth exceeds 3.59 μm (over 150 min), even with small alterations in surface roughness, a notable reduction in LIDT is observed. A phenomenon manifests where the negative impacts of reaction product deposits start to affect LIDT; additionally, excessive HF etching will also cause the severe deterioration of surface roughness. Both factors hinder the practical performance improvement of optical elements in high-power laser systems. In that sense, single HF etching alone is insufficient as a complete precursor treatment for fused silica optical elements to obtain the desired defect-less surface. For further LIDT enhancement, the implementation of additional processes such as IBF [32] and KOH etching [33] would be highly beneficial, particularly in mitigating residual deposits.

5. Conclusions

To explore the evolution of chemical structural defects and micron- to nano-sized secondary contaminative deposition during HF etching, the authors conducted a set of experimental research, and the relevant conclusions are as follows:

(1) During etching, the surface roughness deteriorates steadily; as the etching continues, white mottled gel-like depositions emerge. The roughness deterioration rate illustrates that the concentration of the hydrolysis layer diminishes with the increase in layer depth, and the depositions also add to further surface roughness deterioration;

(2) For chemical structure defects elimination, HF exhibits a significant suppressive effect on ODC defects for fused silica-based optical elements;

(3) The white mottled depositions consist of Na, K, Ca, Cl and C elements originating from either substrate or deionized water. The precisely shared shape between deposition and impure elements distributions illustrate that during etching, the impure ions are absorbed and deposited with the reaction product. For the thorough removal of impure elements, additional processes are suggested.

Author Contributions: Conceptualization, X.S.; methodology, X.S., S.Q. and X.P.; software, X.S.; validation, X.S., X.P. and Y.X.; formal analysis, X.S. and S.Q.; investigation, X.S.; resources, X.S.; data curation, X.S.; writing—original draft preparation, X.S.; writing—review and editing, X.S. and S.Q.; visualization, X.S.; supervision, F.S.; project administration, F.S.; funding acquisition, F.S. All authors have read and agreed to the published version of the manuscript.

Funding: This work is supported by the Strategic Priority Research Program of the Chinese Academy of Sciences (No. XDA25020317) and the Science and Technology Innovation Program of Hunan Province (2022RC1138, 2023JJ30079).

Institutional Review Board Statement: Not applicable.

Informed Consent Statement: Not applicable.

Data Availability Statement: The original contributions presented in the study are included in the article, further inquiries can be directed to the corresponding author.

Conflicts of Interest: The authors declare that there are no conflicts of interest.

References

1. Spaeth, M.L.; Manes, K.R.; Kalantar, D.H.; Miller, P.E.; Heebner, J.E.; Bliss, E.S.; Spec, D.R.; Parham, T.G.; Whitman, P.K.; Wegner, P.J.; et al. Description of the NIF laser. *Fusion Sci. Technol.* **2016**, *69*, 25–145. [CrossRef]
2. Moses, E.I.; Campbell, H.; Stolz, C.; Wuest, C.R. The national ignition facility: The world's largest optics and laser system, High-Power Lasers and Applications. *Int. Soc. Opt. Photonics* **2003**, *5001*, 413–418.
3. Besnard, D. Fusion with the mega joule laser. *J. Phys. Conf. Ser. IOP Publ.* **2008**, *112*, 012004. [CrossRef]

4. Merkle, L.D.; Koumvakalis, N.; Bass, M. Laser-induced bulk damage in SiO_2 at 1.064, 0.532, and 0.355 μm. *J. Appl. Phys.* **1984**, *55*, 772–775. [CrossRef]
5. Yong, J.; Yuan, X.; Zu, X. Research Progress on Subsurface Defects in Fused Silica. *Mater. Rep.* **2009**, *22*, 102–106. [CrossRef]
6. Miller, P.E.; Suratwala, T.I.; Bude, D. Laser damage precursors in fused silica. *Proc. SPIE-Int. Soc. Opt. Eng.* **2009**, *7504*, 318–331. [CrossRef]
7. Wang, H.; Wang, C.; Zhang, M.; Zheng, E.; Hou, J.; Chen, X. Investigation of subsurface damage density and morphology impact on the laser-induced damage threshold of fused silica. *Appl. Opt.* **2019**, *58*, 9839–9845. [CrossRef] [PubMed]
8. He, X.; Cai, C.; Wang, G.; Zhao, H.; Xie, L.; Huang, Y.; Yan, D.; Ma, P. Effect of pad elastic modulus on polishing-induced subsurface damages distribution and laser-induced damage performance of fused silica optics. *Opt. Express* **2019**, *27*, 265–281. [CrossRef] [PubMed]
9. Suratwala, T.I.; Miller, P.E.; Ehrmann, P.R.; Steele, R.A. Polishing slurry induced surface haze on phosphate laser glasses. *J. Non-Cryst. Solids* **2004**, *351*, 2091–2101. [CrossRef]
10. Suratwala, T.; Steele, W.; Wong, L.; Feit, M.D.; Miller, P.E.; Dylla-Spears, R.; Shen, N.; Desjardin, R. Chemistry and Formation of the Beilby Layer During Polishing of Fused Silica Glass. *J. Am. Ceram. Soc.* **2015**, *98*, 2395–2402. [CrossRef]
11. Pfiffer, M.; Longuet, J.L.; Labrugère, C.; Fargin, E.; Bousquet, B.; Dussauze, M.; Lambert, S.; Cormont, P.; Néauport, J. Characterization of the Polishing-Induced Contamination of Fused Silica Optics. *J. Am. Ceram. Soc.* **2017**, *100*, 96–107. [CrossRef]
12. Bude, J.; Miller, P.E.; Shen, N.; Suratwala, T.; Laurence, T.; Steele, W.; Baxamusa, S.; Wong, L.; Carr, W.; Cross, D.; et al. Silica laser damage mechanisms, precursors and their mitigation. *Proc. Laser-Induc. Damage Opt. Mater.* **2014**, *9237*, 108–120.
13. Bude, J.; Miller, P.; Baxamusa, S.; Shen, N.; Laurence, T.; Steele, W.; Suratwala, T.; Wong, L.; Carr, W.; Cross, D.; et al. High fluence laser damage precursors and their mitigation in fused silica. *Opt. Express* **2014**, *22*, 5839–5851. [CrossRef] [PubMed]
14. Ye, H.; Li, Y.; Zhang, Q.; Wang, W.; Yuan, Z.; Wang, J.; Xu, Q. Post-processing of fused silica and its effects on damage resistance to nanosecond pulsed UV lasers. *Appl. Opt.* **2016**, *55*, 3017–3025. [CrossRef] [PubMed]
15. Suratwala, T.I.; Miller, P.E.; Bude, J.D.; Steele, W.A.; Shen, N.; Monticelli, M.V.; Feit, M.D.; Laurence, T.A.; Norton, M.A.; Carr, C.W.; et al. HF-Based Etching Processes for Improving Laser Damage Resistance of Fused Silica Optical Surfaces. *J. Am. Ceram. Soc.* **2011**, *94*, 416–428. [CrossRef]
16. Wong, L.; Suratwala, T.; Feit, M.D.; Miller, P.E.; Steele, R. The effect of HF/NH4F etching on the morphology of surface fractures on fused silica. *J. Non-Cryst. Solids* **2009**, *355*, 797–810. [CrossRef]
17. Ye, X.; Huang, J.; Liu, H.; Geng, F.; Sun, L.; Jiang, X.; Wu, W.; Qiao, L.; Zu, X.; Zheng, W. Advanced Mitigation Process (AMP) for Improving Laser Damage Threshold of Fused Silica Optics. *Sci. Rep.* **2016**, *6*, 31111. [CrossRef] [PubMed]
18. Bude, J.; Carr, C.W.; Miller, P.E.; Parham, T.; Whitman, P.; Monticelli, M.; Raman, R.; Cross, D.; Welday, B.; Ravizza, F.; et al. Particle damage sources for fused silica optics and their mitigation on high energy laser systems. *Opt. Express* **2017**, *25*, 11414–11435. [CrossRef]
19. Zheng, Z.; Zu, X.; Jiang, X.; Xiang, X.; Huang, J.; Zhou, X.; Li, C.; Zheng, W.; Li, L. Effect of HF etching on the surface quality and laser-induced damage of fused silica. *Opt. Laser Technol.* **2012**, *44*, 1039–1042. [CrossRef]
20. Knotter, D.M. Etching mechanism of vitreous silicon dioxide in HFBased solutions. *J. Am. Ceram. Soc.* **2000**, *122*, 4345–4351.
21. Tian, Y. The Key Technology Research of High Precision and Low-Defect Fabrication for Silicon Reflectors. Doctor's Thesis, University of National Defense, Changsha, China, 2018.
22. Judge, S. A Study of the Dissolution of SiO2 in Acidic Fluoride Solutions. *J. Electrochem. Soc.* **1971**, *118*, 1772. [CrossRef]
23. Van Grieken, R.; Markowicz, A. *Handbook of X-ray Spectrometry*, 2nd ed.; Marcel Dekker, Inc.: New York, NY, USA, 2002; pp. 47–48.
24. Neauport, J.; Lamaignere, L.; Bercegol, H.; Pilon, F.; Birolleau, J.C. Polishing-induced contamination of fused silica optics and laser induced damage density at 351 nm. *Opt. Express* **2005**, *13*, 10163–10171. [CrossRef] [PubMed]
25. Kohli, R.; Mittal, K.L. *Methods for Removal of Particle Contaminants, in Developments in Surface Contamination and Cleaning*; Elsevier Inc.: Oxford, UK, 2011.
26. Chmel, A.; Eronko, S.B. Optical strength of glasses implanted with argon ions. *Glass Technol.* **1998**, *39*, 32–34.
27. Feit, M.D.; Rubenchik, A.M. Influence of subsurface cracks on laser-induced surface damage. In *Laser-Induced Damage in Optical Materials*; SPIE: Bellingham, WA, USA, 2004.
28. Salh, R. Silicon Nanocluster in Silicon Dioxide: Cathodoluminescence, Energy Dispersive X-Ray Analysis, Infrared Spectroscopy Studies. In *Crystalline Silicon Properties and Uses*; InTech: Atyrau, Republic of Kazakhstan, 2011.
29. Shu, Y. Study on etching process of fused silica with concentrated HF. *Optik* **2018**, *178*, 544–549. [CrossRef]
30. Jian, C.; Jinghe, W.; Peiyue, Z.; Lei, Z. Experimental study on HF etching of fused silica optical elements. *High Power Laser Part. Beams* **2017**, *29*, 7.
31. Quarzglas, H. *Quartz Glass for Optics Data and Properties*; Heraeus Quarzglas GmbH & Co. KG: Hanau, Germany, 2004.
32. Zhong, Y.; Shi, F.; Tian, Y.; Dai, Y.; Song, C.; Zhang, W.; Lin, Z. Detailed near-surface nanoscale damage precursor measurement and characterization of fused silica optics assisted by ion beam etching. *Opt. Express* **2019**, *27*, 10826–10838. [CrossRef]
33. Zhong, Y.; Dai, Y.; Tian, Y.; Shi, F. Effect on nanoscale damage precursors of fused silica with wet etching in KOH solutions. *Opt. Mater. Express* **2021**, *11*, 884–894. [CrossRef]

photonics MDPI

Article

The Application of Optical Sensors with Built-in Anchor-like Cavities in the Detection of Hemoglobin Concentration

Wen Jiang [1,2,3], Shubin Yan [2,3,*], Yiru Su [2], Chong Wang [2,3], Taiquan Wu [2,3], Yang Cui [2,3], Chuanhui Zhu [2,3], Yi Zhang [2,3], Xiangyang Mu [1] and Guowang Gao [1]

1 School of Electrical Engineering, Xi'an Shiyou University, Xi'an 710065, China; 22212030531@stumail.xsyu.edu.cn (W.J.); muyou98@xsyu.edu.cn (X.M.); wwgao@xsyu.edu.cn (G.G.)
2 School of Electrical Engineering, Zhejiang University of Water Resources and Electric Power, Hangzhou 310018, China; 2022b03001@stu.zjweu.edu.cn (Y.S.); wangchong@zjweu.edu.cn (C.W.); wutq@zjweu.edu.cn (T.W.); cuiy@zjweu.edu.cn (Y.C.); zhchh@zjweu.edu.cn (C.Z.); zhangyi@zjweu.edu.cn (Y.Z.)
3 Zhejiang-Belarus Joint Laboratory of Intelligent Equipment and System for Water Conservancy and Hydropower Safety Monitoring, Hangzhou 310018, China
* Correspondence: yanshb@zjweu.edu.cn; Tel.: +86-186-3611-2255

Abstract: This paper introduces a refractive index sensor based on Fano resonance, utilizing a metal–insulator–metal (MIM) waveguide structure with an Anchor-like cavity. This study utilizes the finite element method (FEM) for analyzing the propagation characteristics of the structure. The evaluation concentrated on assessing how the refractive index and the structure's geometric parameters affect its sensing characteristics. The designed structure demonstrates optimum performance, achieving a maximum sensitivity of 2440 nm/RIU and an FOM of 63. Given its high sensitivity, this nanoscale refractive index sensor is ideal for detecting hemoglobin concentrations in blood, and the sensor's sensitivity is 0.6 nm·g/L, aiding in clinical prevention and treatment.

Keywords: refractive index sensor; MIM; Fano resonance; concentration detection

Citation: Jiang, W.; Yan, S.; Su, Y.; Wang, C.; Wu, T.; Cui, Y.; Zhu, C.; Zhang, Y.; Mu, X.; Gao, G. The Application of Optical Sensors with Built-in Anchor-like Cavities in the Detection of Hemoglobin Concentration. *Photonics* **2024**, *11*, 402. https://doi.org/10.3390/photonics11050402

Received: 26 March 2024
Revised: 17 April 2024
Accepted: 23 April 2024
Published: 26 April 2024

1. Introduction

Common optical refractive index sensors include optical fiber optic sensors [1], micro-ring resonator sensors [2], photonic crystal sensors [3], and surface plasmon resonance (SPR) sensors. Photonic devices excel over electronic devices in operating bandwidth; however, their miniaturization and integration are constrained by the optical diffraction limit, which in turn limits the development of optical refractive index sensor sizes [4]. However, surface plasmon polaritons (SPPs) are capable of exceeding the diffraction limit and can precisely manipulate light on the nanoscale [5]. Consequently, an increasing number of scholars are focusing on SPPs.

Surface plasmon polaritons (SPPs) are electromagnetic waves are formed by the intricate interplay between the free electrons of the metal and incident photons [6], enabling the over-coming of diffraction limits and allowing for the precise manipulation of light at the nanoscale [7]. Therefore, SPPs have immense promise for applications in integrated optical circuits and subwavelength photonic devices, ushering in a new era of transformative possibilities in nanophotonics. This electromagnetic wave can propagate across the interfaces and confine the field energy to a small region around the interfaces, resulting in a rapid decrease in intensity in the direction perpendicular to the interfaces [8]. SPP (surface plasmon polariton) waveguides provide significant advantages in photonics, chiefly through their ability to overcome the physical limitations of the diffraction limit that constrain conventional dielectric waveguides. This property enables the control and manipulation of light at sub-wavelength dimensions, thus facilitating the development of ultra-compact and highly efficient photonic devices. Due to their unmatched nanoscale light manipulation capabilities, SPPs are at the forefront of next-generation photonic technology, serving as the cornerstone for developing highly integrated chip-scale photonic devices. In addition, the potential for the seamless integration of SPPs with existing semiconductor platforms heralds new possibilities for advancements in optical computing and communications, positioning them as key innovations in optoelectronics and nanotechnology [9,10].

Surface plasmon excitations markedly enhance the electromagnetic field at the metal–dielectric interface, a phenomenon of critical importance for optical sensing and information processing. Firstly, the composite structure of micro- and nanomaterials reacts to these excitations through optical nonlinear effects, rendering the system highly sensitive to the local structure of the metal surface and to refractive index changes in the medium. This sensitivity not only aids in detecting changes in the surface morphology and refractive index but also enables the fabrication of devices such as surface plasmon resonance sensors [11–13], all-optical switches [14], and optical modulators [15]. These devices find widespread use in various fields, including biomolecular detection, chemical analysis, environmental monitoring, and optical communication, offering key technologies for optical sensing, super-resolution imaging, and information processing and transmission. These functions are crucial in the fields of nanophotonics, optical sensing, photonic chips, and surface-enhanced spectroscopy and serve as the foundation for developing novel nanophotonic devices, underscoring their significant application value in modern optoelectronics [16].

Surface plasmon polaritons (SPPs) can not only be generated at metal–dielectric interfaces but can also be excited through various waveguide structures, including metal– dielectric–metal (MIM) [17], dielectric–metal–dielectric [18], and dielectric–dielectric–metal [19] waveguides. In particular, MIM waveguides have attracted attention due to their excellent SPP confinement, low ohmic losses, and long transmission distances [20]. These waveguide systems can be designed with various structures to tune broadband and narrowband modes, achieving precise control of Fano resonances. Traditional MIM waveguide structures typically include straight waveguides and circular resonators to produce Fano resonances, but this method has certain performance limitations. Research has found that introducing defects into waveguides or resonators can effectively induce Fano resonances by disrupting propagation symmetry, enhancing performance. For example, by introducing a conical defect in a circular resonator, Rahmatiyar developed a new design for a nanoscale refractive index sensor, combining a ring resonator with an MIM waveguide [21]. Further research explored plasmonic systems with defective silver nanoribbons, discovering that these systems can produce two Fano resonances. In these systems, quadrupole modes, which are not directly excitable on complete nanoribbons, become feasible through the introduction of defects. These defective silver nanoribbons support super radiant dipole and quadrupole modes, thereby producing two asymmetric Fano tilts in the transmission spectrum [22]. Fei Hu's work further demonstrated how adjusting the geometric parameters of silver nanorod defects, such as radius and the height or base length of an isosceles triangle resonator, can independently adjust the shape of the Fano resonances. These studies highlight the importance and potential of structural defects in designing nanophotonic devices with specific optical properties [23].

Additionally, Yang and colleagues proposed a metal–insulator–metal (MIM) waveguide structure that includes a cross-shaped cavity and baffles. This structure achieved a sensitivity of 1075 nm/RIU [24]. Wang and colleagues developed a structure that couples a metal–insulator–metal (MIM) waveguide with a semicircular cavity and two silver baffles, reaching a maximum sensitivity of 2200 nm/RIU [25]. Pang and his team designed a novel structure using an MIM waveguide and an oblique rectangular cavity [26]. Zhou presents a waveguide design incorporating two short stub resonators and a ring resonator. It achieves a peak sensitivity of 1650 nm/RIU and a figure of merit (FOM) of 117.8 [27]. Tathfif reported a sensor composed of a straight waveguide, a groove, and a square ring resonator (SRR), with a maximum sensitivity of 1074.88 nm/RIU and a quality factor of 32.4 [28]. Tang and colleagues proposed a sensor with a sensitivity of 1125 nm/RIU [29]. Chen designed a structure with a semi-elliptical and rectangular ring resonator, achieving an optimal sensitivity of 1384 nm/RIU and a quality factor of 28.4 [30]. This study proposes and investigates a new structure of a nanoscale refractive index sensor utilizing MIM waveguides, composed of an MIM bus waveguide and an internal Anchor-like resonator. It features relatively high sensitivity and quality factors, and the structure is simple, with a relatively small cavity area and low loss. Subsequently, we modified various structural parameters and the refractive index, observed changes in the transmission spectrum, and analyzed the impact on system sensing performance using the finite element method (FEM). Finally, it was applied in detecting hemoglobin concentrations in blood, and the sensor's sensitivity was 0.6 nm·g/L, providing a theoretical basis for the field of nanomedicine.

2. Materials and Methods

This study utilized the finite element method to examine the relationship between the Fano resonance and changes in refractive index. We observed how the resonance frequency and intensity changed with the refractive index and analyzed the magnetic field distribution. By adjusting the structure's geometrical parameters, including size and shape, we investigated their effects on Fano resonance. Given that the designed structure's thickness is negligible compared to the spp surface depth, simulating the 3D structure requires more demanding hardware configuration and meshing. However, as the magnetic field characteristics of the 3D structure are largely similar to those of the 2D structure, we opted for the 2D model for simplicity in calculations [31]. Figure 1 illustrates a 2D schematic of the symmetrical sensor structure along the centerline. As illustrated in Figure 1, the system was modeled in COMSOL, with the Anchor-like structure comprising a circle, a rectangular waveguide, and a segment of a circular arc. Definitions are provided in the figure: a represents the radius of the upper circle, O_1 denotes the center of the upper circle, d is the width of the rectangular bar, r is the radius of the inner circle of the following arc, R represents the radius of the outer circle where R-r equals d, and the arc's angle is 2φ. Here, g signifies the gap between the inverted rectangular waveguide and the Anchor-like, while w denotes the width of the waveguide. Additionally, l represents the distance between the lowest point of the upper circle and the lowest point of the inner circle of the arc, and the height of each individual rectangular bar is denoted by h.

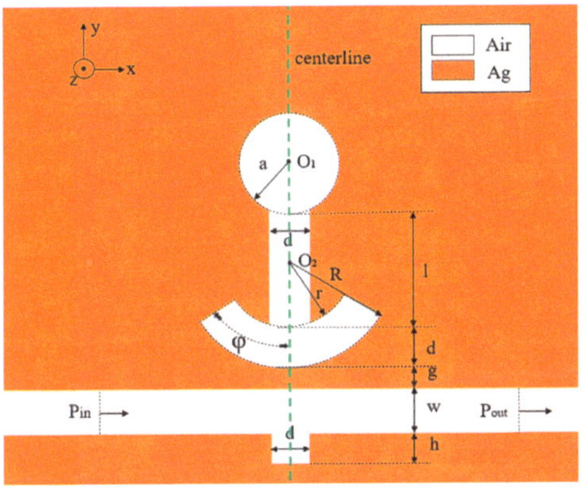

Figure 1. Two-dimensional plan of the overall structure.

Electromagnetic waves propagating along the metal–dielectric interface can be categorized into the two following basic modes according to the orientation of the electric and magnetic fields relative to the direction of propagation: transverse electric (TE) and transverse magnetic (TM) modes. In the TE mode, the electric field vector is completely transverse and perpendicular to both the direction of propagation and the metal–dielectric interface. Given that the electric field lacks a component parallel to the metal surface, it fails to effectively interact with the metal's free electrons, thereby failing to excite surface plasmon polaritons (SPPs). Conversely, in the TM mode, the magnetic field vector is transverse to the direction of propagation, and the electric field includes a component along the interface. This electric field component can interact with the metal surface's free electrons, inducing their collective oscillations and thereby exciting SPPs [32,33]. Therefore, our analysis concentrates on the TM mode, specifically investigating the relationship between surface plasmon polaritons (SPPs) and dispersion within this context [34]:

$$\tanh(k\omega) = -\frac{2k\alpha_c}{k^2 + p^2\alpha_c} \tag{1}$$

where k is the wave vector in the waveguide, $P = \varepsilon_{in}/\varepsilon_m \alpha_c = \left[k_0{}^2 \times (\varepsilon_{in} - \varepsilon_m) + k\right]^{1/2}$, ε_{in} and ε_m are the dielectric constants of the dielectric and the metal, respectively, and p is the ratio of the permittivity ε_{in} to the metal's permittivity, and $k_0 = 2\pi/\lambda_0$ is the free space wave vector.

A key design factor for metal–insulator–metal (MIM) waveguides involve the width of the dielectric layer (d). This dimension crucially affects the supported mode types and their levels of confinement. By matching the dielectric layer's width with the waveguide's (w), specifically at 50 nanometers, the design optimizes the transverse magnetic (TM) mode propagation conditions. It also aims to reduce or eliminate the transverse electric (TE) mode.

In Figure 1, the yellow area represents silver, while the white part represents air. Silver was chosen as the filler metal for the following reasons: among all metallic elements, silver (Ag) has the smallest damping constant and is the best choice for performance at optical frequencies [35], and silver is characterized by a high field strength and low system power; the properties of nanostructured silver make it the most suitable for next-generation plasmonic. We can prepare silver by achieving full deposition on the substrate using the chemical vapor deposition (CVD) and obtain the desired structure through electron beam etching [36]. Considering that the dielectric constant of metallic materials varies with incident frequency, the dielectric constant of silver is modeled using the Drude–Debye approach (assuming a dielectric constant of 1 for air):

$$\varepsilon(\omega) = \varepsilon_\infty + \frac{\varepsilon_s - \varepsilon_\infty}{1 + i\omega\tau} + \frac{\sigma}{i\omega\varepsilon_0} \tag{2}$$

The parameter values are defined as follows: infinite frequency relative permittivity ε_∞ equals 3.8344, static dielectric strength ε_s is -9530.5, relaxation time τ is set at 7.4×10^{-15} s, conductivity σ amounts to 1.1486×10^7 s/m, and ω represents the angular frequency of light at 1.38×10^{16} rad/s.

This study will evaluate the sensor's performance using three parameters: FWHM, sensitivity (S), and figure of merit (FOM). Sensitivity is crucial for assessing sensor performance. Fano resonance-based MIM waveguide structures detect changes in desired physical quantities by observing variations in wavelength position. In general, sensitivity is defined as follows [37]:

$$S = \frac{\Delta\lambda}{\Delta n} \tag{3}$$

Here, $\Delta\lambda$ represents changes in refractive index, while Δn denotes changes in wavelength. Fano resonance manifests a distinctive feature in the transmission spectrum; the sharper the curve, the better the performance. We employ the full width at half maximum (FWHM) to characterize the spectrum's sharpness, specifically, the width of the spectrum at half the height of the resonance peak, which is expressed as follows [38]:

$$\text{FWHM} = \lambda_1 - \lambda_2 \tag{4}$$

Here, Transmittance at λ_1 and λ_2 is $(Tmax + Tmin)/2$. $Tmax$ and $Tmin$ represent the maximum and minimum transmittance in the transmission spectrum. The figure of merit

(FOM) is a crucial parameter for assessing sensors based on MIM waveguide structures and is defined as follows [39]:

$$FOM = S/FWHM. \tag{5}$$

3. Simulation Results and Analysis

We employ COMSOL Multiphysics 5.4a software and a 2D modeling and analysis approach to investigate the frequency domain behavior of electromagnetic waves within the RF module. We focus on frequency domain analysis and boundary mode analysis as the subjects of our study. The 2D model is analyzed through simulation, and the geometric model of the designed sensor system is constructed. We establish the boundary absorption condition as a perfectly matched layer and utilize a free-profile triangular mesh along with hyper-refinement for mesh refinement in the waveguide and Anchor-like structure regions. Mesh refinement for other regions is performed using regularization. This strategy minimizes superfluous computations and upholds calculation precision. Initially, we begin by defining the various parameters of the sensor, the r = 140, R = 190, a = 75, d = h = w = 50 nm, l = 195 nm, g = 10 nm, and $\varphi = 60°$. Figure 2 displays the spectrograms of a single rectangular stub, all systems, and a single Anchor-like structure. These are represented by black, red, and blue curves, respectively. The wavelength range spans from 1700 nm to 2500 nm with 1 nm increments.

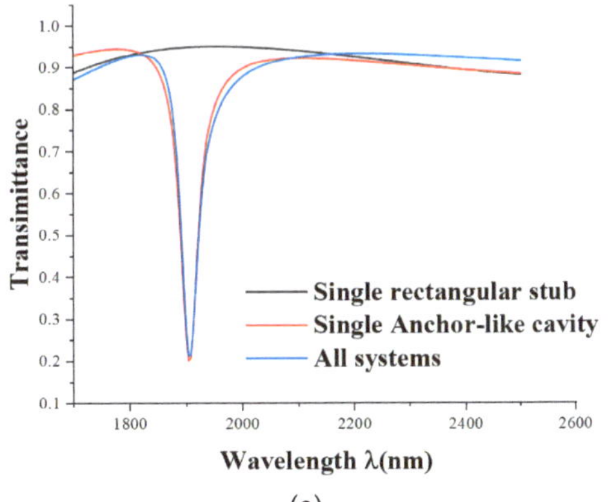

(a)

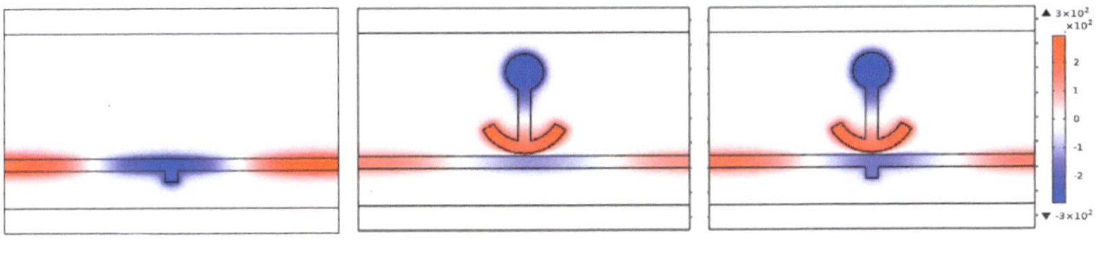

(b)

Figure 2. (**a**) Transmission spectra of single rectangular stub, all systems, and a single Anchor-like structure. (**b**) Magnetic field distribution for different structures.

Spectral analysis reveals that the single anchor structure demonstrates low transmittance and a relatively symmetric shape, indicating a narrow-band mode directly excited by the incident light. Conversely, the transmission spectral profile of the single reverse waveguide structure displays a positive slope with high transmittance, implying a continuous broadband mode. Similarly, the transmission spectrum of a single rectangular stub, with its slightly upward-sloping curve and high transmittance for most optical signal frequencies, exhibits continuous broadband transmission characteristics, this phenomenon is classified as a continuous broadband mode. The structure reveals an asymmetric curve, signifying the clear presence of a Fano resonance. This phenomenon arises from the interaction between the continuous broadband states generated by the rectangular waveguide and the discrete narrowband states associated with each anchor-like structure, resulting in a significant Fano resonance. This phenomenon arises from the interaction between the anchor-like structures. Additionally, we investigated the effect of refractive index variation on the sensing performance of the designed structure.

Refractive indices ranging from n = 1.00 to n = 1.05 RIUs were simulated with constant structural parameters, as illustrated in Figure 3a. With an increase in refractive index, the transmission spectrum exhibits an almost equidistant red shift. However, the change in refractive index does not alter the waveform of the transmission spectrum; it only shifts its position. Figure 3b demonstrates a linear correlation between the wavelength and the refractive index, indicating the potential use of the proposed structure as a refractive index sensor. Through linear regression analysis to calculate the slope, the sensitivity of this sensor is determined to be 2440 nm/RIU, and the FOM is 63.

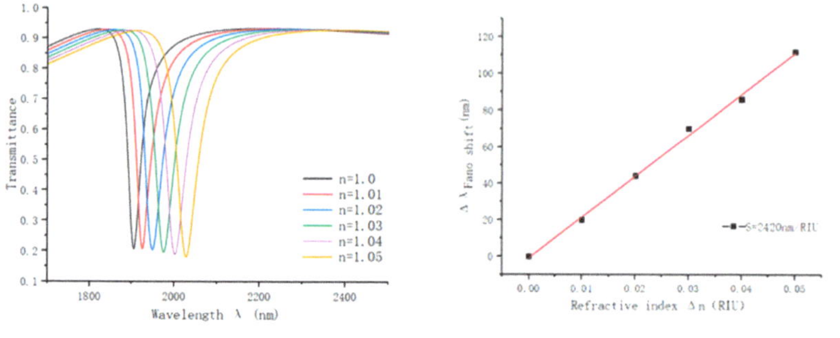

(**a**) (**b**)

Figure 3. (**a**) Transmittance spectrum at different refractive indices and (**b**) sensitivity fitting curve.

We discovered that the refractive index is only one factor influencing the Fano resonance, while the specific parameters of the structure can also significantly affect it as well. To delve deeper into the impact of these specific parameters, we initially adjusted the radius size, denoted as 'a', while keeping the center of circle O_1 fixed. The radius range was set from 75 to 95 nm in steps of 5 nm, while all other structural parameters remained constant. The transmission spectra and sensitivity curves are illustrated in Figure 4. Subsequently, we explored the influence of another parameter, 'l', on the sensor's propagation performance. Maintaining the distance between the lower arc and the waveguide constant, we shifted the upper circle relatively upward. The parameter 'l' varied from 195 nm to 235 nm with a step size of 10 nm, and the other parameters remained unchanged. Figure 5 shows the simulation results.

The results above indicate a continuous redshift in the transmission spectrum as the radii a and l increase, accompanied by minor fluctuations in transmittance and FWHM. Additionally, we conducted linear fitting on resonance curves for various radii. As depicted in Figure 4b, sensor sensitivity increases with larger radii a and l. Notably, the impact of a and l on sensor sensitivity is essentially identical. The maximum sensitivities are 2840 and 2980, respectively, for a = 95 and l = 195 nm.

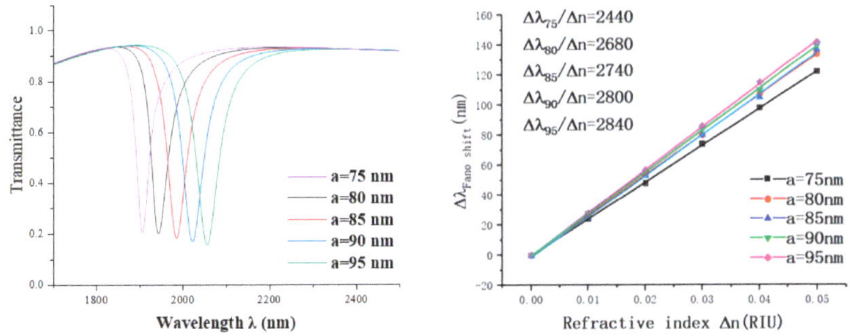

(**a**) (**b**)

Figure 4. (**a**) The transmittance spectra for radii ranging from 75 to 95 nm. (**b**) The sensitivity fitting curves for different radii a.

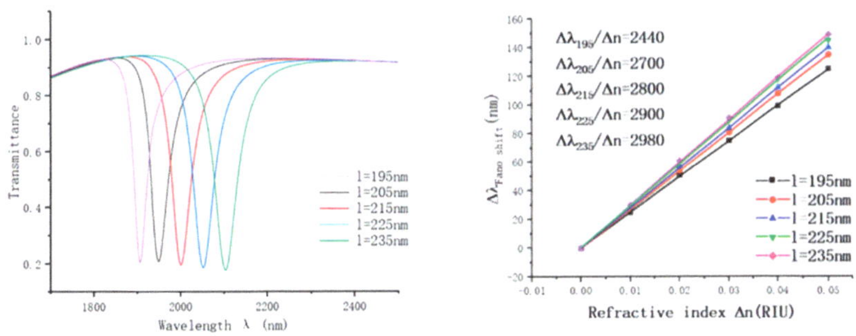

(**a**) (**b**)

Figure 5. (**a**) The transmittance spectra for l ranging from195 to 235 nm. (**b**) The sensitivity fitting curves for different l.

To delve deeper into the factors impacting sensor accuracy and to analyze the impact of the coupling gap on sensor performance, we vary the gap between the system and the anchor-like structure. We incrementally increased the coupling gap from 10 nm to 50 nm. As depicted in Figure 6a, the curve noticeably blueshifts, and in Figure 6c, we can see that FWHM significantly narrows with an increasing g. Additionally, Fano resonance curves are presented in Figure 6a. The increasing transmittance of the Fano resonance curve indicates a notable deterioration in coupling performance, leading to a sharp decline in sensor performance. The sensitivity fitted line for various coupling gaps is illustrated in Figure 6b. The most substantial sensitivity decrease occurs as g increases from 10 to 20. When g exceeds 10, the rate of change in sensor sensitivity decreases. However, when g is less than or equal to zero, implying direct contact between the two structures, the coupling effect becomes exceedingly strong, potentially inducing highly nonlinear behavior. Thus, g = 10 nm emerges as the optimal parameter for this design, corresponding to a sensitivity of 2440 nm/RIU and an FOM of 63.

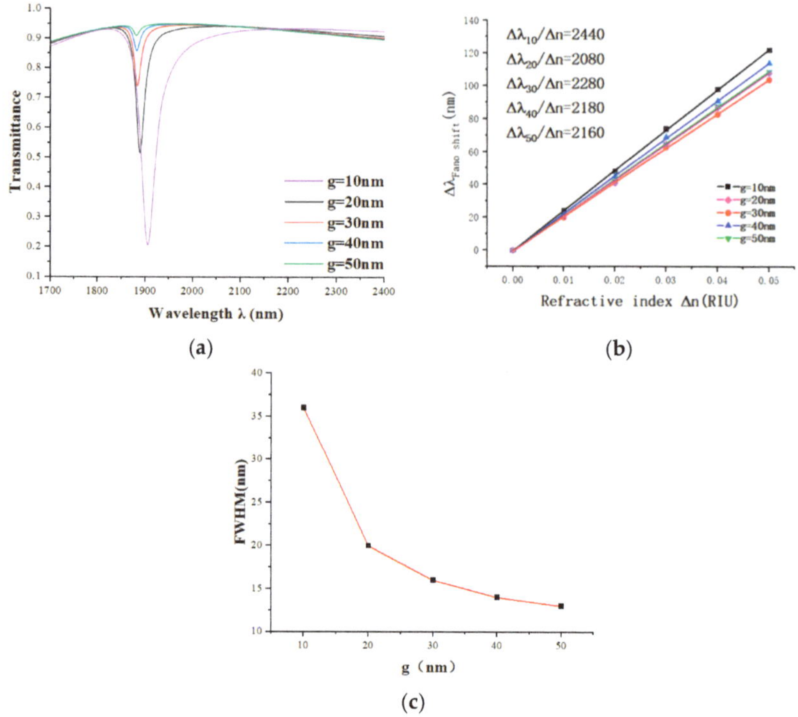

Figure 6. (**a**) The transmittance spectra for g ranging from10 to 50 nm, (**b**) the sensitivity fitting curves for different g, and (**c**) the trend of FWHM change.

To assess the impact of changing the height of the reverse rectangular baffle on the sensor's transmission performance, we incrementally raise it from 45 nm to 85 nm in 10 nm increments while keeping other parameters constant. The simulation results are depicted in Figure 7.

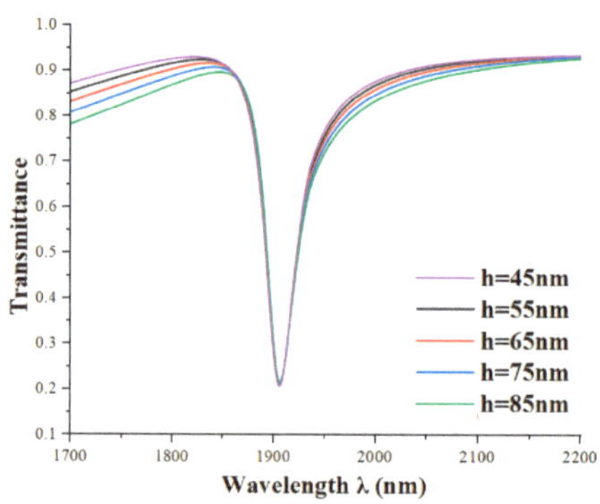

Figure 7. The transmittance spectra for h ranging from 45 to 85 nm.

Regardless of the variation in the height of the inverse rectangle, the position and transmittance of the Fano resonance remain largely stable. Furthermore, the Fano resonance curve exhibits distinct asymmetric features that become more pronounced by increasing the baffle height of the inverse rectangle. Finally, we explore the impact of the arc angle on the sensor's transmission characteristics, maintaining all other parameters constant while incrementally increasing the arc angle φ from 60° to 80° and 2φ from 120° to 160°.

As shown in Figure 8a, as the angle 2φ increases from 120° to 160°, a redshift occurs in the transmission spectrum, signifying an increase in the wavelength of transmitted light. At the same time, as shown in Figure 8b, the sensor's sensitivity undergoes a notable improvement. This indicates that within this angle range, enhancing the angle directly improves the sensor's performance. However, upon exceeding 120°, the rate of improvement starts to decline, showing that beyond a specific threshold angle, further angle increases result in a gradual reduction in the enhancement of sensor sensitivity.

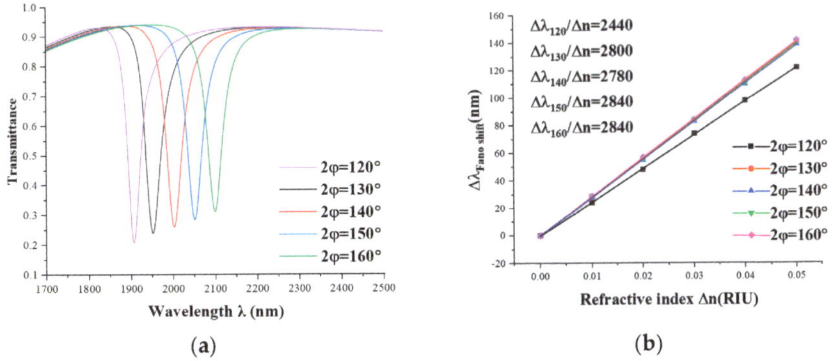

Figure 8. (**a**) The transmittance spectra for 2φ ranging from 120° to 160°, (**b**) the sensitivity fitting curves for different 2φ.

Subsequently, we explored the effects of adjusting the angle 2φ to 360° and setting l to 55 to form a ring. This was compared to merely adjusting l to 55. According to Figure 9a, adjusting l alone results in a smooth curve and exhibits high transmission. When l is set to 55, forming a ring, and compared to only adjusting l to 55, the figure shows that adjusting l alone results in a transmission spectrum that becomes a smooth curve with high transmittance. This study further reveals that decreasing the length of l reduces the sensor's sensitivity. However, with l at 55 and arc 2φ at 360°, forming a circle that encompasses the inner circle, transmittance decreases. Nonetheless, the arc's completeness partially compensates for the reduction in transmittance. From the figure, we can also see that the magnetic field distributions of the two different structures are opposite. From Figure 9b,c, we can also see that the magnetic field resonances of the two different structures are opposite. This suffices to demonstrate that the arc's completeness is a significant factor influencing transmittance at a wavelength of 2082 nm.

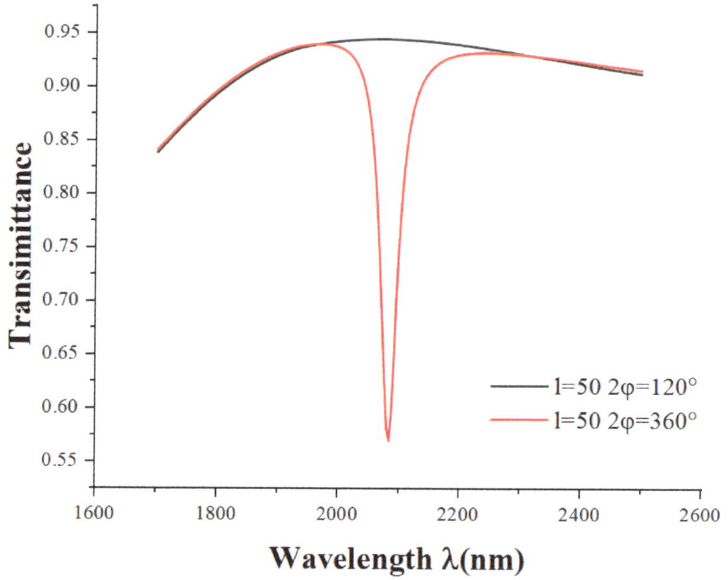

(a)

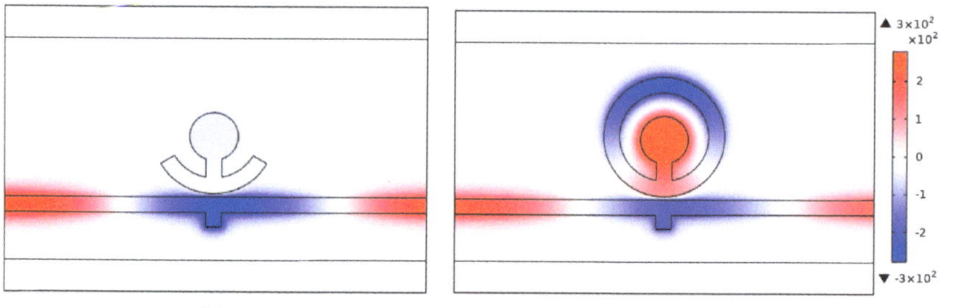

(b) (c)

Figure 9. (**a**) The transmission spectra for l being the same, with 2φ values of 120°and 160°. (**b**) The distribution of the magnetic field at a wavelength of 2082 nm when 2φ is 120°. (**c**) The distribution of the magnetic field at a wavelength of 2082 nm when 2φ is 160°.

4. The Application of this Structure in Hemoglobin Concentration

Red blood cells (RBCs) are often crucial indicators of related disorders. Erythrocytes, major blood cells, carry oxygen and carbon dioxide. In clinical practice, they serve as key indicators in routine blood tests, reflecting the body's RBC production and aiding in diagnosing blood disorders. Hemoglobin (Hb), a special protein in RBCs, transports oxygen. Hemoglobin reference values differ by gender and age: men: 120–165 g/L, women: 110–150 g/L, newborns: 180–190 g/L, and children: 120–140 g/L (12–14 g/dL). Hemoglobin levels in newborns decrease with age, approaching adult levels. Slightly lower hemoglobin is normal in the elderly, during menstruation, or in mid-to-late pregnancy. Chemotherapy, HIV treatment, and conditions such as iron deficiency anemia, aplastic anemia, or thalassemia can reduce red blood cell production or increase loss or destruction, leading to low hemoglobin. Pathological factors such as severe vomiting, diarrhea, extensive burns, and conditions like chronic hypoadrenocorticism, uremia, or renal and hepatic cancer can result in high hemoglobin levels. Monitoring hemoglobin concentration allows for the immediate diagnosis of blood diseases and timely treatment, holding great clinical significance.

The relationship between Hb concentration and RI is defined as follows [40]:

$$n = 1.38 + \frac{\mathrm{H} - \mathrm{H_{normal}}}{5766.5} \tag{6}$$

where $\mathrm{H_{normal}}$ is 140 g/L, H represents the concentration to be measured, the test range spans from 0 to 300 g/L, and the step size is set at 60 g/L. The detection process entails injecting blood with varied hemoglobin concentrations into the sensor chamber, altering the medium's refractive index. This results in spectral displacement. Sensor sensitivity is determined by the following equation:

$$\mathrm{S_H} = \frac{\Delta\lambda}{\Delta\mathrm{H}} \tag{7}$$

where $\Delta\lambda$ represents the spectrum's shift and ΔH indicates changes in the concentration of blood red cells. Output spectra at various concentrations are depicted in Figure 10a. The resonance trough's wavelength changes in proportion to Hb concentration increases, with resonance wavelength spacing being 36 nm. The sensor sensitivity fitting is shown in Figure 10b. The sensor sensitivity's linear fit correlation coefficient exceeds 0.99, the sensor's sensitivity is 0.6 nm·g/L, and the commercial spectral analyzer's minimum resolution is 0.001 nm. The sensor provides more accurate concentration measurements than traditional spectroscopic or colorimeter methods. Compared with demand-labeled assay techniques (e.g., enzyme-linked immunosorbent assay—ELISA), a label-free assay is provided with fewer preprocessing steps [41], demonstrating significant potential for advancing high-precision nanoscale hemoglobin concentration monitoring. However, surface plasma sensors are sensitive to ambient temperatures. Prolonged use in a clinical setting may place higher demands on the device's durability and stability. Ensuring that the device maintains high performance and accuracy over repeated use may be a substantial obstacle for applications requiring large-scale screening, such as routine hemoglobin testing [42].

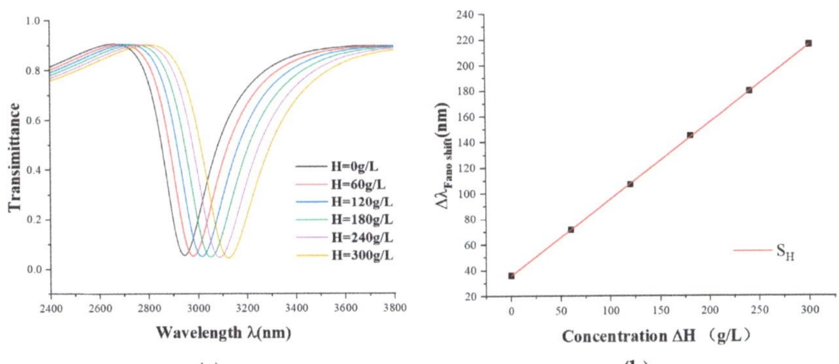

Figure 10. (**a**) The transmittance spectra for H ranging from 0 to 300 g/L. (**b**) The sensitivity fitting curve of hemoglobin concentration changes.

5. Conclusions

This study introduces a nanoscale refractive index sensor that leverages the unique properties of optical Fano resonance and plasma waveguide coupling. The sensor design combines a MIM waveguide structure with an anchor-like configuration, maximizing the benefits of optical Fano resonance and plasma waveguide coupling. Through finite element analysis, the propagation properties of the composite structure were explored, leading to the development of a novel nanoscale refractive index sensor. Subsequently, the anchor-like structure's parameters were adjusted to examine their impact on sensor performance, particularly in relation to spectrum transmittance and refractive index changes. By increasing radius a and length l, it is possible to alter the Fano resonance wavelength, thereby enhancing the sensor's performance. Furthermore, enlarging the coupling gap induces a blue shift, which further refines performance. Consequently, this also boosts the Fano resonance curve's transmittance. Modifying the rectangular baffle's height alters the Fano resonance curve's shape without affecting the trough's wavelength. The arc angle φ significantly influences the transmittance. Adjusting this structure's variables may be essential for specific applications to identify the sensor's optimal manufacturing parameters. Additionally, the sensor can detect blood's red blood cell concentration, and the sensor's sensitivity is 0.6 nm·g/L, which can provide some basis for clinical medicine. Moreover, the varying concentration of red blood cells in different blood types enables the differentiation between them, which is foundational for nanosensor research in biomedical and clinical medicine.

Author Contributions: Conceptualization, S.Y. and W.J.; methodology, W.J.; software, Y.C.; validation, S.Y., Y.S. and C.Z.; formal analysis, W.J. and Y.Z.; investigation, W.J.; data curation, W.J. and T.W.; writing—original draft preparation, W.J.; writing—review and editing, Y.S. and C.W.; visualization, G.G. and X.M.; supervision, S.Y.; project administration and funding acquisition, S.Y. All authors have read and agreed to the published version of the manuscript.

Funding: The author(s) declare financial support was received for the research, authorship, and/or publication of this article. This work was supported in part by the National Natural Science Foundation of China under Grant No. 62374148, in part by the Zhejiang Provincial Natural Science Foundation of China under Grant No. LD21F050001, the Key Research Project by Department of Water Resources of Zhejiang Province under Grant No. RA2101, the Key Research and Development Project of Zhejiang Province under Grant No. 2021C03019, and the Funds for Special Projects of the Central Government in Guidance of Local Science and Technology Development under Grant No. YDZJSX20231A031.

Institutional Review Board Statement: Not applicable.

Informed Consent Statement: Not applicable.

Data Availability Statement: The data provided in this study are available upon request from the corresponding author.

Acknowledgments: The authors are grateful to their other colleagues in the laboratory for their understanding and help. They also thank the affiliates for providing the research platform and the sponsors for financial support.

Conflicts of Interest: The authors declare no conflicts of interest.

References

1. Zhao, Y.; Shen, J.-C.; Liu, Q.; Zhu, C.-L. Optical fiber sensor based on helical Fibers: A review. *Measurement* **2022**, *188*, 110400. [CrossRef]
2. Hu, C.; Shi, Y.; Zhou, T.; Xu, C.; Zhu, A. A small size on-chip temperature sensor based on a microring resonator. *Silicon* **2021**, *14*, 5847–5854. [CrossRef]
3. Hu, Y.; Tian, Z.; Ma, D.; Qi, C.; Yang, D.; Huang, S. Smart colloidal photonic crystal sensors. *Adv. Colloid Interface Sci.* **2024**, *324*, 103089. [CrossRef]
4. Zhu, J.; Wu, C. Optical refractive index sensor with Fano resonance based on original MIM waveguide structure. *Results Phys.* **2021**, *21*, 103858. [CrossRef]
5. Chen, J.-H.; Wang, L.; Feng, Y.-T.; Chen, J.-X.; Zeng, Y.-P. A nano-refractive index sensor based on a MIM waveguide with a semicircular ring rectangular resonator. *Micro Nanostructures* **2024**, *186*, 207730. [CrossRef]
6. Shan, H.; Zu, S.; Fang, Z. Research Progress in Ultrafast Dynamics of Surface Plasmon Hot Electrons. *Laser Optoelectron. Prog.* **2017**, *54*, 30002. [CrossRef]
7. Zhang, J.; Zhang, L.; Xu, W. Surface plasmon polaritons: Physics and applications. *J. Phys. D Appl. Phys.* **2012**, *45*, 113001. [CrossRef]
8. Li, T.; Yan, S.; Liu, P.; Zhang, X.; Zhang, Y.; Shen, L.; Ren, Y.; Hua, E. A nanoscale structure based on an MIM waveguide coupled with a Q resonator for monitoring trace element concentration in the human body. *Micromachines* **2021**, *12*, 1384. [CrossRef] [PubMed]
9. Schuller, J.A.; Barnard, E.S.; Cai, W.; Jun, Y.C.; White, J.S.; Brongersma, M.L. Plasmonics for extreme light concentration and manipulation. *Nat. Mater.* **2010**, *9*, 193–204. [CrossRef]
10. Barnes, W.L.; Dereux, A.; Ebbesen, T.W. Surface plasmon subwavelength optics. *Nature* **2003**, *424*, 824–830. [CrossRef]

11. Pérez-Ocón, F.; Pozo, A.M.; Cortina, J.; Rabaza, O. Design of a Surface Plasmon Resonance CO Sensor. *Sensors* **2022**, *22*, 3299. [CrossRef] [PubMed]
12. Vahedi, A.; Kouhi, M. Liquid crystal-based surface plasmon resonance biosensor. *Plasmonics* **2020**, *15*, 61–71. [CrossRef]
13. Mahmoudpour, M.; Dolatabadi, J.E.N.; Torbati, M.; Homayouni-Rad, A. Nanomaterials based surface plasmon resonance signal enhancement for detection of environmental pollutions. *Biosens. Bioelectron.* **2019**, *127*, 72–84. [CrossRef] [PubMed]
14. Fang, M.; Shi, F.; Chen, Y. Unidirectional all-optical absorption switch based on optical Tamm state in nonlinear plasmonic waveguide. *Plasmonics* **2016**, *11*, 197–203. [CrossRef]
15. Wei, D.; Hu, C.; Chen, M.; Shi, J.; Luo, J.; Zhang, X.; Wang, H.; Xie, C. Optical modulator based on the coupling effect of different surface plasmon modes excited on the metasurface. *Opt. Mater. Express* **2020**, *10*, 105–118. [CrossRef]
16. Fang, Y.; Sun, M. Nanoplasmonic waveguides: Towards applications in integrated nanophotonic circuits. *Light. Sci. Appl.* **2015**, *4*, e294. [CrossRef]
17. Chandran, A.; Barnard, E.S.; White, J.S.; Brongersma, M.L. Metal-dielectric-metal surface plasmon-polariton resonators. *Phys. Rev. B* **2012**, *85*, 085416. [CrossRef]
18. Dai, M.; Chen, Z.; Zhao, Y.; Gandhi, M.S.A.; Li, Q.; Fu, H. State-of-the-art optical microfiber coupler sensors for physical and biochemical sensing applications. *Biosensors* **2020**, *10*, 179. [CrossRef]
19. Li, X.; Gu, Y.; Luo, R.; Wang, L.; Gong, Q. Effects of dielectric anisotropy on surface plasmon polaritons in three-layer plasmonic nanostructures. *Plasmonics* **2013**, *8*, 1043–1049. [CrossRef]
20. Veronis, G.; Fan, S. Crosstalk between three-dimensional plasmonic slot waveguides. *Opt. Express* **2008**, *16*, 2129–2140. [CrossRef]
21. Rahmatiyar, M.; Afsahi, M.; Danaie, M. Design of a refractive index plasmonic sensor based on a ring resonator coupled to a MIM waveguide containing tapered defects. *Plasmonics* **2020**, *15*, 2169–2176. [CrossRef]
22. Qin, M.; Zhai, X.; Wang, L.; Li, H.; Xia, S.; Lin, Q.; Zhang, B. Double Fano resonances excited in a compact structure by introducing a defect. *EPL* **2016**, *114*, 57006. [CrossRef]
23. Hu, F.; Chen, F.; Zhang, H.; Sun, L.; Yu, C. Sensor based on multiple Fano resonances in MIM waveguide resonator system with silver nanorod-defect. *Optik* **2021**, *229*, 166237. [CrossRef]
24. Yang, Q.; Liu, X.; Guo, F.; Bai, H.; Zhang, B.; Li, X.; Tan, Y.; Zhang, Z. Multiple Fano resonance in MIM waveguide system with cross-shaped cavity. *Optik* **2020**, *220*, 165163. [CrossRef]
25. Wang, Y.; Huo, Y.; Cui, P.; Song, M.; Zhao, C.; Liao, Z.; Liu, T.; Zhang, Z.; Xie, Y. Ultrahigh FOM and multiple Fano resonances in MIM waveguide systems with half-ring and rectangular cavities. *J. Comput. Electron.* **2023**, *22*, 839–848. [CrossRef]
26. Pang, S.; Huo, Y.; Xie, Y.; Hao, L. Fano resonance in MIM waveguide structure with oblique rectangular cavity and its application in sensor. *Opt. Commun.* **2016**, *381*, 409–413. [CrossRef]
27. Zhou, C.; Huo, Y.; Guo, Y.; Niu, Q. Tunable multiple fano resonances and stable plasmonic band-stop filter based on a metal-insulator-metal waveguide. *Plasmonics* **2021**, *16*, 1735–1743. [CrossRef]
28. Tathfif, I.; Rashid, K.S.; Yaseer, A.A.; Sagor, R.H. Alternative material titanium nitride based refractive index sensor embedded with defects: An emerging solution in sensing arena. *Results Phys.* **2021**, *29*, 104795. [CrossRef]
29. Tang, Y.; Zhang, Z.; Wang, R.; Hai, Z.; Xue, C.; Zhang, W.; Yan, S. Refractive index sensor based on Fano resonances in metal-insulator-metal waveguides coupled with resonators. *Sensors* **2017**, *17*, 784. [CrossRef]
30. Chen, J.-H.; Chen, J.-H.; Wang, L.; Wang, L.; Chen, J.-X.; Chen, J.-X.; Zeng, Y.-P.; Zeng, Y.-P.; Liang, Q.-Q.; Liang, Q.-Q.; et al. A refractive index sensor based on the MIM waveguide with a semi-elliptical and a rectangular ring resonant cavity. *Mod. Phys. Lett. B* **2023**, *37*, 2350141. [CrossRef]
31. Wang, L.; Zeng, Y.-P.; Wang, Z.-Y.; Xia, X.-P.; Liang, Q.-Q. A refractive index sensor based on an analogy T shaped metal–insulator–metal waveguide. *Optik* **2018**, *172*, 1199–1204. [CrossRef]
32. Chen, A.T.C.; Rakib Uddin, M. Transverse Electric (TE) and Transverse Magnetic (TM) Modes Dependent Effective Index Analysis for a Nano-scale Silicon Waveguide. In Proceedings of the 2022 24th International Conference on Advanced Communication Technology (ICACT), Pyeongchang, Republic of Korea, 13–16 February 2022.
33. Lee, S.-Y.; Lee, I.-M.; Park, J.; Oh, S.; Lee, W.; Kim, K.-Y.; Lee, B. Role of magnetic induction currents in nanoslit excitation of surface plasmon polaritons. *Phys. Rev. Lett.* **2012**, *108*, 213907. [CrossRef] [PubMed]
34. Choi, B.B.; Kim, B.; Bice, J.; Taylor, C.; Jiang, P. Inverse DVD-R grating structured SPR sensor platform with high sensitivity and figure of merit. *J. Ind. Eng. Chem.* **2022**, *116*, 321–330. [CrossRef]
35. Rakhshani, M.R.; Mansouri-Birjandi, M.A. High sensitivity plasmonic refractive index sensing and its application for human blood group identification. *Sensors Actuators B Chem.* **2017**, *249*, 168–176. [CrossRef]
36. Höflich, K.; Jurczyk, J.; Zhang, Y.; dos Santos, M.V.P.; Götz, M.; Guerra-Nuñez, C.; Best, J.P.; Kapusta, C.; Utke, I. Direct electron beam writing of silver-based nanostructures. *ACS Appl. Mater. Interfaces* **2017**, *9*, 24071–24077. [CrossRef] [PubMed]
37. Liu, N.; Weiss, T.; Mesch, M.; Langguth, L.; Eigenthaler, U.; Hirscher, M.; Sönnichsen, C.; Giessen, H. Planar metamaterial analogue of electromagnetically induced transparency for plasmonic sensing. *Nano Lett.* **2010**, *10*, 1103–1107. [CrossRef] [PubMed]
38. Liu, H.; Gao, Y.; Zhu, B.; Ren, G.; Jian, S. A T-shaped high resolution plasmonic demultiplexer based on perturbations of two nanoresonators. *Opt. Commun.* **2015**, *334*, 164–169. [CrossRef]
39. Kekatpure, R.D.; Hryciw, A.C.; Barnard, E.S.; Brongersma, M.L. Solving dielectric and plasmonic waveguide dispersion relations on a pocket calculator. *Opt. Express* **2009**, *17*, 24112–24129. [CrossRef] [PubMed]

40. Yadav, G.K.; Metya, S.K.; Zafar, R.; Garg, A.K. High sensitivity plasmonic refractive index sensor for early anemia detection. *Photonics Nanostructures-Fundam. Appl.* **2024**, *58*, 101235. [CrossRef]
41. Jaria, G.; Calisto, V.; Otero, M.; Esteves, V.I. Monitoring pharmaceuticals in the aquatic environment using enzyme-linked immunosorbent assay (ELISA)—A practical overview. *Anal. Bioanal. Chem.* **2020**, *412*, 3983–4008. [CrossRef]
42. You, K.E.; Uddin, N.; Kim, T.H.; Fan, Q.H.; Yoon, H.J. Highly sensitive detection of biological substances using microfluidic enhanced Fabry-Perot etalon-based optical biosensors. *Sensors Actuators B Chem.* **2018**, *277*, 62–68. [CrossRef]

photonics MDPI

Article

A Nanosensor Based on Optical Principles for Temperature Detection Using a Gear Ring Model

Lei Li [1,2], Shubin Yan [2,3,*], Yang Cui [2,3], Taiquan Wu [2,3], Chuanhui Zhu [2,3], Yi Zhang [2,3], Yiru Su [3], Qizhi Zhang [1] and Guowang Gao [1]

1 School of Electrical Engineering, Xi'an Shiyou University, Xi'an 710065, China; 22212030534@stumail.xsyu.edu.cn (L.L.); zhangqz@xsyu.edu.cn (Q.Z.); wwgao@xsyu.edu.cn (G.G.)
2 Zhejiang-Belarus Joint Laboratory of Intelligent Equipment and System for Water Conservancy and Hydropower Safety Monitoring, Hangzhou 310018, China; cuiy@zjweu.edu.cn (Y.C.); wutq@zjweu.edu.cn (T.W.); zhchh@zjweu.edu.cn (C.Z.); zhangyi@zjweu.edu.cn (Y.Z.)
3 School of Electrical Engineering, Zhejiang University of Water Resources and Electric Power, Hangzhou 310018, China; 2022b03001@stu.zjweu.edu.cn
* Correspondence: yanshb@zjweu.edu.cn; Tel.: +86-186-3611-2255

Abstract: Based on the characteristics of plasmonic waveguides and resonators, we propose a refractive index (RI) sensor that couples a gear ring with a metal–insulator–metal (MIM) waveguide. Using the finite element method (FEM), we conduct extensive spectral analysis of the sensor's properties in the near-infrared spectrum. Furthermore, we investigate the structural parameters affecting the refractive index sensing characteristics. This study reveals that the complexity of the ring cavity edge can significantly enhance the sensitivity of the nanosensor. Optimal structural performance parameters are selected when the number of gears is six, resulting in a sensitivity of 3102 nm/RIU and a Figure of Merit (FOM) of 57.4 for the sensing characteristics of the gear ring. It possesses the advantages of small size and high sensitivity. This nanoscale sensor design demonstrates high sensitivity in the field of industrial material temperature detection.

Keywords: refractive index sensor; MIM; Fano resonance; temperature detection

Citation: Li, L.; Yan, S.; Cui, Y.; Wu, T.; Zhu, C.; Zhang, Y.; Su, Y.; Zhang, Q.; Gao, G. A Nanosensor Based on Optical Principles for Temperature Detection Using a Gear Ring Model. *Photonics* **2024**, *11*, 311. https://doi.org/10.3390/photonics11040311

Received: 6 March 2024
Revised: 18 March 2024
Accepted: 20 March 2024
Published: 28 March 2024

1. Introduction

Surface Plasmon Polaritons (SPPs) are electromagnetic waves that exist at the interface between a metal and dielectrics [1]. When light illuminates the interface, it interacts with the free electrons in the metal, giving rise to these electromagnetic waves. SPPs exhibit unique spectral characteristics, such as the ability to localize optical fields at the nanoscale, surpassing classical diffraction limits. Due to their ability to concentrate light in subwavelength structures, resulting in enhanced electric fields, SPPs are particularly suitable for nanoscale integration and find widespread applications in sensing, optical switches, and other fields [2,3]. They concentrate light within structures smaller than the wavelength of light. In these structures, light is confined to extremely small volumes, leading to a significant enhancement in the electric field. Field enhancement occurs when SPPs confine the light beam at the metal–dielectric interface at the nanoscale, resulting in the electromagnetic field intensity far exceeding the original electromagnetic field intensity of the incident light. This enhanced field is crucial for the performance of sensors, as it significantly improves the sensor's responsiveness to target substances (such as biomolecules or chemicals) [4,5].

Moreover, the suitability of SPPs for nanoscale integration means that smaller, more precise sensors and optical components can be manufactured [6]. This capability is particularly advantageous for designing and fabricating systems with size constraints, achieving efficient light control and sensing in very small spaces [7,8]. For instance, in biosensors, SPPs can detect the presence of biomolecules, such as DNA hybridization or blood components. Thus, these properties of SPPs not only provide new tools for research but also give new possibilities for industrial and technological applications, especially in the field of precision measurement and high-sensitivity detection [9,10].

In optical devices, plasma sensors work when the refractive index of the filled media changes, allowing for the detection of even minor refractive index changes [11]. Therefore, enhancing the sensing behavior of plasma sensors requires adjusting the design parameters, optimizing structures, and using new materials. For example, silicon membrane primitive surfaces are a novel type of optical material structure capable of effectively controlling the propagation and scattering of incident light by designing micro- and nanoscale structures on the surface of a silicon membrane. When the surface of the silicon primitive membrane is exposed to optical signals, it can influence the intensity of the incident light. By monitoring this intensity change, variations in the refractive index of the surrounding medium can be inferred. The intensity detection technology based on silicon membrane primitive surfaces can also find applications in the field of communication. For example, it can be used to achieve modulation and demodulation of optical signals in photonic integrated circuits, as well as data transmission and processing in optical sensor networks [12,13].

The overall structure is Au-SiO_2–graphene, which can be realized by tuning the Fermi energy level of graphene. The sensing performance of the absorber in an aqueous solution was simulated to discover the application of high refractive index sensitivity. The absorbers have an important role in high-sensitivity sensors, photothermal detection, and thermal radiation [14,15].

Metal–insulator–metal (MIM) waveguides are a variant of SPP waveguides, composed of two layers of metal sandwiching a layer of insulating material. They possess unique advantages, enabling the realization of complex optical structures in small dimensions with high integration. When light bends within the waveguide, it maintains high localization inside the waveguide, minimizing energy loss. MIM waveguides are also capable of deeply constraining light at subwavelength scales, thereby increasing the efficiency of light interaction with the medium [16–18].

Many studies have investigated the unique designs of RI sensors based on MIM waveguides. For instance, Butt explored a refractive index sensor based on a square ring-shaped MIM structure and utilized the finite element method (FEM) to model the transmission spectra and electric field distribution of the nanosensing structure. Their study achieved a sensitivity of 1200 nm/RIU and a FOM of 19.7 [19]. Zhu established an MIM waveguide with a U-shaped resonator, achieving a sensitivity of 825 nm/RIU and an FOM of 21.54. The superior sensing structure is capable of accurately identifying five types of pure edible vegetable oils at room temperature [20]. Rohimah designed an r-shaped resonator for a refractive index sensor in a MIM waveguide coupling system, achieving a sensitivity of 1333 nm/RIU [21]. Kazanskiy demonstrated a semi-ring resonator cavity by means of decorating nanodots with a sensitivity of 1084.21 nm/RIU [22].

Numerical results have been obtained by applying the designed sensing structure to different fields. For example, Sagor R H et al. designed a coupled cavity based on a metal–insulator–metal (MIM) waveguide combined with two unequal vertical rectangular cavities. This sensing was finally obtained with a maximum sensitivity of 2625.87 nm/RIU and a Figure of Merit (FOM) of 26.04. Human blood type was determined by utilizing the refractive index modeling of different human blood types including A, B, and O. In addition, as a temperature sensor, it had a sensing factor of -1.04 nm/°C [23].

Tavana S et al. proposed a metal–insulator–metal (MIM) structure consisting of a semicircular resonant cavity (SCRC) and a circular split ring resonator (SRR). Because of the interaction between the narrowband modes of the SRR and the broadband modes of the SCRC, this optical sensor could generate and support a double Fano resonance. Its maximum sensitivity and Figure of Merit (FOM) were 579 nm/RIU and 12.46, respectively [24]. Shen S et al. explored the transmission characteristics of a MIM waveguide structure based on a square ring resonator for SRS in a waveguide system. The optimization of the structure yielded a sensitivity of 6000 nm/RIU and a Figure of Merit (FOM) of 17 RIU^{-1}. Also, it had a temperature sensing coefficient of -20.3 nm/°C [25].

However, the sensitivity of these RI sensors is generally limited, restricting their application scenarios. Therefore, this paper proposes a simple yet highly sensitive plasmonic sensor. Building upon the foundation of a ring-coupled cavity and a non-rectangular barrier, this paper further explores the impact of ring edge complexity on sensor sensitivity and the limiting capability of the rectangular barrier to light. When the measured substance is exposed to contact with the sensor, the refractive index changes, leading to alterations in the wavelength, intensity, or phase of the spectral curve. Finally, the RI sensor is composed of a gear ring coupled with a MIM waveguide. The complexity of the ring-shaped cavity edge contributes to enhancing sensor sensitivity, while the addition of the rectangular barrier aids in amplifying resonant SPP modes. With a typical sample sensitivity of 3102 nm/RIU and a FOM of 57.4, the device delivers excellent performance. This highly sensitive sensor design can detect variations in the refractive index of industrial material temperatures, obtaining propagation curves at different temperatures with minimal error.

2. Materials and Methods

The designed structure consists of a gear ring coupled to a MIM waveguide with a rectangular stub, as shown in Figure 1.

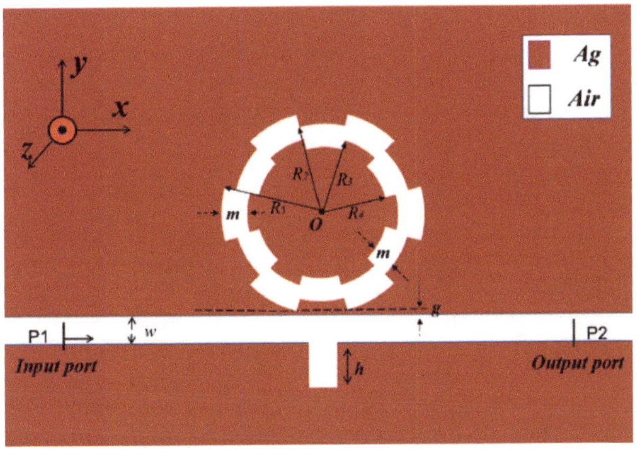

Figure 1. Two-dimensional plan of the overall structure.

In the two-dimensional planar design of the model, the red and white areas represent silver and air, respectively. In the near-infrared region, silver has the minimum imaginary part of the relative dielectric constant. This implies that silver exhibits high sensitivity to light in this wavelength range, making it suitable for optical sensors (Table 1).

Table 1. The parameters of the gear ring model.

Parameter	Definition
R_1, R_2, R_3, R_4	Inner and outer radii of single gear ring circles
m	The dielectric layer of the gear ring thickness
w	The waveguide breadth
g	Gap at the distance between the gear ring and waveguide
h	The rectangular stub height
Ts	The edge complexity of the gear ring
O	The center of the gear ring's circle
P1, P2	Input port and output port

This study uses a two-dimensional (2D) model to emulate the magnetic properties of a three-dimensional (3D) system [26,27]. On the one hand, this choice is motivated by the uniform nature of material properties, where the 3D effect has almost no influence on the magnetic domain profile. On the other hand, this method reduces the computational complexity, providing results that are nearly identical to those obtained in a real 3D scenario.

Simultaneously, the waveguide allows the SPP transmission by importing an isolating layer between the two metallic layers. The reason is that the odd symmetric modes are strongly coupled and attenuated in the insulating layer, leading to much greater energy loss and shorter transmission distances [28]. In contrast, the transmission distance of even-numbered symmetric modes is longer with less energy loss. This guarantees that there is only one even-symmetric transmission mode, and also effectively improves the transmission efficiency of even-symmetric mode and reduce energy losses. The dielectric layer thickness is chosen to be 50 nm, which can meet the requirements [29].

In Figure 1, the red area is the substrate with metallic silver material, which has a low loss of light; the white part stands for the air medium. The relative permittivity of the underlying silver material is defined [30]:

$$\varepsilon(\omega) = \varepsilon_\infty + \frac{\varepsilon_s - \varepsilon_\infty}{1 + i\tau\omega} + \frac{\sigma}{i\omega\varepsilon_0} \tag{1}$$

where ε_∞ is the infinite frequency relative dielectric constant, ε_d is the air relative dielectric constant, ε_s is the static dielectric constant, τ is the relaxation time, and σ is the conductivity.

The fundamental transverse magnetic mode of the MIM waveguide can be expressed as [31]:

$$\tan(kw) = -\frac{2k\alpha_c}{k^2 + p^2\alpha_c} \tag{2}$$

where $k = 2\pi/\lambda$ is the wave vector in the waveguide, w is the width of the waveguide, $p = \varepsilon_i/\varepsilon_m$, and $\alpha_c = \sqrt{[k_0(\varepsilon_i - \varepsilon_m) + k^2]}$. ($\varepsilon_i$ and ε_m are the diameters of the insulator, respectively).

In this article, the properties of the plasma sensor are measured by means of three different parameters as follows: *FWHM* denotes the sharpness of the spectral lines and *FOM* and sensitivity signify the index of component performance or sensing behavior. These three key parameters are essential. The formula is as follows [32]:

$$S = \frac{\Delta\lambda}{\Delta n} \tag{3}$$

$$FOM = \frac{S}{FWHM} \tag{4}$$

where Δn and $\Delta\lambda$ show the changes in the refractive index and wavelength, respectively. After confirming the required conditions, COMSOL Multiphysics 5.4a (COMSOL Inc., Stockholm, Sweden) is used to build a geometrical model of the plasma sensor. An ultra-fine triangular mesh section is chosen for the whole coupling structure. The reason is that ultra-fine triangular meshes can better capture the details of complex structures, especially for nanostructures or devices with tiny features. By increasing the density of the mesh, numerical errors can be reduced, making the simulation results more reliable and accurate. At the same time, ultra-fine meshes can capture the details of the structure more efficiently, reducing the need for large-scale computation.

Following the preliminary simulations, the wavelength range is chosen to be 1700–2800 nm in a 1 nm step. The initial parameters of the gear structure are given in the following table (Table 2).

Table 2. Initial structural parameters.

Parameter	Value (nm)
R_1	220
R_2	195
R_3	170
R_4	145
m	50
g	10
w	100
h	120

3. Simulation Results and Analysis

In this section, we choose a gear ring with six gear edges for further analysis.

In Figure 2, a comprehensive representation is provided. The black line represents the transmission spectrum of a single waveguide with a rectangular barrier, depicted as a slightly upward-sloping line. This line signifies that most frequencies of light signals exhibit high transmittance, demonstrating a continuous broadband transmission characteristic.

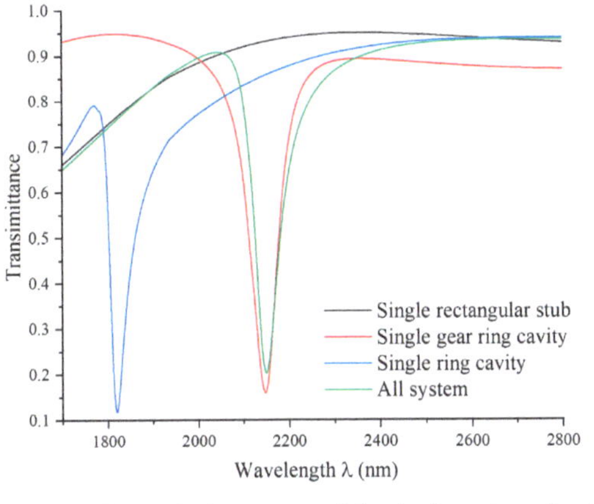

Figure 2. Transmission spectra of the single rectangular stub (black line), single gear ring cavity (red line), single ring cavity (blue line), and the whole system (green line).

The red, blue, and green lines correspond to the single gear ring cavity, the single ring cavity with a smooth surface, and the gear ring cavity with six gear edges, respectively. These cavities all exhibit Fano resonance, which occurs when light passes through a cavity with a constant refractive index, leading to sharp peaks in the spectral graph within a very narrow wavelength range. To further investigate the implications of the cavity geometry parameters on the Fano resonance, we focus on the rectangular stub height and edges of the gear cavity complexity.

In Figure 3, it can be noticed that when the SPPs propagate through the waveguide and couple at the junction with the cavity, the resonance phenomenon is produced. In comparison, for the structure in Figure 3a, the magnetic field distribution associated with the waveguide is more concentrated within the cavity. The inclusion of a rectangular stub seems to restrict the propagation of SPPs in the waveguide, potentially making them more prone to coupling into the cavity. Thus, at specific wavelengths, the propagation of light in the structure is constrained, leading to enhanced resonance phenomena.

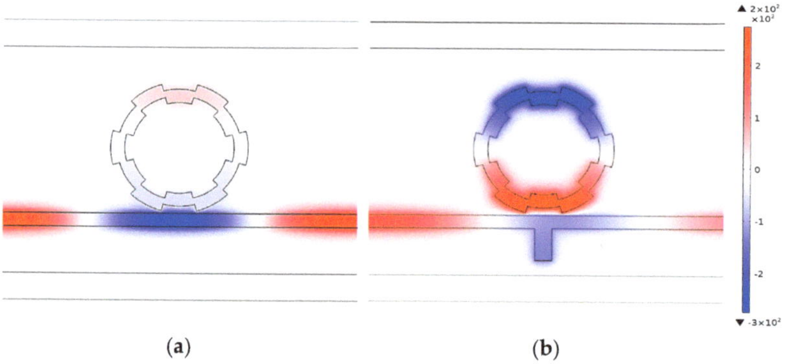

Figure 3. (**a**) The magnetic field distribution of the single gear ring cavity at λ = 2150 nm. (**b**) The magnetic field distribution of the whole system at λ = 2150 nm.

In Figure 4, the complexity of the cavity edge does not impact the resonance phenomenon. However, the complexity of the gear ring cavity edge results in different ring-shaped magnetic field distributions for the two structures, which may lead to variations in light confinement capabilities. This difference ultimately results in distinct plasmonic sensing performances. As can be seen from the transmission curves shown by the blue and green lines in Figure 2, the complexity of the cavity edges improves the sensing performance, especially in terms of sensitivity.

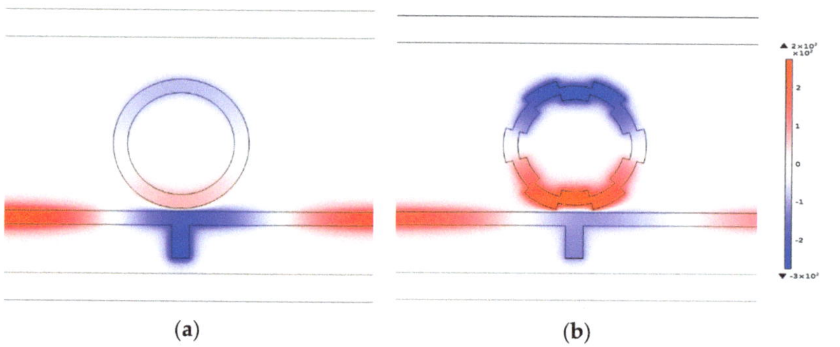

Figure 4. (**a**) The magnetic field distribution of the ring cavity with a smooth surface at λ = 2150 nm. (**b**) The magnetic field distribution of the gear ring cavity with six gear edges at λ = 2150 nm.

According to the above analysis under different refractive indices, the sensing performance is initially analyzed when the number of gears is six and other parameters are fixed. In Figure 5a, various transmittance spectra are plotted for refractive indices ranging from 1.00 to 1.05. When the refractive index gradually increases, the appropriate wavelengths increase, leading to a redshift in the resonant wavelength. The waveform of the transmittance spectrum stays unaffected by changes in the refractive index, and they move equidistant, showing a certain stability. The RI sensor can be designed from this property. Figure 5b illustrates the linear relationship between the resonant wavelength and refractive index. Through linear fitting, the sensitivity of the gear ring nanosensor to the refractive index can reach 3102 nm/RIU with an FOM of 57.4.

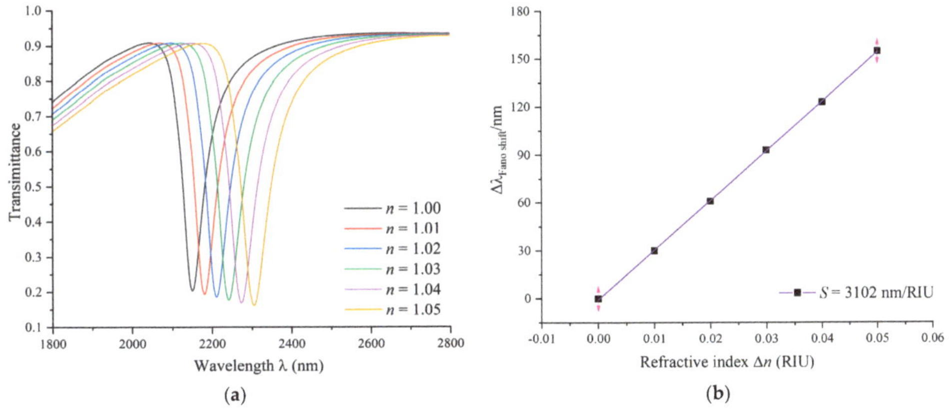

Figure 5. (**a**) Transmission spectra of multiple refractive indices. (**b**) Fitted line for the variation in the refractive index.

Next, we analyze the complexity of the edges of the gear ring cavity.

The primary structural values remain fixed. We compare the transmission spectra for different numbers of gear edges and delve into the impact of gear ring edge complexity on the Fano resonance. The number of gear edges is set to 0, 2, 4, 6, 8, and 10. When the number of gear edges is 0, it presents a ring cavity with a smooth surface.

Figure 6a illustrates the transmittance spectra for different numbers of gear edges. As the number of gear edges increases, the wavelength also increases, leading to a redshift in the resonant wavelength. The difference between adjacent peaks increases, and when the number of gear edges is 10, an asymmetric transmission curve with a larger wavelength is observed, indicating a highly sensitive Fano resonance with a value of 3600 nm/RIU.

However, as the number of gear edges increases, the FWHM of the obtained transmission curves widens. By calculation, we find a value of only 41.9. Although the sensitivity increases to some extent, the decrease in FOM suggests that the coupling cavity is more susceptible to environmental changes, resulting in a broader peak in the electromagnetic field and some loss of resolution.

Therefore, under the optimal geometric parameters of the cavity, namely, with six gear edges, we further investigate the size of the overall model.

Similarly, the initial values of other parameters remain invariant, the outer radius R_1 is varied in steps of 5 nm from 210 nm to 230 nm, and the different circular radii of the ring are varied to guarantee that the structure is enlarged or reduced. The transmittance spectra for coupling cavities of different sizes are shown in Figure 7a, where the size of the gear ring increases with the radius, and the resonant wavelength exhibits a noticeable redshift.

Figure 7b depicts the sensitivity values for gear ring cavities of different sizes obtained using a mathematical linear method. The sensitivity goes from 2900 nm/RIU to 3362 nm/RIU as the increment in the size of the gear ring increases. We find that the control of the trough wavelength and the sensitivity can be enhanced by adapting the size of the overall structure. Therefore, in the design of optical devices, utilizing different sensitivities to choose the appropriate device size is feasible.

Subsequently, the influence of the rectangular stub height in a coupling structure is investigated. The height values are varied from 100 nm to 140 nm. As shown in Figure 8, the variation in the height of the barrier has no obvious effect on the angle position of the Fano resonance. When the rectangular barrier height becomes larger, the shape of the Fano resonance becomes more uneven, and the asymmetry becomes more pronounced. The introduction of the rectangular barrier, being a unique model, results in an asymmetric shape of the Fano resonance curve, which becomes more pronounced with increasing barrier height. This structure also induces a continuous broadband state that affects the line profile of the Fano resonance but does not change the wavelength at the dip position.

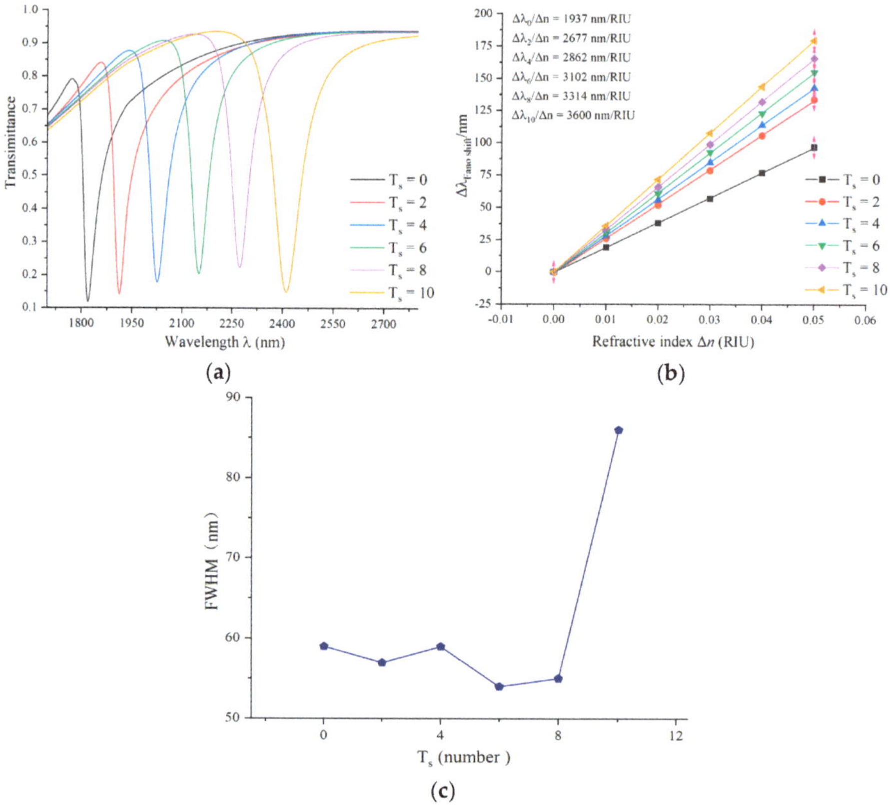

Figure 6. (**a**) Transmission spectra of edge complexity of gear rings. (**b**) Fitting lines of sensitivity at different complexities at the edge of gear rings. (**c**) Variation in FWHM values for different complexities of gear rings.

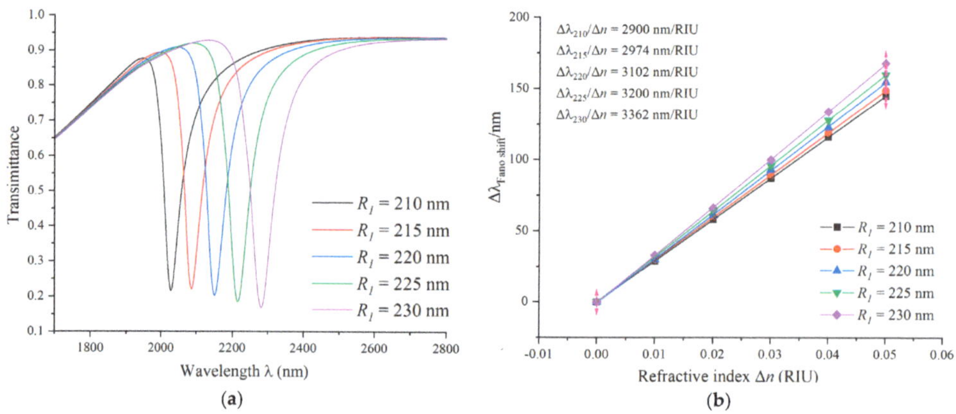

Figure 7. (**a**) Transmission spectra of different sizes of gear rings. (**b**) Fitting lines of sensitivity for different sizes of gear rings.

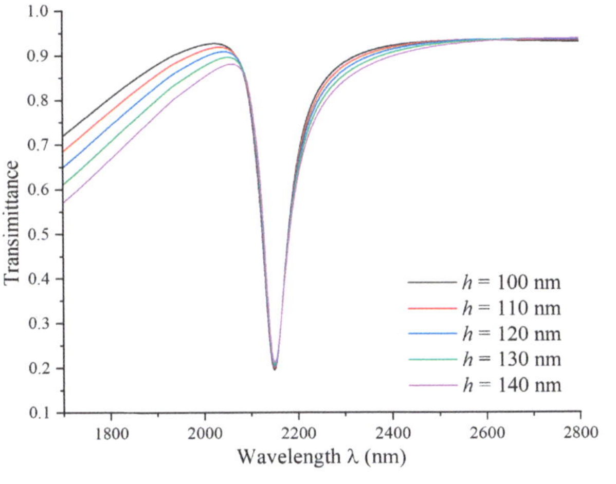

Figure 8. Transmission spectra of rectangular stubs at different heights.

Finally, the influence of the coupling gap at the distance between the waveguide and the gear ring structure on the sensing performance is assessed. There is no change in the initial values, but the coupling gap g varies in the range of 10 nm to 30 nm in 5 nm steps. As shown in Figure 9a, by altering the amount of g, a blue shift in the spectral curve is observed, indicating a shift towards shorter wavelengths. In addition, the FWHM of the curve becomes diminished, and the transmittance increases. This suggests that the coupling becomes more challenging between SPP and the gear ring structure, and with an increase in the coupling gap, the electric field intensity gradually weakens.

The sensitivity fitting curve in Figure 9b shows a decrease in sensitivity from 3102 nm/RIU to 2900 nm/RIU as the coupling gap increases. So, the sensitivity can reach a maximum of 3102 nm/RIU when g = 10 nm is chosen. In a nutshell, the variation in the coupling gap has some effects on the optical properties, including a blue shift, changes in FWHM, and an increase in transmittance. However, within a certain range, changes in the coupling gap do not significantly alter the sensitivity of the sensor, suggesting that these variations may be negligible in sensor applications.

The preparation of nanoscale refractive index sensors typically involves precise nanomanufacturing techniques. For smaller-sized plasmonic sensors, advanced lithography and nanomanufacturing technologies may be necessary. The process begins with substrate preparation, where standard semiconductor processing techniques are employed to select a suitable substrate material, such as silicon, as the foundational layer for the sensor.

Subsequently, a layer of photoresist is coated onto the substrate. Typically, this is a photosensitive polymer that, through exposure and development, defines the desired pattern. Using a photolithography machine, the photoresist is exposed to project the desired sensor structure pattern onto it. For features smaller than 20 nanometers, high-resolution lithography techniques like electron beam or near-field lithography are often required.

After exposure, the developed photoresist removes material from unexposed areas, creating the desired pattern and essentially forming a template or mask. Following this, nanomanufacturing techniques such as electron beam etching (EBE) or ion beam etching (IBE) are utilized to etch the substrate based on the template, creating nanoscale structures.

Plasma deposition techniques are then employed to deposit the necessary sensing layer onto the nanostructures. This layer is typically made of materials like silver metal, which are suitable for measuring refractive index changes. Finally, the process concludes with necessary steps like cleaning, annealing, or other post-processing to ensure the performance and stability of the sensor.

In practical fabrication, the specific steps and techniques may vary depending on the sensor design and desired characteristics. It is crucial to note that nanoscale fabrication requires highly precise equipment and techniques.

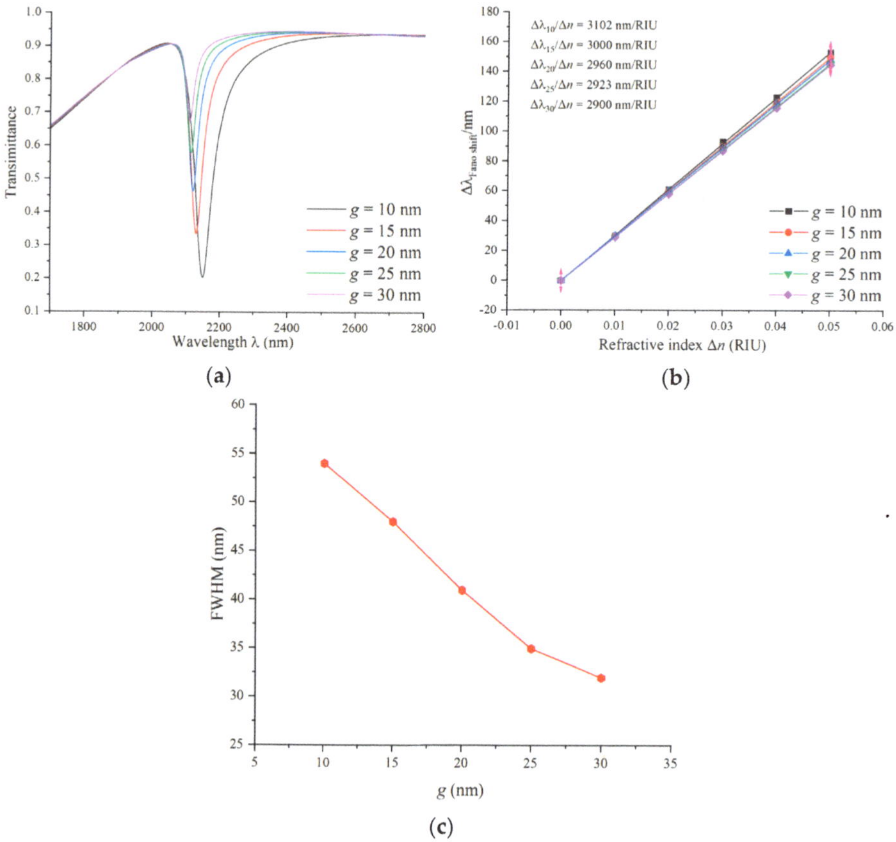

Figure 9. (**a**) Transmission spectra of a range of coupling gaps. (**b**) Sensitivity-fitted lines with different coupling gaps. (**c**) Variation in FWHM values at the different coupling gaps.

The manufacturing process flow for the specific steps is shown in Figure 10.

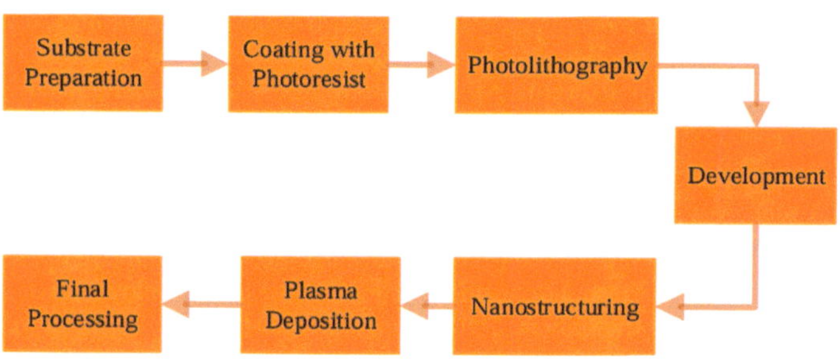

Figure 10. Nanosensing manufacturing process flow steps.

Many studies have undertaken the design of this plasmonic sensor (Table 3).

Table 3. Summary of the other literature sources.

Model Type	Sensitivity (nm/RIU)	Figure of Merit (FOM)	Wavelength Range
Square ring resonator [19]	1200	19.7	1300 < λ < 1900
U-shaped cavities [20]	825	21.54	600 < λ < 1400
An r-shaped resonator [21]	1333	58.76	600 < λ < 2000
NDs decorated semi-ring [22]	1084.21	57.06	800 < λ < 1200
Two unequal vertical rectangular cavities [23]	2625.87	26.04	1200 < λ < 2000
A SCR cavity and a circular split ring resonator [24]	579	12.64	800 < λ < 1700
A gear ring cavity	3102	57.4	1700 < λ < 2800

4. Application

The sensing feature of the gear ring structure mentioned above brings the advantages of high sensitivity, easy miniaturization and integration, and it is also applicable to the temperature detection of industrial materials. Ethanol, being an excellent temperature-sensing medium, is picked to be the filling material. The melting point and boiling point of ethanol are −114.1 °C and 80 °C, respectively [33].

The refractive index temperature parameters of the different media in the sensing structure are 3.94×10^{-4} (°C^{-1}) for ethanol, 9.30×10^{-6} (°C^{-1}) for metallic silver, and 8.60×10^{-6} (°C^{-1}) for a quartz substrate. Since the refractive index temperature sensing parameters of metallic silver and quartz are much smaller than that of an ethanol medium, they are negligible in the study of the effect of sensing characteristics.

When ethanol is used as a medium, the equation for the relationship between its refractive index and temperature is as follows [34]:

$$n = 1.3605 - 3.94 \times 10^{-4}(T - T_0) \tag{5}$$

where T and T_0 represent the temperature of industrial materials and room temperature, respectively. Room temperature is set at 20 °C.

The geometric parameters of the initial structure remain unchanged, with a gear ring consisting of six gears designed as the temperature sensing structure. The temperature detection range is set between −75 °C and 75 °C. After placing the nanosensor in ethanol, upon contact, air is expelled from the groove and replaced by the liquid. The refractive index of ethanol in the groove varies with temperature. The nanosensor measures the refractive index by monitoring light transmission or reflection, generating a Fano resonance curve. As the refractive index of ethanol changes with temperature, it influences the Fano resonance curve. The nanosensor detects and compares these changes with a predetermined temperature–refractive index relationship, thereby inferring the external temperature. Calibration against known temperatures ensures accurate measurements. The sensitivity relationship for temperature sensing is expressed as follows [29]:

$$S_T = \frac{\Delta\lambda}{\Delta n} \tag{6}$$

where $\Delta\lambda$ and Δn represent the changes in wavelength and temperature, respectively. When the temperature increases in increments of 20 °C, ranging from −55 °C to 45 °C, the refractive indices are calculated as 1.38995, 1.38207, 1.37419, 1.36631, 1.36237, and 1.35015. The simulated transmission curves are shown in Figure 11a, where an increase in temperature leads to a blue shift in the spectral curve. Due to the excellent performance of modern optical instruments in precision measurement and wavelength accuracy, nanoscale wavelengths can be accurately detected. The trough in the transmission curve moves from 3540 nm to 3377 nm, indicating $\Delta\lambda$ = 163 nm. Through linear fitting, the sensitivity is obtained, as shown in Figure 11b. The temperature sensor has a sensitivity of 1.664 nm/°C for good temperature sensing performance.

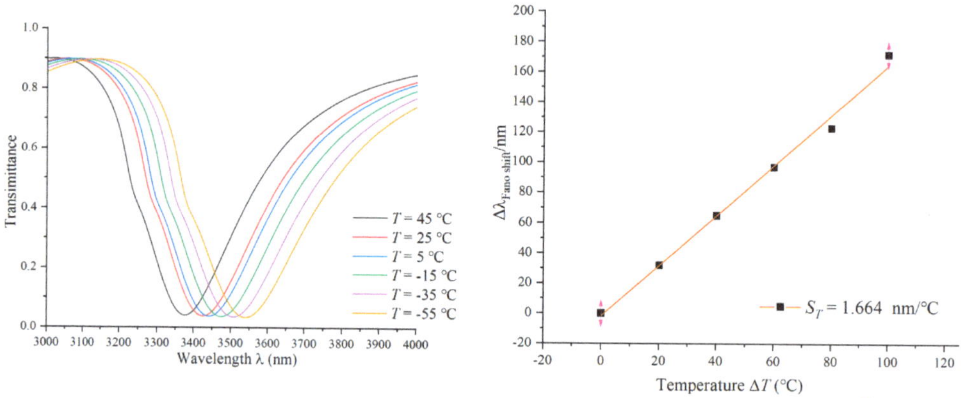

(a) (b)

Figure 11. (**a**) Transmission spectra at gradually increasing temperatures. (**b**) Sensitivity-fitted lines with different temperatures.

5. Conclusions

By utilizing the peculiar properties in optical Fano resonance, a highly precise nanoscale refractive index sensor has been successfully developed with a MIM waveguide and a gear ring. By analyzing the different geometric parameters of the sensing behavior, factors are considered, such as gear ring edge complexity, the inner and outer radii of the gear ring, and the gap at the distance between the gear ring and rectangular waveguide. Particularly, emphasis is given to the presentation of a rectangular stub and the influence of gear ring edge complexity on the sensing performance.

In addition, considering the complexity of the cavity edge, it is found to significantly enhance the sensitivity performance of the sensor. By comparing different numbers of gears, a gear ring with six gears is ultimately selected as the optimal configuration for the cavity.

In the end, upon optimizing the geometrical parameters, we managed to obtain a novel nanosensor. This gear ring cavity sensor exhibits outstanding performance, with a sensitivity of 3102 nm/RIU and an FOM of 57.4. The nanosensor not only performs well under industrial materials temperatures but also reliably monitors changes in temperature. The experimental data indicate that the refractive index temperature sensor has a sensitivity of 1.664 nm/°C, demonstrating excellent temperature sensing performance. This study provides strong support for the further application of nanoscale sensors.

Author Contributions: Conceptualization, S.Y. and L.L.; methodology, L.L.; software, Y.C.; validation, S.Y. and C.Z.; formal analysis, L.L. and Y.Z.; investigation, L.L.; resources, T.W. and Y.S.; data curation, L.L.; writing—original draft preparation, L.L.; writing—review and editing, Q.Z.; visualization, G.G. and Q.Z.; supervision, S.Y.; project administration and funding acquisition, S.Y. All authors have read and agreed to the published version of the manuscript.

Funding: This research was funded by the National Natural Science Foundation of China, grant numbers 62374148, 62003315, and 61975189; by the Zhejiang Provincial Natural Science Foundation of China, grant number LD21F050001; by the Key Research Project by the Department of Water Resources of Zhejiang Province, grant number RA2101; by the Key Research and Development Project of Zhejiang Province, grant number 2021C03019; and by the Scientific Research Foundation of Zhejiang University of Water Resources and Electric Power, grant number xky2022032.

Data Availability Statement: The data provided in this study are available upon request from the corresponding author.

Acknowledgments: The authors are grateful to other colleagues in the laboratory for their understanding and help. They also thank their affiliates for providing the research platform and their sponsors for financial support.

Conflicts of Interest: The authors declare no conflicts of interest.

References

1. Terekhin, P.; Benhayoun, O.; Weber, S.; Ivanov, D.; Garcia, M.; Rethfeld, B. Influence of surface plasmon polaritons on laser energy absorption and structuring of surface. *Appl. Surf. Sci.* **2020**, *512*, 144420. [CrossRef]
2. Han, X.X.; Rodriguez, R.S.; Haynes, C.L.; Ozaki, Y.; Zhao, B. Surface-enhanced Raman spectroscopy. *Nat. Rev. Methods Prim.* **2021**, *1*, 87. [CrossRef]
3. Akhavan, A.; Ghafoorifard, H.; Abdolhosseini, S.; Habibiyan, H. Metal–insulator–metal waveguide-coupled asymmetric resonators for sensing and slow light applications. *IET Optoelectron.* **2018**, *12*, 220–227. [CrossRef]
4. Butt, M.A. Metal-insulator-metal waveguide-based plasmonic sensors: Fantasy or truth—A critical review. *Appl. Res.* **2013**, *2*, e202200099. [CrossRef]
5. Kazanskiy, N.L.; Khonina, S.N.; Butt, M.A. Plasmonic sensors based on Metal-insulator-metal waveguides for refractive index sensing applications: A brief review. *Phys. E Low-Dimens. Syst. Nanostruct.* **2020**, *117*, 113798. [CrossRef]
6. Chou Chau, Y.F.; Chou Chao, C.T.; Huang, H.J.; Kumara, N.T.R.N.; Lim, C.M.; Chiang, H.P. Ultra-high refractive index sensing structure based on a metal-insulator-metal waveguide-coupled T-shape cavity with metal nanorod defects. *Nanomaterials* **2019**, *9*, 1433. [CrossRef] [PubMed]
7. Blau, Y.; Gilad, T.; Hanein, Y.; Boag, A.; Scheuer, J. High efficiency coupling to metal-insulator-metal plasmonic waveguides. *Opt. Express* **2022**, *30*, 13757–13764. [CrossRef] [PubMed]
8. Maier, S.A. Surface Plasmon Polaritons at Metal/Insulator Interfaces. In *Plasmonics: Fundamentals and Applications*; Springer: New York, NY, USA, 2007; pp. 21–37.
9. Tan, Y.M.; Chao, C.-T.C.; Kooh, M.R.R.; Huang, H.J.; Thotagamuge, R.; Lim, C.M.; Chiang, H.-P.; Chau, Y.-F.C. Mid infrared sensing structure based on a metal–insulator–metal waveguides with a triangular-shaped resonator. *Opt. Commun.* **2022**, *516*, 128282. [CrossRef]
10. Rashid, K.S.; Hassan, F.; Yaseer, A.A.; Tathfif, I.; Sagor, R.H. Gas-sensing and label-free detection of biomaterials employing multiple rings structured plasmonic nanosensor. *Sens. Bio-Sens. Res.* **2021**, *33*, 100440. [CrossRef]
11. Paul, S.; Ray, M. Multispectral switching using Fano resonance and plasmon-induced transparency in a plasmonic waveguide-coupled resonator system. *Plasmonics* **2019**, *14*, 1113–1122. [CrossRef]
12. Tognazzi, A.; Rocco, D.; Gandolfi, M.; Locatelli, A.; Carletti, L.; De Angelis, C. High quality factor silicon membrane metasurface for intensity-based refractive index sensing. *Optics* **2021**, *3*, 193–199. [CrossRef]
13. Sherif, S.M.; Swillam, M.A. Silicon-based mid infrared on-chip gas sensor using Fano resonance of coupled plasmonic microcavities. *Sci. Rep.* **2023**, *1*, 12311. [CrossRef] [PubMed]
14. Shangguan, Q.; Zhao, Y.; Song, Z.; Wang, J.; Yang, H.; Chen, J.; Liu, C.; Cheng, S.; Yang, W.; Yi, Z. High sensitivity active adjustable graphene absorber for refractive index sensing applications. *Diam. Relat. Mater.* **2022**, *128*, 109273. [CrossRef]
15. Ma, J.; Wu, P.; Li, W.; Liang, S.; Shangguan, Q.; Cheng, S.; Tian, Y.; Fu, J.; Zhang, L. A five-peaks graphene absorber with multiple adjustable and high sensitivity in the far infrared band. *Diam. Relat. Mater.* **2023**, *136*, 109960. [CrossRef]
16. Ravindran, N.; Kumar, S.; S, R.; CA, M.; Thirunavookarasu S, N.; CK, S. Recent advances in Surface Plasmon Resonance (SPR) biosensors for food analysis: A review. *Crit. Rev. Food Sci. Nutr.* **2023**, *63*, 1055–1077. [CrossRef] [PubMed]
17. Yesudasu, V.; Pradhan, H.S.; Pandya, R.J. Recent progress in surface plasmon resonance based sensors: A comprehensive review. *Heliyon* **2021**, *7*, e06321. [CrossRef] [PubMed]
18. Pommier, D.; Hufschmitt, Z.; Zhang, C.; Lai, Y.; Dujardin, G.; Le Moal, E.; Sauvan, C.; Greffet, J.J.; Wang, J.; Boer-Duchemin, E. Nanoscale Electrical Excitation of Surface Plasmon Polaritons with a Nanoantenna Tunneling Junction. *ACS Photonics* **2023**, *10*, 2641–2649. [CrossRef]
19. Butt, M.A.; Khonina, S.N.; Kazanskiy, N.L. Plasmonic refractive index sensor based on MIM square ring resonator. In Proceedings of the 2018 International Conference on Computing, Quetta, Pakistan, 12–13 November 2018.
20. Zhu, J.; Wu, C. Optical refractive index sensor with Fano resonance based on original MIM waveguide structure. *Results Phys.* **2021**, *21*, 103858. [CrossRef]
21. Rohimah, S.; Tian, H.; Wang, J.; Chen, J.; Li, J.; Liu, X.; Cui, J.; Xu, Q.; Hao, Y. Fano resonance in the plasmonic structure of MIM waveguide with r-shaped resonator for refractive index sensor. *Plasmonics* **2022**, *17*, 1681–1689. [CrossRef]
22. Kazanskiy, N.; Butt, M.; Khonina, S. Nanodots decorated MIM semi-ring resonator cavity for biochemical sensing applications. *Photonics Nanostruct.-Fundam. Appl.* **2022**, *42*, 100836. [CrossRef]
23. Sagor, R.H.; Hassan, M.F.; Yaseer, A.A.; Surid, E.; Ahmed, M.I. Highly sensitive refractive index sensor optimized for blood group sensing utilizing the Fano resonance. *Appl. Nanosci.* **2021**, *11*, 521–534. [CrossRef]
24. Tavana, S.; Bahadori-Haghighi, S. Visible-range double fano resonance metal–insulator–metal plasmonic waveguide for optical refractive index sensing. *Plasmonics* **2022**, *6*, 2441–2449. [CrossRef]

25. Shen, S.; She, S.; Wang, Z.; Tan, Q.; Xiong, J.; Zhang, W. MIM waveguide structure consisting of two triangle stubs, side-coupled with an eight-like resonant cavity. *Opt. Commun.* **2021**, *495*, 127087. [CrossRef]
26. Bell, S.E.J.; Charron, G.; Cortés, E.; Kneipp, J.; De La Chapelle, M.L.; Langer, J.; Procházka, M.; Tran, V.; Schlücker, S. Towards reliable and quantitative surface-enhanced Raman scattering (SERS): From key parameters to good analytical practice. *Angew. Chem. Int. Ed.* **2020**, *59*, 5454–5462. [CrossRef] [PubMed]
27. Stone, J.M.; Hawley, J.F.; Gammie, C.F.; Balbus, S.A. Three-dimensional magnetohydrodynamical simulations of vertically stratified accretion disks. *Astrophys. J.* **1996**, *463*, 656. [CrossRef]
28. Zhu, J.; Li, N. MIM waveguide structure consisting of a semicircular resonant cavity coupled with a key-shaped resonant cavity. *Opt. Express* **2020**, *28*, 19978–19987. [CrossRef]
29. Bian, Z.-Y.; Liang, R.-S.; Zhang, Y.-J.; Yi, L.-X.; Lai, G.; Zhao, R.-T. Multifunctional disk device for optical switch and temperature sensor. *Chin. Phys. B* **2015**, *24*, 107801. [CrossRef]
30. Yang, Z.; Wang, J.; Shao, Y.; Jin, Y.; Yi, M. Studying corrosion of silver thin film by surface plasmon resonance technique. *Opt. Quantum Electron.* **2020**, *52*, 31. [CrossRef]
31. Zhang, C.; Ji, C.; Park, Y.B.; Guo, L.J. Thin-Metal-Film-Based Transparent Conductors: Material Preparation, Optical Design, and Device Applications. *Adv. Opt. Mater.* **2021**, *9*, 2001298. [CrossRef]
32. Liu, L.; Luo, Y.; Zhao, Z.; Zhang, W.; Gao, G.; Zeng, B.; Wang, C.; Luo, X. Large area and deep sub-wavelength interference lithography employing odd surface plasmon modes. *Sci. Rep.* **2016**, *6*, 30450. [CrossRef]
33. Gao, Z.; Chen, H.; Feng, Y.; Ullah, S.; Li, H.; Jing, X.; Li, S. Ultra-wide range and high-sensitivity temperature sensor based on a simple SPR system. *Infrared Phys. Technol.* **2023**, *131*, 104676. [CrossRef]
34. Xu, D.; Yan, S.; Yang, X.; Su, H.; Wu, X.; Hua, E. A nanoscale structure based on a ring with matchstick-shape cavity for glucose concentration and temperature detection. *IEEE Sens. J.* **2020**, *21*, 4442–4450. [CrossRef]

MDPI AG
Grosspeteranlage 5
4052 Basel
Switzerland
Tel.: +41 61 683 77 34

Photonics Editorial Office
E-mail: photonics@mdpi.com
www.mdpi.com/journal/photonics

www.ingramcontent.com/pod-product-compliance
Lightning Source LLC
Chambersburg PA
CBHW041351210926
51850CB00013B/3051

* 9 7 8 3 7 2 5 8 3 3 4 4 3 *